LOW-TEMPERATURE X-RAY DIFFRACTION

Apparatus and Techniques

MONOGRAPHS IN LOW-TEMPERATURE PHYSICS

Edited by John G. Daunt

Stevens Institute of Technology
Hoboken, New Jersey

and

K. Mendelssohn

The University of Oxford
Oxford, England

Bernard Yates — Thermal Expansion

Reuben Rudman — Low-Temperature X-Ray Diffraction

LOW-TEMPERATURE X-RAY DIFFRACTION

Apparatus and Techniques

Reuben Rudman

Department of Chemistry
Adelphi University
Garden City, New York

Plenum Press · New York and London

Library of Congress Cataloging in Publication Data

Rudman, Reuben.

Low-temperature X-ray diffraction.

(Monographs in low-temperature physics)
Bibliography: p.
Includes index.
1. X-ray crystallography. 2. X-rays—Diffraction. I. Title. II. Series

| QD945.R77 | 548'.83 | 76-23259 |

© 1976 Plenum Press, New York
Softcover reprint of the hardcover 1st edition 1976

A Division of Plenum Publishing Corporation
227 West 17th Street, New York, N.Y. 10011

ISBN 978-1-4615-8773-6   ISBN 978-1-4615-8771-2 (eBook)
DOI 10.1007/978-1-4615-8771-2

This book is dedicated to the memory of those Israelis who fell in the Yom Kippur War — 5734.

ה י״ד

Preface

Low-temperature X-ray diffraction (LTXRD) investigations offer many challenges to the diffractionist, not all of which are technical or scientific in nature.

LTXRD studies can be *frustrating:* There are at least two reports of investigations ruined by the loss of crystals (grown with extreme difficulty) because of the widespread power failure and blackout in the northeastern United States in late 1965.

LTXRD studies can cause *discomfort:* In several instances, "low temperatures" have been attained by opening all the windows in the X-ray laboratory.

LTXRD studies can be *dangerous:* It was once reported that a crystal was lost because a laboratory assistant fell down a flight of stairs and lay unconscious for about an hour on his way to refilling a liquid-nitrogen (LN_2) dewar. This last report indicated the disposition of the crystal but not that of the laboratory assistant.

However, in general, the results of low-temperature X-ray diffraction investigations cannot be obtained in any other manner, and one is well compensated for the effort expended in constructing and maintaining a low-temperature system. Crystal-structure analyses of solidified liquids and gases, phase transformation investigations, accurate crystal-structure analyses and electron-density maps, thermal expansion measurements, and defect structure studies are a few of the many important applications of LTXRD.

In this book, the results of nearly sixty years of research in LTXRD instrumentation are summarized, classified, and presented for both the novice and the expert. The necessary information required to initiate simple, exploratory low-temperature investigations is presented along with detailed discussions of the basic design principles which have been used to construct modern, sophisticated low-temperature systems compatible with modern, sophisticated X-ray diffraction apparatus. In addition, the techniques used to prepare samples for LTXRD investigations are presented in a systematic manner for the first time.

The primary purpose of this book is to present the principles of low-temperature experimentation, in general, and these techniques as applied to X-ray diffraction, in particular, for the use of the X-ray diffractionist who wishes to expand his experimental abilities in this direction.

Although most of the references deal with LTXRD, a few were taken from the closely related fields of neutron diffraction, X-ray spectroscopy, and X-ray study of liquids if they could be adapted easily to LTXRD.

During the course of writing this book and classifying the various instruments, a computerized apparatus classification scheme was developed. All LTXRD instruments have been assigned a code number and have been sorted according to this code number. In this manner, it is possible to determine immediately whether or not a particular type of LTXRD apparatus or technique has been described. Details of this classification scheme are found in the introduction to the Bibliography.

I have found that over the years there has been a great deal of duplication in the development of LTXRD apparatus and techniques. This is due to the lack of any comprehensive review of the field and the difficulties encountered in abstracting and indexing LTXRD material. Many of the early devices were described in the "experimental" section of a larger paper and were easily overlooked by later experimentalists.

Suggestions for the proper method of reporting LTXRD apparatus are presented in Appendix 13.

It is very difficult, if not impossible, to describe an ideal LTXRD system. Throughout the book I have tried to offer suggestions as to what system or technique is proper for a given application or X-ray instrument.

Critical comments have been kept to a minimum; what is best for one experimenter or laboratory may not work for another. In many cases several alternative methods are described in the literature. I have attempted to present the various options and allow the reader to decide what course he will follow.

The text consists of a brief introduction to the apparatus and applications of LTXRD, followed by the major portions of the book—sections on apparatus and on technique. The last part contains a series of short appendices (in which general low-temperature techniques are described) and the Bibliography with the computer-sorted apparatus and technique code numbers.

I am sure that many more unpublished low-temperature systems, adapters, and techniques exist. With the hope of publishing an updated bibliography from time to time, I would be interested in receiving descriptions of any LTXRD apparatus and techniques that I have overlooked or that will be developed in the future.

With regard to the units of temperature, I have, in general, used kelvin units for temperatures lower than 195 K and degrees Celsius for temperatures greater than $-80°C$.

I am deeply grateful to Prof. Ben Post for having introduced me to the field of LTXRD and to the late Dr. Walter C. Hamilton, who afforded me the opportunity to pursue these interests.

This book was written while I was a Visiting Professor of Chemistry at the Hebrew University of Jerusalem, Israel, during the very difficult year 1973–1974. I thank Prof. Saul Patai, Chairman of the Institute of Chemistry, and Prof. Henry Selig, Chairman of the Department of Inorganic and Analytical Chemistry, for their kind invitation and assistance. A particular debt of gratitude is owed Prof. Yitzchak Mayer for his gracious hospitality and constant encouragement during my stay at the Hebrew University.

By the time I had finished this book, the list of individuals who assisted me in this task had become quite formidable. I would like to thank all of them for their gracious help in providing reprints, preprints, data, translations, and helpful discussions. Dr. Z. Barnea and Dr. F. Hirshfeld critically reviewed the early versions of the computerized Bibliography. Frank Ely of the Adelphi University computer center was extremely helpful in the preparation of the final version of the Bibliography. Dr. S. C. Abrahams, Dr. R. J. Feldmann, Prof. Y. Okaya, Mr. M. Paunovic, Dr. B. Penfold, Dr. J. Przedmojski, and Prof. W. Streib assisted me in obtaining and/or translating many references that were otherwise unobtainable.

In particular, I would like to thank Profs. P. Coppens, H. W. King, G. Petsko, and B. Post for having critically read the manuscript. Their penetrating comments and helpful suggestions were incorporated in the final text. Of course, the author takes full responsibility for all errors of omission and commission.

A special word of acknowledgment is due Mr. Robert Lippman for his many helpful comments and his assistance in preparing the figures for publication.

I sincerely thank the librarian of the Adelphi University Science Library, Mrs. L. Matzka, for her untiring efforts in helping me track down and acquire many of the references. I am deeply grateful to my indefatigable secretary, Mrs. Sue Goddard, for her always gracious assistance in typing the several versions of this manuscript.

I finally understand why all authors thank their families for their patience and understanding. The myriad of details that engulf an author as he completes the manuscript really do encroach on his personal and professional life. I thank my dear wife, Idelle, and our children, Zave, Rachel, Benjamin, Ephraim, and Sara, for their patience, understanding, and encouragement during the preparation of this book.

תושלב״ע

Reuben Rudman

Jerusalem
June 17, 1974
27 Sivan, 5734

New York
May 1, 1976

Contents

PART II: Low-Temperature X-Ray Diffraction Apparatus

PART III: Low-Temperature X-Ray Diffraction Techniques

PART I

Introduction and Applications

CHAPTER 1

Introduction

1.1. Historical Background

As methods for achieving low temperatures were developed at the beginning of this century, scientific investigators began using these techniques to study the low-temperature properties of materials. The earliest reports of low-temperature crystallography were the low-temperature microscopy investigations of Goldschmidt (1912), Wahl (1913), and Rinne (1914).

Some of these methods were later adapted for use in low-temperature X-ray diffraction (LTXRD) studies. Rinne (1917) (Figure 1–1) and St. John (1918) initiated their research on the crystal structure of ice as early

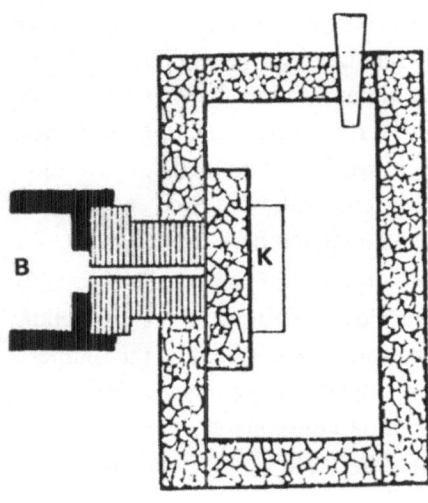

Fig. 1–1. Schematic diagram of earliest reported low-temperature device for X-ray diffraction (Rinne, 1917). The crystal (K) was held in a cork box and cooled with liquid air. The X-ray beam entered through the collimator (B) and passed through a layer of cork before reaching the crystal.

as 1916, only three years after the first crystal-structure analysis. By 1922, the structure of mercury at 160 K was reported (McKeehan and Cioffi), followed soon after by argon at 40 K (Simon and Simson, 1924a) and para-hydrogen at 1.65 K (Keesom et al., 1930). More recently, a temperature of 30 mK has been reported (Simmons, 1976).

These early measurements were often made in specially designed cameras and involved complex cooling techniques. The difficulties encountered were quite formidable and simple techniques for accurate investigations were not available. As a result, low-temperature studies were limited to only a few laboratories.

During the late 1940s and early 1950s, cooling devices that could be assembled from common laboratory items and the techniques to use them were developed and popularized by several groups. These improvements led to the renaissance of low-temperature X-ray diffraction. Although excellent results can still be obtained with the basic cooling systems described in the early 1950s, one now encounters descriptions of such sophisticated items as a liquid-helium cryostat mounted on an automatic single-crystal four-circle diffractometer or a mechanically refrigerated back-reflection powder camera that can be cooled to 1.4 K.

1.2. Cooling Apparatus

All X-ray diffraction instruments consist of three basic components: a source of X-rays, a sample-positioning device, and a means of detecting the diffracted beam. A low-temperature investigation requires the addition of a device for cooling the sample while causing *minimal* loss of data (e.g., by restricting the number of reflections that can be measured or by reducing the intensity of the beam). The term *low-temperature,* as used in this book, includes all temperatures from just below ambient to just above absolute zero. However, any given low-temperature device is generally designed for optimum operation within a specific temperature range.

The low-temperature system itself consists of several components, which can be classified into three categories according to their function: provision for cooling the sample, control and measurement of the temperature, and frost prevention. A number of different approaches have been described in the literature for each of these components, and many successful modern low-temperature systems are based upon a combination of contributions from several sources.

1.2.1. Provision for Cooling the Sample

There are three general methods for cooling a sample.

1.2.1.1. Gas Stream

A stream of cold gas (e.g., air, nitrogen, or helium), generated by boiling a liquefied gas or by passing warm dry gas through a heat exchanger, is directed over the sample.

This system has the advantages of being relatively easy to set up and having minimum absorption problems. It suffers from the disadvantages of being most susceptible to frost formation and using relatively large amounts of cryogen for generating the cold gas stream. However, the latter disadvantage is largely offset by recently developed mechanically refrigerated dewars and probes which can be used to replace the classical cryogens in many applications (see Section 3.4.1.1).

1.2.1.2. Conduction

The sample and sample chamber are cooled by means of a good thermal conductor in contact with a cooling unit. The cooling unit may be a cold bath, mechanical refrigerator, thermoelectric cooler, or Joule–Thomson expansion device.

Advantages of this system include minimal use of cryogen, relatively good thermal equilibrium, and frost-free operation. Disadvantages are the need for a cooling chamber, difficulty in observing the sample, and errors introduced by absorption of the X-ray beam by the cooling-chamber windows. The possibility of thermal gradients in the sample (due to poor conduction properties) must also be considered.

1.2.1.3. Immersion

It is possible to cool the sample by immersing the entire camera in a cold liquid or by dripping a cold liquid over the sample. This technique, although not often used, has the advantage of maintaining a constant temperature, which is usually that of the cold liquid. Absorption problems and a lack of fine temperature control are some of the difficulties to be considered. A variation of this method is to place the entire apparatus in a cold room or cold chamber, with the temperature normally limited to a low of $-50°C$.

Interestingly, in one of the earliest LTXRD investigations reported,

St. John (1918) appears to have used prototype versions of all three methods. He initially cooled his sample by opening the windows and allowing the cold air to flow over his sample (gas-stream method). When this was not satisfactory, he enclosed the camera in a chamber and placed cans of ice and salt inside the chamber (cold-room immersion technique). Finally, he built a special spectrometer with a small ammonia refrigerating machine which was mounted in a well-insulated refrigerator box. No further details are given, but this may well have operated on the conduction principle.

1.2.2. Control and Measurement of the Sample Temperature

The temperature is generally monitored with a thermocouple which may be attached to a recording potentiometer. In some applications it is necessary to know the temperature to $0.01°$, while in others $±5°$ accuracy is sufficient.

The temperature can be controlled by mixing a warm gas stream with the cold gas stream or by using a heater in the gas stream or near the conducting cold finger.

1.2.3. Frost Prevention

Atmospheric gases, most notably water and carbon dioxide, should not be allowed to condense on the sample and all other points along the path of the X-ray beam from source to detector. The most common approach for gas-stream systems is to use an outer concentric stream of warm, dry gas which acts as a sheath around the inner cold gas stream, thus preventing atmospheric gases from reaching the cold sample. In the conduction method, a dry or evacuated chamber protects the sample. However, if vacuum insulation is not used, frost is prevented by directing a flow of warm, dry air over the cryostat windows or movable parts of the apparatus. For the immersion technique, the steady flow of cold liquid over the sample is generally sufficient to prevent condensation.

Finally, in many instances the construction of a dry box or plastic tent over the diffraction apparatus prevents moist air from coming in contact with the sample.

1.3. Problem Areas in Designing LTXRD Apparatus

When one designs and/or uses LTXRD apparatus, there are several areas that require special attention. These are outlined briefly in the following paragraphs and are discussed in greater detail in later chapters.

1.3.1. Absorption of X-Radiation

The passage of X-radiation through any material reduces the intensity of the beam in accordance with the well known formula $I = I_o e^{-\mu t}$, where I, the intensity of the transmitted beam, is a function of I_o, the incident beam, μ, the linear absorption coefficient of the absorber, and t, the thickness of the absorber.

A low-temperature device must be designed with a minimum amount of material in the paths of the incident and diffracted beams. This requirement places severe restraints on the design of cooling chambers, in which the use of very thin beryllium or Mylar windows is strongly suggested. This problem is not encountered when gas-stream cooling is used.

1.3.2. Size of Instrument

The size of the cooling apparatus, or at least that portion of the apparatus which is in close proximity to the sample, is determined by the size of the diffraction apparatus. In many cases, the relative sizes of the diffraction and cooling instruments, as well as the minimum temperature desired, control the quantity of data that can be collected (by determining maximum and minimum values of 2θ and by limiting the sample orientation with respect to the X-ray beam).

While it is fine to consider an ideal low-temperature device, i.e., one in which any desired temperature can be reached and any desired reflection measured, the effort needed to design, construct, and operate such an instrument is not necessarily worthwhile. Several low-temperature devices, each adapted to a specific need, are more efficient in both design and construction efforts and in operating characteristics than a single "universal" unit. It is worthwhile to remember that, all other factors being equal, the larger the system (dewar or cryostat), the greater will be the consumption of cryogen. Also, if the cooling device is required to cool too large a volume, unnecessarily large amounts of cryogen will be consumed and critical parts of the diffraction apparatus may be cooled, resulting in misalignment of the instrument, rusting of sensitive parts of the apparatus, or icing up of movable joints.

1.3.3. Frost Formation

As was mentioned previously, cooling the sample below room temperature can result in the condensation of various vapors on the sample and cold portions of the instrument. This condensation is a serious problem on many LTXRD devices, especially when the humidity is high. Aside from the adverse effects on the diffraction instrument (mentioned in the preceding paragraph), frost formation can affect the data collection by leading to errors in measuring relative intensities as a function of time (due to increasing amounts of frost) and as a function of d^* (due to uneven formation of frost on the sample). Finally, excessive amounts of frost on the sample or on the sample support can move the crystal, thus affecting its alignment. In extreme cases, the weight of this frost on a small single crystal and its fiber support can move the sample out of the beam.

1.3.4. Temperature Stability

For obvious reasons, a low-temperature device should be capable of maintaining a constant temperature during the investigation. Acceptable limits of long-range stability vary from $\pm 5°$ to better than $0.01°$.

Equally important, but sometimes overlooked during the design of the instrument, is the fact that the apparatus must often maintain a stable temperature during cryogen-refill periods. This problem is of greatest concern in gas-stream devices where the gas stream is obtained from a boiling cryogenic liquid.

1.3.5. Temperature Calibration

In general, the position of the thermocouple does not coincide with that of the sample, for if the thermocouple is too close to the sample, it may diffract the X-ray beam. Therefore, in many instruments, it is necessary to calibrate the temperature of the sample vs. the thermocouple reading.

A second potential source of error is the possibility of thermal gradients in the sample, particularly if the sample is large (as in some powder-diffraction studies).

1.3.6. Misalignment

The diffraction apparatus and/or the sample may become misaligned as the temperature is altered. This fact, which must be considered when designing or operating low-temperature instruments, is particularly bothersome when cryostats are used.

1.3.7. Sample

There are several potential sources of difficulty that must be considered when low-temperature techniques are employed, among which are the following.

The *mounting adhesive* used for samples that are solids at room temperature must not contract at a rate too different from that of the sample and mounting pin. Otherwise, the crystal may fall off the pin or become misaligned, and perhaps shattered, by the mechanical strains which develop as the temperature is lowered.

Solids with high vapor pressures, liquids, and *gases* must be contained within a glass tube or other sample holder with thin walls, in order to reduce absorption of the X-ray beam. These substances may also present problems in growing single crystals or in preparing randomly distributed polycrystalline samples, since preferred orientation, twinning, and partial crystallinity can all occur. Special attention is required if counter diffractometer, rather than film, techniques are used.

Finally, it is possible that a *phase transition* may occur when the sample is cooled. A preliminary investigation of the effects of cooling a particular sample is always in order.

CHAPTER 2

Applications

The results of low-temperature X-ray diffraction investigations have been the subject of several critical reviews, starting with Keesom (1924). The literature prior to 1937 was reviewed by Ruhemann and Ruhemann (1937), while the specific area of metallurgical applications was covered rather extensively by Barrett (1957). However, because of the popularization of LTXRD techniques in the 1950s, the number of low-temperature investigations reported in the literature increased so greatly that no comprehensive critical review has appeared in the past twenty-five years. A selected bibliography was prepared in 1964 (Post), while several short reviews have appeared on specific subjects (see Code Number 20 in the Bibliography).*

The topics covered in this chapter are representative of the areas in which LTXRD techniques have been applied. In some instances, it is impossible to obtain the desired results without cooling the sample, while in others, the use of low-temperature techniques serves to improve the quality of data which could also be obtained at room temperature. A comprehensive review of LTXRD applications and results is beyond the scope of this book.

2.1. Solidified Liquids and Gases

It is quite apparent that materials which are normally liquids or gases at room temperature must be cooled if their crystal structures are to be

---

*In the Bibliography, all LTXRD instruments have been classified by code numbers assigned to them. This classification system is explained in the Introduction to the Bibliography and in Table B-1, which for convenience also appears on a foldout page in the back of the book.

studied. Methods have been developed for preparing single-crystal or poly-crystalline samples of these materials directly on the diffraction apparatus or for growing them elsewhere and transferring them to the apparatus. Details of sample preparation are found in Chapters 6 and 7.

Although such studies are somewhat tedious, requiring skill in sample preparation as well as a carefully constructed and regulated apparatus, the data obtained from these investigations cannot be acquired by any other means. Important information concerning molecular structures, interatomic distances, and electron-density distributions in many simple molecules, fundamental to a proper understanding of molecular bonding theories, has been obtained in this way.

Examples include the crystal-structure determinations of mercury (McKeehan and Cioffi, 1922), ice (Barnes, 1929), neon (de Smedt et al., 1930), hydrogen chloride (Natta, 1933), tetranitromethane (Oda et al., 1943), chlorine (Collin, 1952), pentaborane (Dulmage and Lipscomb, 1952), diborane (Smith and Lipscomb, 1965), helium (Schuch and Mills, 1962), and methylchloroform (Silver and Rudman, 1972). Numerous other examples are found in the Bibliography.

2.2. Crystal-Structure Analysis

As the temperature of a sample is lowered, the thermal motions of the atoms within the sample are reduced. This factor is responsible for the preferred use of low-temperature data over room-temperature data in crys-tal-structure analyses, regardless of the physical state of the sample at room temperature. More data will be obtained, the peak-to-background ratio will be improved (Figure 2–1), the estimated standard deviations (esd) of the atomic coordinates will be lowered, and more accurate bond lengths and angles can be calculated.

An analysis of the decrease in the esd's when low-temperature data are used was first presented by Burbank (1953), while Cruickshank (1956, 1960) discussed the decrease in molecular vibrations as the tem-perature is lowered (see also Section 8.1.1.).

An early description of the improvement in accuracy as a result of using low-temperature methods was also given by Hirshfeld and Schmidt (1956). They pointed out that the increased number of reflections that were measured at low temperatures improved the two-dimensional Fourier map to the point where atoms that were poorly resolved when room-tem-

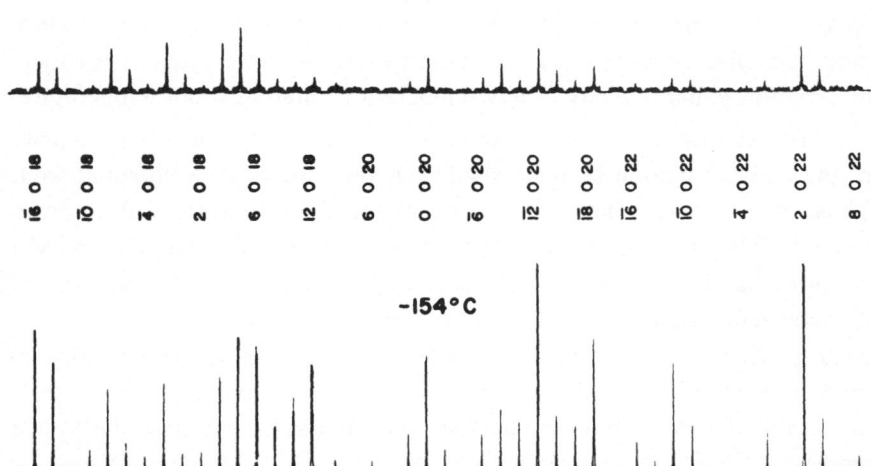

Fig. 2–1. Comparison of intensities at 23 and −154°C for an organic phosphate (Abowitz and Ladell, 1968). (Reproduced with the permission of the Institute of Physics.)

perature data were used became accurately resolved with low-temperature data. This resulted in a marked increase in accuracy of the atomic coordinates.

They also found that the peripheral atoms tended to have lower peak densities than those near the molecular centers. This was correctly attributed to the thermal motion of the molecules.

Later development of libration corrections resulted in improved accuracy of bond distances, providing that the angles of libration were not too large. Further improvement was afforded by the use of the rigid-body treatment. A lattice-dynamical interpretation of thermal motion in molecular crystals has recently been published (Filippini et al., 1974).

In all cases, these correction factors are approximations which are most accurate for small librations. Thus, if a molecule has a large amplitude of libration at room temperature, it is necessary to collect low-temperature data rather than attempt to correct room-temperature data if accurate results are desired. Filippini et al. (1974) show that their approach works best for a translational rms amplitude of 0.15 Å and rotational rms amplitude of 2.5°. Thus, for benzene, they find that the data must be collected below 135 K.

The advantages of using low-temperature data in crystal-structure analyses have been summarized by Coppens (1972). He shows that, in spite of the improved accuracy of the correction term for low-temperature data, it is also possible for a molecule to show a more pronounced thermal-motion anharmonicity at low temperatures than at room temperature.

This follows from the following reasoning: In general, anharmonicity of the thermal motion is much smaller at the bottom of a potential well. Thus, those modes which are excited at room temperature will conform more closely to the harmonic approximation at a lower temperature. On the other hand, many of the high-frequency internal modes are not excited at room temperature and will thus become relatively more important on cooling. Therefore, molecules behave less like rigid bodies at low temperatures and, in principle, the anharmonicity could be more pronounced at low temperatures *if* the internal modes are less harmonic than the lattice modes.

Several reports have appeared recently in which the crystal structure of a given compound has been determined at several different temperatures. For example, Spencer and Lundgren (1973) studied the $CF_3SO_3^-$ ion at 298 K and 83 K (Figure 2–2). Note that the differences in the three chemically equivalent $C-F$ and $S-O$ bond lengths are considerably greater at 298 K than at 83 K. This is due to the inability of the correction factor to fully correct for the large atomic librations that are present at room temperature. Similar results have been obtained for *p*-nitropyridine-*N*-oxide at 300 K and 30 K (Figure 2–3) by Wang et al. (1976).

There is an apparent shortening of the bond lengths at the higher temperature due to the insufficiency of the thermal-motion correction factors. A more sophisticated treatment of thermal motion (e.g., higher-order cumulants) can improve the results obtained from room-temperature data. However, the use of low temperatures negates the need for these complex corrections and offers the advantages of improved quality of the data and subsequent crystal-structure determination.

2.3. Electron-Density Distribution

The electron distributions of atoms participating in chemical bonds differ from those of the free atoms. Theoretical evaluations of these differences show that the calculated changes depend on the assumptions made for the wave functions of the molecules. Direct experimental information can, in principle, be obtained by X-ray diffraction as by this method the

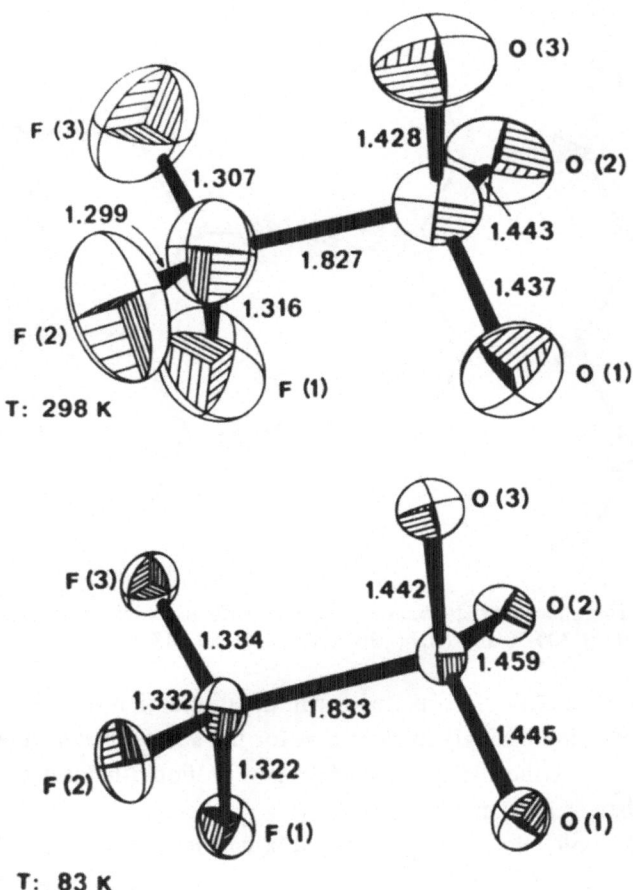

Fig. 2–2. Bond lengths (Å) in the $CF_3SO_3^-$ ion at 298 and 83 K. The thermal ellipsoids are scaled to enclose 50% probability (Spencer and Lundgren, 1973). (Reproduced with the permission of the International Union of Crystallography.)

one-electron density function in a molecule can be determined (e.g., Verschoor and Keulen, 1971; Coppens and Vos, 1971). Recent results have been critically reviewed by Coppens (1975).

The requirements of accuracy of the experimental observations in charge-density studies are quite severe. Symmetry-equivalent reflections should be measured and should agree to better than 2% in the intensities. Systematic effects such as multiple reflections, absorption of X-rays in the crystal, and thermal diffuse scattering should be considered, and elimi-

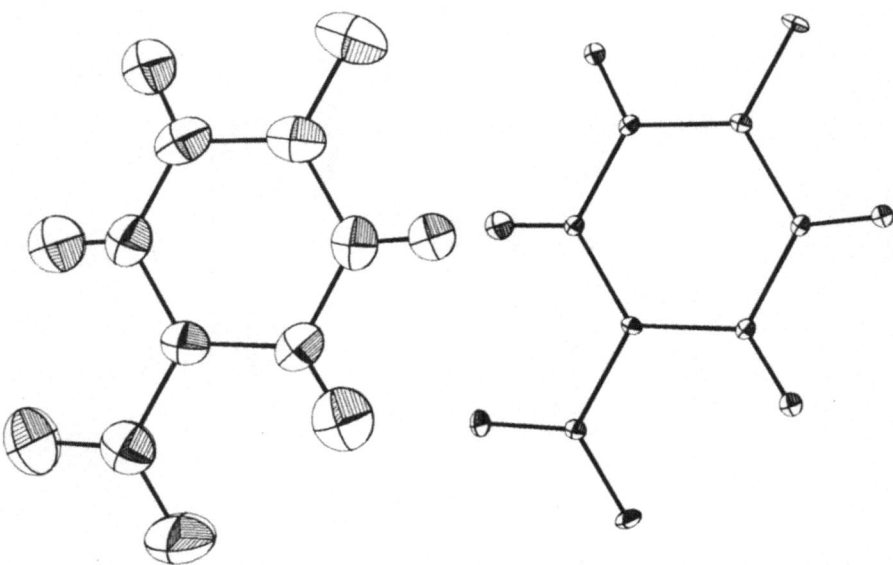

Fig. 2–3.   Thermal ellipsoids (scaled to enclose 50% probability) for *p*-nitropyridine-*N*-oxide at (left) 300 K and (right) 30 K (Wang et al., 1976).

nated experimentally or corrected for in data reduction or refinement. Furthermore, it is increasingly evident that for the study of molecular crystals, data should be collected at liquid-nitrogen or liquid-helium temperatures for a number of reasons:

1. The collection of high-order data will lead to an increase of resolution in the electron-density maps and allow a more accurate determination of the positional and thermal parameters.

2. A reduction in temperature will reduce the anharmonicity of the external molecular vibrations so that introduction of a more complex model in the refinement can be avoided.

3. The effect of thermal diffuse scattering decreases relative to the intensity of the Bragg reflections when the temperature is lowered (Figure 2–4), thus reducing one of the sources of error in the temperature parameters.

Although this field of investigation is relatively new and the number of compounds that have been studied is limited, the results are quite impressive. The electron density is determined by calculating a Fourier difference density map, where the difference term represents the difference between the experimental and "high-angle" X-ray structure factors. The

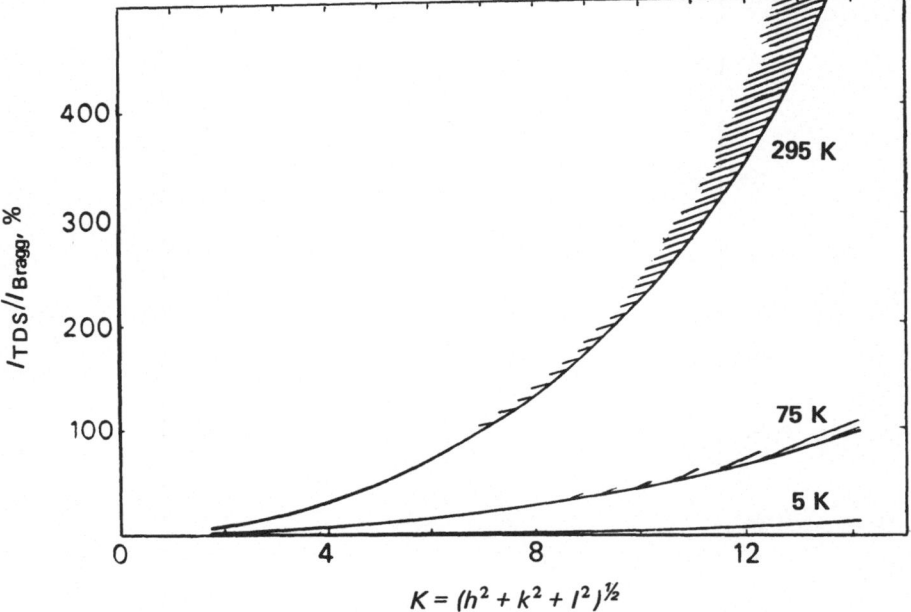

Fig. 2–4. (a) NaCl. The ratio I_{TDS}/I_{Bragg} expressed as a percentage for a range of scattering lengths K (in units of $2\pi/d$) and for the temperatures 295, 75, and 5 K. The thermal diffuse scattering is integrated over a sphere in reciprocal space whose radius extends 0.124 to the (100) zone boundary. The shaded region represents the anisotropy that arises from true multiphonon processes which exactly sum to a reciprocal-lattice vector.

"high-angle" structure factors are the structure factors *calculated* from the parameters obtained by a least-square refinement in which only high-angle data are used. Typical difference maps show excess density in the bonding and lone-pair regions, and an electron deficiency near the nuclear positions (Figure 2–5).

2.4. Radiation Damage

The interaction of X-radiation with a crystal lattice often results in irreversible damage to the crystal. A steady decline in the intensity of the standard reflections measured during the course of data collection usually indicates that the crystal has decomposed. Often this is the result of the exposure of the crystal to X-rays. The exact effect of the decomposition products on the measurement of relative intensities is not predictable. Some reflections may even have their intensities increased. Although many

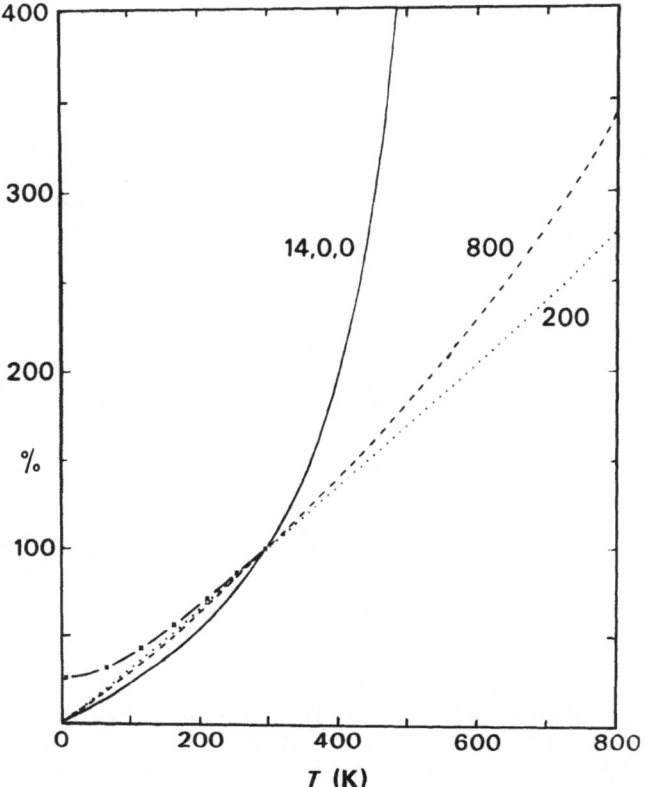

Fig. 2–4. (b) The temperature dependence of the ratio
I_{TDS}/I_{Bragg} for the reflections 14,0,0, 800, and 200 calculated under the conditions given in (a). Each ratio is expressed as a percentage of its value at 295 K. The line of
crosses shows in a similar way the low-temperature variation
of the Debye–Waller B terms (Reid, 1973). (Reproduced
with the permission of the International Union of Crystallography.)

investigators have calculated correction factors for room-temperature data,
the most accurate analyses require data collected at low temperatures. As
the temperature is lowered, the damage caused by exposure to X-radiation
is reduced. Elimination of this damage is extremely important when crystals that are difficult to replace are used. Thus, a large effort has been
expended in developing low-temperature techniques for protein crystallography (see Sections 2.5 and 6.1.3).

Guttormson and Robertson (1973), in a study of a spiro compound
containing the maximum possible number of substituents ($C_8H_6Cl_4$), noted

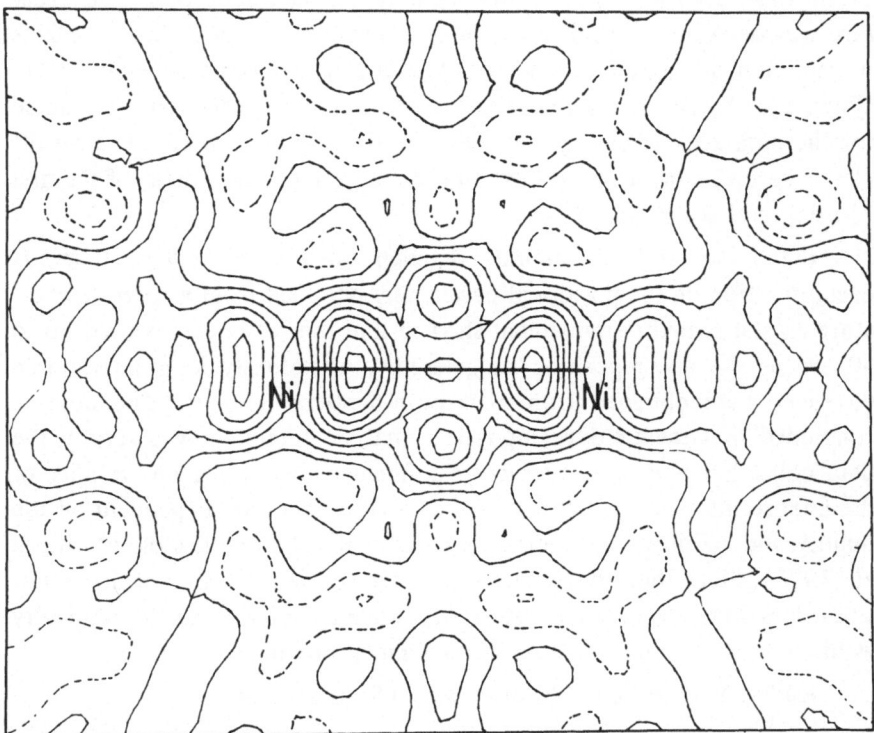

Fig. 2–5. Deformation density in the Ni–Ni Bond in μ-acetylene-dicyclopentadienyl-dinickel at 77 K. Plane parallel to the acetylene C–C bond; contours at 0.1 e Å^{-1} (negative contours are dotted). (Wang and Coppens, 1976.)

that the crystals deteriorated in the X-ray beam at ambient temperatures too quickly for any meaningful collection of data. The melting point of the compound is 36–37°C, and at room temperature the white crystals have a thin liquid layer on their surface. In the X-ray beam, the fraction of material in the liquid state increases, allowing the remaining solid material to change its orientation as the crystal is rotated. The data were collected at −65°C, and 1979 independent reflections were successfully refined to a residual of 0.072.

During the collection of data from $[Cr(tn)_3][Ni(CN)_5] \cdot 2H_2O$ at room temperature, the crystal decomposed in the X-ray beam with the result that the standard reflections were reduced to 14% of their initial intensities. To compensate for the decomposition, a correction factor was calculated and applied to obtain the values of the intensity and standard deviation that would have been measured at the beginning of data collection. The structure was determined from this data set. However, to check on the effects of radiation damage, a second set of data were collected at $-80°C$. No systematic change in the intensities of the standard reflections was noted for the data set, indicating that radiation damage was absent. The results of this second data set were in substantial agreement with those of the first (Jurnak and Raymond, 1974).

On the other hand, another detailed study of the effect of crystal decomposition on the positional parameters obtained from a room-temperature crystal-structure refinement showed that the parameters varied up to 50 esd as the decomposition progressed. Anomalously long interatomic distances were obtained from the room-temperature data. The authors concluded that their study emphasizes that no matter how carefully the data collection and crystal-structure refinement are done, the validity of the final results and the inferences drawn from them are dependent on the quality and stability of the crystals used in the data collection (Nolte et al., 1975). Since their investigation also showed that the rate of decomposition is a function of temperature, the obvious solution to this difficulty would have been to collect a set of low-temperature data.

Radiation damage is also discussed in Section 8.3.4.

2.5. Protein Crystallography

There are many reasons for studying protein crystals at subzero temperatures (Petsko, 1975). The rate of radiation damage is markedly reduced. The rates of motion and exchange of loosely held groups are reduced, thus providing an improved image of "floppy" areas of the protein, and revealing details of conformational flexibility of backbone and side chains. In favorable cases, protein crystallography may also provide a clear view of "bound" solvent molecules, giving valuable information about protein–water interactions and liquid structure. Another extremely important potential use for low-temperature protein crystallography is the direct observation of enzyme–substrate complexes and unstable intermediates. Such complexes can be stabilized in solution at subzero temperatures, and their lifetimes are long enough for high-resolution X-ray data collection. Recent improvements in the techniques used to cool protein crystals to subzero temperatures are discussed in Section 6.1.3.

2.6. Defect Structures

In many cases, it is necessary to lower the temperature to retain stacking faults and point defects for study with diffraction techniques.

Simmons (Losee and Simmons, 1968; Balzer and Simmons, 1974) has shown that information about thermally created defects in a solid can be obtained by comparing the change of macroscopic volume V and of X-ray cell volume Ω for different temperatures by means of the relation

$$\frac{\Delta V}{V} - \frac{\Delta \Omega}{\Omega} = c$$

where c is the net concentration of vacancy-type defects.

2.7. Solid–Solid Phase Transition

A number of materials undergo transitions from one crystal structure to another as the temperature and/or pressure are changed. Many of these phases are unstable at room temperature and can be studied only at low temperatures. Studies of this sort lead to an understanding of the mechanism of transition as well as a knowledge of the various structures. Techniques and apparatus for studying these phenomena are described in Sections 6.1.4, 7.2.2.4, and 7.3.

2.7.1. Plastic Crystals

Crystals consisting of molecules which are approximately spherical often undergo phase transitions in the solid state and exhibit abnormally low entropies of fusion and unusually high melting points. X-ray studies of the phases stable just below the melting point generally indicate highly disordered systems, which have been shown to be the result of hindered molecular reorientation.

As the temperature is lowered, crystals of this sort generally undergo one or more transformations to systems of lower symmetry. The soft, waxy, and volatile high-temperature forms transform into relatively hard and brittle modifications. In most cases, the sum of the entropies of the solid-state transitions is considerably greater than the entropy of fusion. The dielectric constants and nmr spectra of the high-temperature phases resemble those of liquids; below the transition temperature, they take on characteristics associated with normal solids.

The high-temperature phases of these solids are often referred to as "plastic crystals" or "orientationally disordered crystals."

Single crystals of the disordered high-temperature modifications are often easily prepared, but it is very difficult to obtain a significant quantity of X-ray diffraction data useful for structure determination from these phases. Methods that have been developed to extract useful structural information from these data are described in Section 8.4.2.

Although sufficient data can be obtained from single crystals of the ordered low-temperature phases (Figure 2–6), the crystals themselves are difficult to form inasmuch as the plastic crystals invariably shatter as they are cooled through the transition.

Single crystals are easily prepared from compounds with ordered phases stable at room temperature (e.g., camphor and carbon tetrabromide). However, many of the most interesting pseudospherical molecules are simple, low-melting compounds such as carbon tetrachloride, *t*-butyl chloride, and neopentane. In these cases, it is necessary to prepare single crystals of the low-temperature phases well below room temperature. To complicate matters further, they must be grown in a manner suitable for study by low-temperature X-ray diffraction techniques. This necessitates a small crystal (smaller than 0.5 mm), grown from the melt and enclosed in some manner so as to facilitate handling and prevent losses due to the high vapor pressure. The temperature must be carefully controlled to remain within the prescribed temperature range for the given crystalline modification. A useful, though not necessary, feature is that the crystal be grown directly on the X-ray diffraction apparatus.

A number of techniques for preparing single crystals of the various

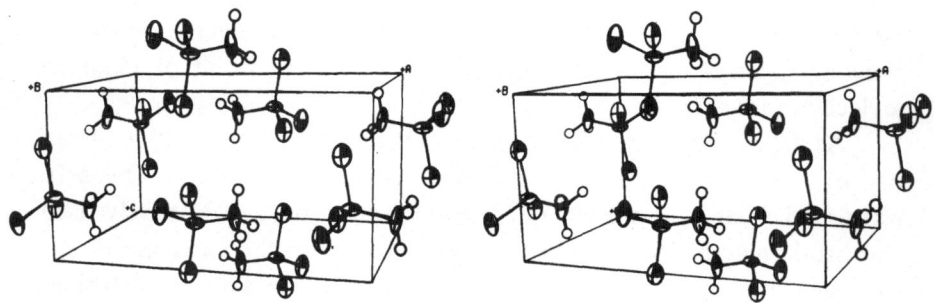

Fig. 2–6. Stereo view of the crystal structure of methylchloroform. The molecules lie on mirror planes perpendicular to b at $y = \frac{1}{4}$ and $y = \frac{3}{4}$ (Silver and Rudman, 1972). (Reproduced with the permission of the American Institute of Physics.)

crystalline modifications of low-melting pseudospherical molecules are discussed in Chapter 7.

2.7.2. Metastable States

Ultrarapid quenching of metallic samples allows one to retain a metastable state long enough for data to be collected (Girt et al., 1974, 1975).

2.7.3. Phase Transitions

Temperature-dependent structure changes are often studied using low-temperature X-ray diffraction methods. Cold-working of metals (Barrett, 1957), order–disorder transitions, and other types of phase transitions have been investigated by determining the structures of both the high- and low-temperature phases.

2.7.4. High-Pressure, Low-Temperature Studies

A number of transitions occurring at high pressures and low temperatures have been examined, using specially designed apparatus (e.g., McDonald et al., 1966; Bronsveld et al., 1973; Gerard and Pernolet, 1973; Balzer and Simmons, 1974). In addition, several regular high-pressure X-ray diffraction cells have been cooled successfully with liquid helium as well as liquid nitrogen (S. Block, personal communication; Morosin and Schirber, 1974) (Figure 2–7).

McDonald et al. (1966) were able to retain high-pressure phases in a metastable state at atmospheric pressure and 4.2 K. Both AgI and gallium were investigated. Mills and Schuch (1974 and earlier papers) have studied the crystal structures of several forms of helium. Other examples of materials studied at high pressures and low temperatures include ethylene hydrate clathrate (Gerard and Pernolet, 1973) and tetramethylammonium manganese (II) trichloride (Morosin and Schirber, 1974).

2.8. Thermal Expansion Coefficients and Precision Lattice Constants

The effect of temperature on the unit cell parameters and volume has been a fertile field of investigation for many years. Both powder and single-crystal samples have been examined. Many devices have been designed

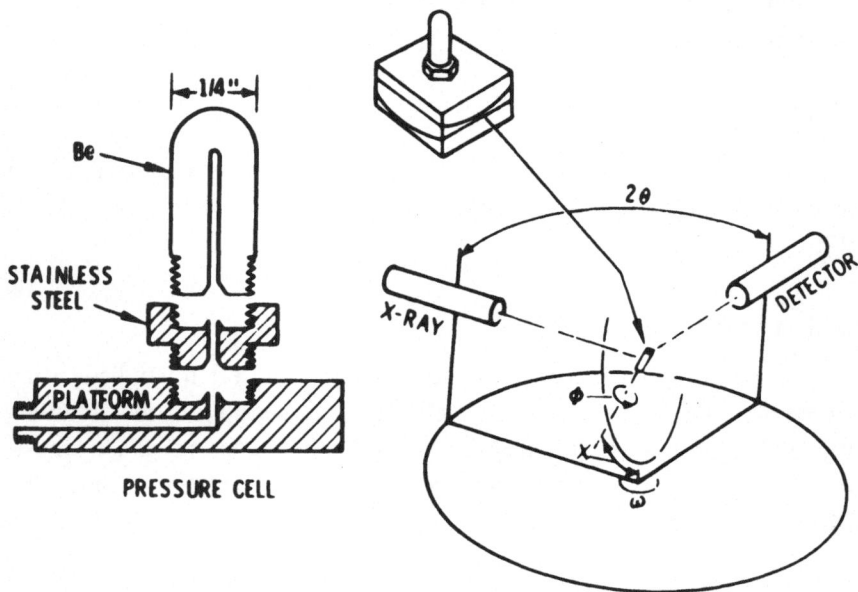

Fig. 2–7. Schematic diagram for high-pressure X-ray cell as used on a diffractometer. The cell is cooled by blowing cold nitrogen gas over the pressure cell or by dripping a steady stream of liquid nitrogen on the cell (Morosin and Schirber, 1974). (Reproduced with the permission of the International Union of Crystallography.)

specifically for this purpose, and often a powder diffractometer or similar instrument is used with a large, platelike single crystal. Data on metastable, as well as high-pressure, modifications have been obtained.

King and Preece (1967) have developed techniques and apparatus for measuring lattice parameters with a reported reproducibility of 1 part in 150,000 at liquid-helium temperatures.

The lattice parameters and thermal expansion coefficients of aluminum, silver, and molybdenum to 30 K were determined using a symmetrical back-reflection vacuum camera with the sample holder attached to a mechanically cooled refrigeration unit (Straumanis and Woodard, 1971).

Eubig and Tomizuka (1972) describe techniques used to measure the lattice parameter changes in quenched aluminum to one part per million at liquid-nitrogen temperatures. Their results are related to studies of radiation damage, cold-working, and the number of defects introduced by quenching.

The accuracy of temperature readout and control must be very high for precision lattice measurements. Axon et al. (1953) calculated that, for many metals, an error of 1°C corresponds to an error in the lattice spacing of 1 part in 40,000.

2.9. Low-Temperature Powder Patterns

If the crystallites in a powder sample are randomly aligned, a pattern which is characteristic of a particular crystalline substance can be quickly obtained. Thus, it is possible to rapidly (1) survey a group of possibly isostructural compounds; (2) determine the occurrence of major phase transitions; and (3) determine lattice parameter changes and coefficients of thermal expansion at low temperatures. Most of these applications have been discussed in previous paragraphs.

Another interesting application is the use of a powder camera in which the film is translated as the temperature is changed. Many materials cannot be prepared as single cystals. As a result, powder data of phases that occur at low temperatures cannot be indexed easily. In these cases, one can follow the shifting of lines in the powder pattern as the temperature changes. Several cameras have been described in the literature (e.g., Simon, 1971) and are available commercially (see Appendix 12). It is also possible to use a standard-low-temperature Weissenberg goniometer to obtain similar results (see end of Section 3.5.4).

2.10. Other X-Ray Diffraction Applications

Goto et al. (1969) modified a low-temperature powder diffractometer to allow the simultaneous measurement of the powder pattern and differential thermal-analysis data. This was used for the study of phase transitions of fats and fatty compounds. The X-ray pattern can be correlated with the thermal data since the two sets of data are obtained under identical conditions.

The range of samples studied by X-ray topography has recently been extended to include low-temperature measurements (Ando, 1973; Coulon et al., 1974). For example, α-iron single crystals have been investigated between 123 K and 296 K up to 20% deformation. Berg–Barrett X-ray reflection topography was used to examine the overall dislocation substructure (Coulon et al., 1974). Below 200 K, the observed dislocation substructure is homogeneous, while above 200 K, it becomes more and more heterogeneous as temperature increases.

Many low-temperature devices can also be used for moderately high-temperature uses. A gas-stream nozzle can be used to reach 100–150°C. The heat exchanger may be immersed in a warm bath or an in-line heater may be used to warm the gas stream. However, excessive heating may cause the dewar tube to fracture. Baun and Renton (1963) suggest the use of a conduction cryostat filled with a warm bath for high-temperature applications. The limiting factor in this case is the heat-resistant property of the solder or epoxy used in constructing the cryostat.

2.11. Application of Low-Temperature Methods to Other Types of Physical Measurements

The applications discussed in this chapter, as well as the apparatus and techniques described in the following chapters, have all been drawn from the area of low-temperature X-ray diffraction. However, often the same apparatus and the same techniques can be adapted for use with other types of physical measurements. In particular, the methods described for preparing samples and single crystals are easily adapted for use with other instruments.

The cryostats described in Chapter 4 usually have windows specially designed for the transmission of X-rays, but the gas-stream technique discussed in Chapter 3 has nearly universal applicability (Figure 2–8). It has been successfully used in low-temperature optical microscopy, nmr, esr, and infrared and neutron diffraction investigations.

Although cooling of specimens is readily accomplished in specially designed cryostats, it is often desirable to cool samples in a manner that does not interfere with visual observation or physical measurements. In addition, the simple gas-stream cooling apparatus described in Section 3.3 can be easily and cheaply constructed from readily available apparatus. Thus, preliminary measurements can be conducted without the necessity of investing in an expensive cryogenic system.

For example, the device described by Albracht (1974, *J. Mag. Res.* **13**, 299–303) is quite similar to the gas-stream devices for LTXRD that have been developed over the past fifty years. A great deal of unnecessary duplication and expenditure of energy could be prevented if investigators

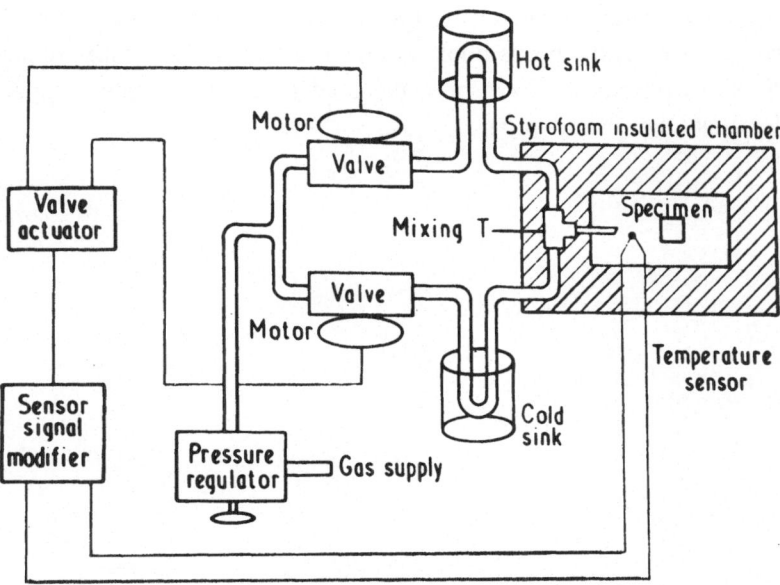

Fig. 2–8. Gas-flow temperature controller designed for cooling specimens in insulated chamber (Huber, 1969). (Reproduced with permission of the Institute of Physics.)

in other disciplines would take advantage of the efforts of nearly sixty years of low-temperature X-ray diffraction studies.

Low-temperature neutron diffraction investigations require special comment because of the similarity between X-ray and neutron diffraction. There are two major differences between low-temperature apparatus designed for these techniques.

Neutron diffraction apparatus is much larger than comparable X-ray diffraction apparatus. Similarly, the typical neutron diffraction sample is considerably larger than an X-ray diffraction sample. As a result, low-temperature apparatus designed for use on a neutron diffractometer has, in general, fewer restrictions on its overall volume and on the size of the components used in its construction.

The second major difference is that aluminum windows can be used in neutron diffraction apparatus, while X-ray diffraction apparatus is limited to beryllium or thin plastic windows. Thus the neutron diffraction devices, in addition to being larger, are also easier to fabricate and are structurally sturdier. A review of neutron diffraction cryostats has been prepared by Atoji (1965).

Low-temperature neutron diffraction apparatus has not been described extensively in this book, although a few devices, with applicability to X-ray diffraction, have been included (see Bibliography, Apparatus Code, First Digit, Number 6).

PART II

Low-Temperature X-Ray Diffraction Apparatus

CHAPTER 3

Gas-Stream Cooling Apparatus

3.1. General Principles

X-ray diffraction samples can be cooled readily by blowing a stream of cold gas over the sample. The systems that have been described in the literature range from crude gas-stream devices, useful for preliminary survey studies, that can be set up from common laboratory items to sophisticated, electronically controlled systems capable of operating unattended for long periods of time.

The first gas-stream cooling devices used in X-ray diffraction studies were reported by Cioffi and Taylor (1922) and by Eastman (1924), but fell into relative disuse until the method was rediscovered twenty-five years later (Kaufman and Fankuchen, 1949; Abrahams et al., 1950). Since then, many modifications and versions of the basic apparatus have been described in the literature (see listing under Apparatus Code, Second Digit, Number 1, in the Bibliography).

A specimen cooled in a flowing gas stream has only two means of raising its temperature above that of the gas stream: conduction through the specimen support and radiation (Robertson, 1960). In the case of a single-crystal specimen, these effects are rather small. A typical glass-fiber support, approximately 0.5 cm long and 0.02 mm in diameter, will conduct heat to the specimen at the rate of 0.002 mcal/s at 10 K. Below 100 K, a crystal of approximately 1 mm^3 volume can receive a maximum of 0.08 mcal/s from its surroundings (at 20° C) by radiation, with this figure reduced for higher operating temperatures (Figure 3–1).

On the other hand, Figure 3–2 shows the "cooling power" of a cold gas stream in terms of the heat capacity per unit volume, i.e., millicalories

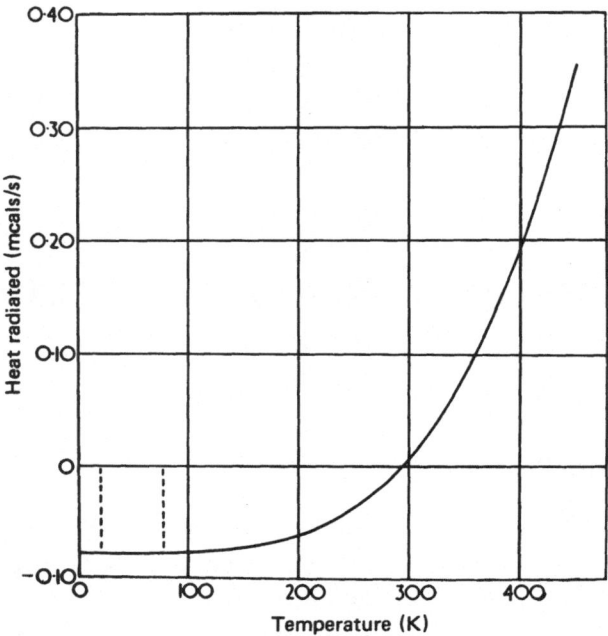

Fig. 3–1.   Heat radiated to a small crystal as a function of its temperature, the environment being at 20°C. Crystal size: height 2 mm, diameter 1 mm (Robertson, 1960). (Reproduced with the permission of the Institute of Physics.)

of heat taken up per cubic centimeter of gas, at atmospheric pressure, when the gas temperature is altered by one degree. It is clear that the amount of heat which could be absorbed by a stream flowing as slowly as 1 cm³/s, and rising only 1° in its temperature, is sufficient to cool a typical single-crystal specimen (even though the heat exchange between crystal and gas stream is not complete or instantaneous). Although powder specimens and sample holders are generally larger, rapidly flowing gas streams and properly designed specimen supports will give equally satisfactory results. Thermal gradients (ΔT) within the sample are not automatically ruled out, even though all parts of the surface are at the same temperature. However, these effects are more severe in heated samples than in cooled samples, and except for the case of large powder samples, an effective ΔT of less than 0.1° may be expected (Young, 1966).

In short, gas-stream cooling techniques have the advantages of simplicity, convenience, dynamic response, and excellent heat-transfer capabilities.

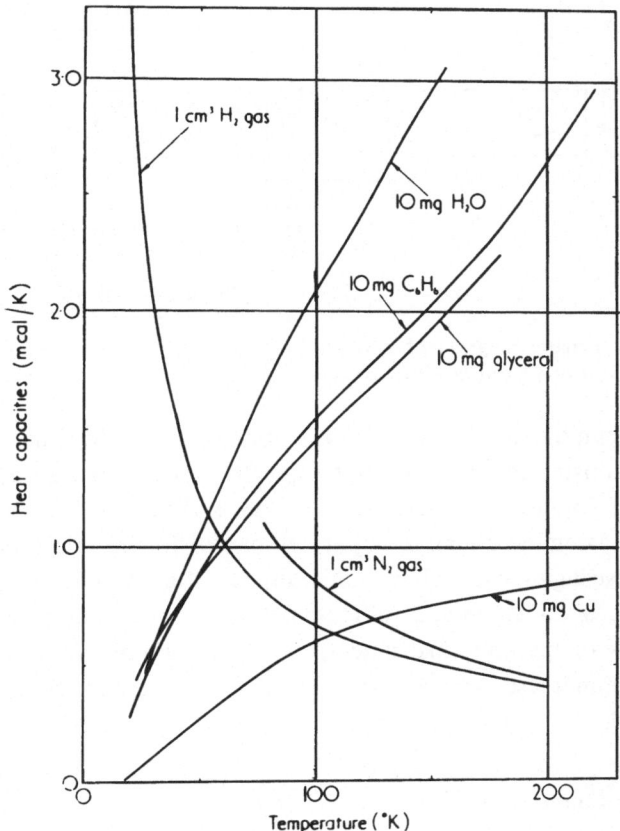

Fig. 3–2.   Heat capacities per unit volume of hydrogen gas
and of nitrogen gas compared with heat capacities of certain
solids at low temperature (Robertson, 1960). (Reproduced
with the permission of the Institute of Physics.)

A complete description of a satisfactory low-temperature gas-stream
instrument (Figure 3–3) must include provision for generating a stream of
cold gas (gas generator), directing it over a specimen (gas-delivery
system), controlling and measuring the specimen temperature, and preventing atmospheric vapors from condensing on the specimen. Several
devices and/or techniques have been described in the literature for each of
these components. Since the experimental difficulties encountered and the
instrumentation used are not necessarily unique to the field of X-ray diffraction, most LTXRD devices are based on contributions from several
sources.

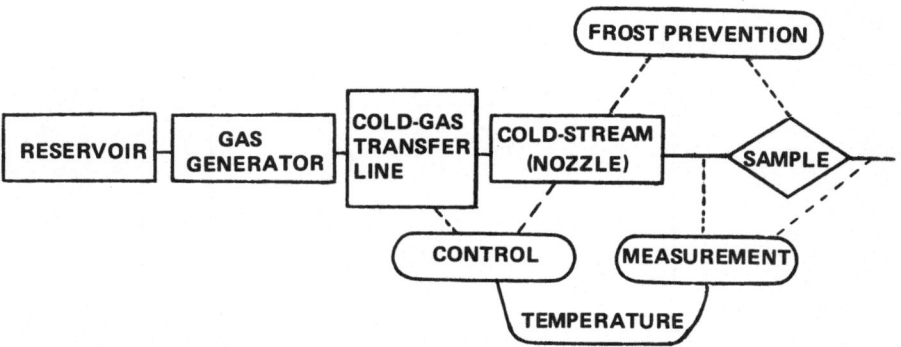

Fig. 3–3. Schematic diagram of a generalized gas-stream cooling apparatus. Components that can be placed at either of two points are indicated by dashed lines.

It is emphasized that the apparatus acquired by any laboratory should be constructed on a level of sophistication commensurate with the experiments for which it is to be used. As is shown in Section 3.3, it is possible to assemble a simple apparatus, useful for a variety of low-temperature investigations, with a minimum of effort and expense. On the other hand, if an automatic diffractometer is to be used in conjunction with the low-temperature instrument, the complexity and cost of the apparatus will be increased considerably. In all cases, the same basic principles are involved.

3.1.1. Gas Generator

The part of the system from which there flows a steady, uninterrupted stream of *cold* gas is called the gas generator.

Water vapor must be absent from the gas stream at all temperatures below 0°C, or the water will freeze in the line and block the gas stream. As the temperature is lowered, higher-boiling components of the gas must be removed. Thus, properly dried air can be used satisfactorily for temperatures above 130 K, while at lower temperatures gaseous nitrogen is recommended. (Liquid oxygen, which forms below 90 K, can spurt from the transfer line and affect the flow of cold gas.) At the lowest end of the temperature range (below 78 K), helium (or hydrogen) should be used. Other gases can be substituted, but they are usually not economical. Gas from a storage tank or compressor may be passed through a heat exchanger immersed in a cryogenic bath, or the gas may be generated by boiling a liquefied gas. In a heat-exchanger system, compressed air passed through a suitable drier (Appendix 10) may be used with the heat exchanger immersed in a dry-ice (CO_2) slurry of, e.g., acetone, propanol, or trichloroethylene (the latter is nonflammable). However, if a liquid-nitrogen bath is used, the most economical gas is dry nitrogen.

The advantages of using an external gas source with a heat exchanger are (a) simplicity of the apparatus and (b) the fact that the cold bath is not coupled to the gas-stream flow. The disadvantages are that one must maintain, i.e., replenish, both the cold bath and the gas stream. In the case of a dry-ice slurry, this entails the handling and crushing of 20–30 pounds of dry ice daily with a lower temperature limit of approximately $-65°C$ at the sample. When a liquid-nitrogen bath is used, the bath can be filled automatically, but the cold, boiling nitrogen is then lost to the atmosphere. In addition, if cylinders of gaseous nitrogen are used for the gas stream, they must be replaced regularly (a cylinder lasts 4–12 hours). These disadvantages are minor for short-term use, but can become bothersome and expensive for extensive, long-term investigations. Recently, mechanical refrigerators of sufficient capacity to maintain the bath at a constant temperature (to as low as 135 K) have become available commercially.

When liquid nitrogen is employed, the most efficient method is to use gaseous nitrogen obtained from boiling liquid nitrogen as the cold gas stream. The difficulty is that the temperature at the crystal varies with the pressure in the generator, and, unless suitable precautions are taken, large pressure fluctuations arise in the generator every time it is refilled with liquid nitrogen. This is caused by the higher pressure in the reservoir *vis à vis* the gas generator and the change in the volume of gas within the generator as liquid nitrogen flows in.

This latter difficulty is perhaps the most perplexing problem facing designers of gas generators. It has been solved in several ways.

The most obvious and effective solution is to use a very short refill cycle, so that the liquid level does not change very greatly, and to refill from a reservoir that is not at a very high pressure relative to the gas generator (Altona, 1964a,b; Renaud and Fourme, 1967; Enraf-Nonius, 1971; Fourme, 1975; Hardy, 1968).

A second approach, which is also very effective, is somewhat more expensive and complicated to design. This entails the use of a cryogenic solenoid relief valve which is wired into the refill controller. When the sensor calls for a refill cycle to begin, the first step is to open the solenoid relief valve, which releases all the gaseous nitrogen that formed in the reservoir-to-generator transfer line between filling cycles. This solenoid is set to close after a predetermined time period and then to actuate the solenoid valve between the reservoir and generator. In this manner, only liquid nitrogen enters the generator. A pop-off valve (set at a pressure only slightly higher than the operating pressure of the gas generator) maintains a reasonably constant pressure within the generator (Abowitz and Ladell, 1968).

A third method, which is often used in conjunction with one of the other methods, is to introduce the liquid nitrogen into the gas generator by means of a tube extending *below* the level of liquid nitrogen already present in the generator (Renaud and Fourme, 1967; Abowitz and Ladell, 1968; Rudman and Godel, 1969; Bolhuis, 1971). The use of a porous phase separator to aid in the smooth flow of liquid nitrogen is also recommended.

Other methods that have been used with success are discussed in Section 3.4.2. They include generating the gas in a separate chamber that is connected to the main chamber by a small opening (Hori and Matsuno, 1972; Burbank, 1973); generating the gas inside an inverted bell submerged in the liquefied gas (Robertson, 1960); warming the gas to room temperature, passing it through a sensitive pressure regulator, and then recooling it in a heat exchanger submerged in the gas generator (Silver and Rudman, 1971); and using an intermediate dewar from which the LN_2 is gravity-fed into the gas generator, thus reducing pressure surges (Thaxton and Jacobson, 1970).

Another problem that arises when LN_2 is used is the presence of ice crystals in the liquid nitrogen. Several investigators have reported that LN_2 obtained from standard commercial sources often contains small ice particles. These particles are carried along with the LN_2 into the generator and from there to the sample, where they are deposited. This manner of ice formation on the sample has been successfully avoided by inserting a wire screen [70-mesh (210-μm) to 2500-mesh (5-μm) screens have been recommended] in the transfer line between reservoir and generator to filter out the ice particles (Abowitz and Ladell, 1968; Frenz et al., 1969; Bolhuis, 1971). Many other investigators do not appear to have been bothered by this phenomenon; it probably depends on the local source of LN_2.

When boiling liquid nitrogen is used, natural heat leaks into the system are not sufficient to generate the required quantities of cold gas.

Therefore, a heater is usually immersed in the liquid nitrogen. This heater may be operated at a fixed setting or it may be controlled automatically by a pressure switch or by feedback from the thermocouple. In the case of liquid hydrogen or helium, a heater is normally not required. At least one investigator has used a battery to power the heater so that the crystal would not be lost in case of an electric power failure (Cruickshank et al., 1956).

To summarize, a properly designed gas generator furnishes a controlled flow of cold gas at a constant temperature and pressure throughout the entire experiment, *including* periods of cryogen replenishment.

3.1.2. Transfer Lines and Nozzle

The transfer line is an insulated tube through which the cold gas is transported from gas generator to crystal. The lowest temperature that can be reached at the specimen will depend to a large extent on the efficiency of the transfer line and is usually proportional to its length. A simple apparatus can be constructed using styrofoam or similar insulation (e.g., Armaflex tubing), while the most efficient systems use vacuum-jacketed metal or glass transfer lines. The glass lines are easier to construct and just as efficient as metal lines. However, the use of flexible metal transfer lines offers some advantages in positioning the line.

A detailed description of special nozzle designs will be found in Section 3.2, while the nozzle's proper location relative to the sample is discussed in Section 3.5.

3.1.3. Temperature Control

The temperature of the cold gas stream can be controlled in one of two ways: mixing the cold gas with a warmer gas or heating it with an in-line heater. In addition, a change in the flow rate will change the temperature.

When a mixer is employed, either a separate source of warm gas is used (Reed and Lipscomb, 1953) or some cold gas is warmed and then mixed with the cold gas in the nozzle (Rudman and Godel, 1969).

Huber (1969) improved upon this method by using the temperature measured by a thermocouple near the sample to control two valve actuators that open or close the warm and cold gas stream valves (Figure 2–8). A "warm sink," as well as a "cold sink," is used; the valves are located in the gas lines before the gas enters the heat exchangers (so that standard, noncryogenic valves can be used).

When a heater is used, it is generally placed in the nozzle, but it can also be located in the transfer line. It must be designed to warm the gas uniformly at the maximum flow rate of the system without disrupting the smooth flow of cold gas.

Henshaw (1957) used two heaters: one for "rough-warming" and one for "fine control." The latter is a Pt-wound heater which also serves as a temperature sensor. Darlington and Megaw (1973) used a temperature-sensing diode at the end of the nozzle. The signal from this device was constantly compared with a preselected voltage, corresponding to some chosen temperature, via two balanced field-effect transistors. The output of this device determined the voltage across the heater used to generate cold gas. In another variation, the on–off operation of the heater is controlled by a pressure switch so as to maintain a constant pressure in the gas generator (Rudman and Godel, 1969).

The temperature at the *end* of the nozzle depends on the efficiency of the generator and transfer lines. If the production and flow rate of cold gas is steady and the transfer lines are suitably insulated, the temperature will be constant to within 0.1°. Reproducible settings may be obtained by using flowmeters in the gas lines (Kreuger, 1955; Harding, 1956).

The temperature difference between the end of the nozzle and the sample depends on the distance from the end of the nozzle to the sample (Figure 3–4), the ambient temperature, the relative positions of the nozzle, collimator, and sample, the flow rate of the gas stream (Figure 3–5), and the presence of air currents in the working area. The normal nozzle-to-sample distance is between 0.3 and 1.0 cm.

In air-conditioned laboratories, air currents can be strong enough to adversely affect the flow of cold gas in the vicinity of the sample. In addition to bringing moist air into contact with the sample, these currents will cause a large temperature gradient between the end of the nozzle and the sample. Techniques to remedy this situation are discussed in Appendix 9.

3.1.4. Temperature Measurement

The fundamentals of temperature measurement are given in Appendix 1. A copper–constantan thermocouple is recommended for use between ambient and liquid nitrogen temperatures, although iron–

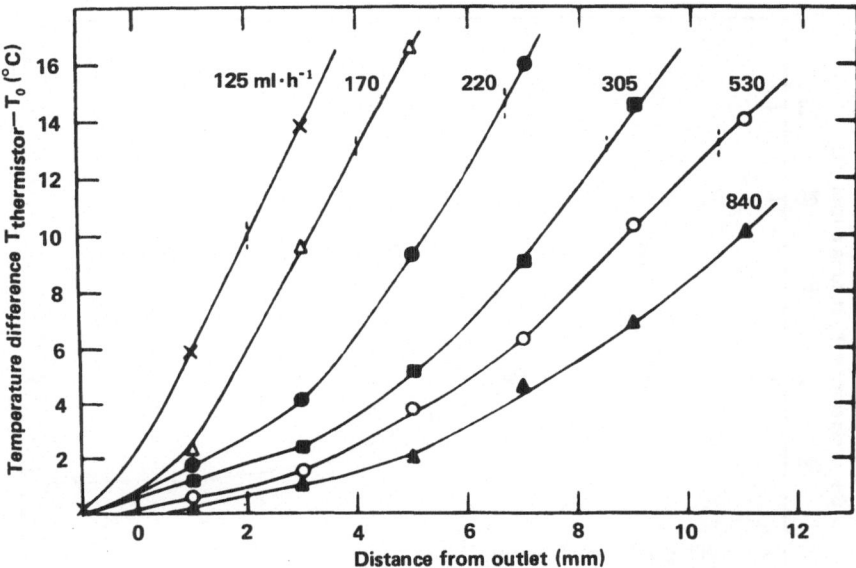

Fig. 3–4. $(T_{\text{thermistor}} - T_0)$ as a function of distance from the tube outlet for various rates of boiling of liquid nitrogen. T_0 is the gas temperature at the outlet of the delivery tube (Altona, 1964a,b). (Reproduced with the permission of the International Union of Crystallography.)

constantan is also linear above 120 K. In order to minimize the effect of heat conduction away from the thermocouple tip, very thin wires (0.08 mm diameter) should be used (Young, 1966). It is also important to place the thermocouple inside the transfer line. Otherwise, especially if the thermocouple wires are larger than 0.08 mm in diameter, temperature readings may be off by as much as 20°. When the thermocouple is not placed in direct contact with the sample (e.g., for a single-crystal specimen), it is necessary to determine the temperature difference between the position of the thermocouple and the crystal (see Section 8.1.2).

LaCour (1974) placed the thermocouple between the crystal and the detector so as to minimize turbulence at the cold gas-stream outlet. Other investigators (e.g., Thewlis and Davey, 1955) use two thermocouples to measure any thermal gradients that may be present. Burbank (1973) calibrated the temperature in terms of the power settings on the three heating elements in his system and assumed a one-to-one correspondence between the calibrated power settings and the desired temperature during the experimental investigations.

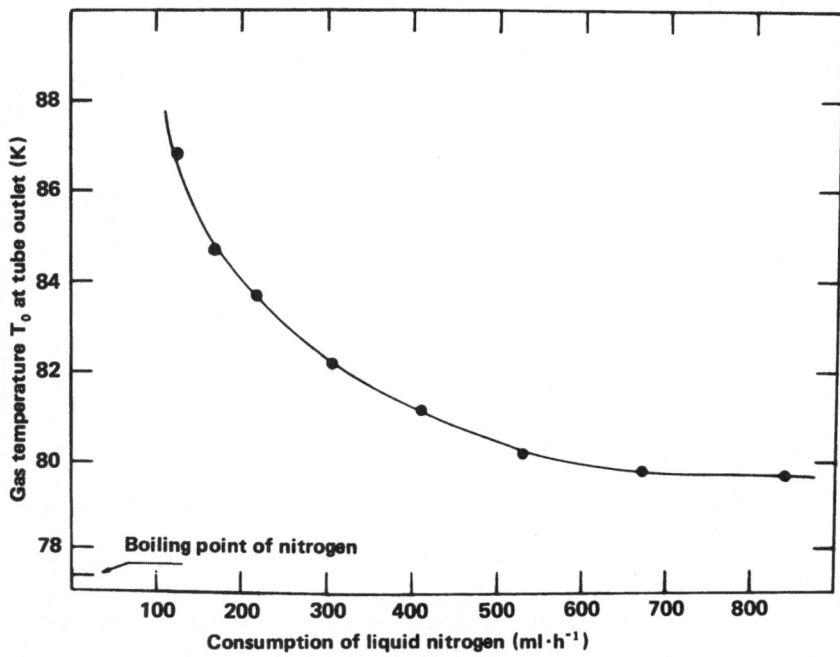

Fig. 3–5.   Gas temperature (T_0) at the outlet of the delivery tube vs. consumption of liquid nitrogen (Altona, 1964a,b). (Reproduced with the permission of the International Union of Crystallography.)

3.1.5. Frost Prevention

Any ice forming on the sample or in the path of the X-ray beam will absorb the diffracted X-rays, give spurious reflections or powder lines, possibly disturb the crystal alignment, and prohibit accurate measurement of relative intensities. The latter will occur because the formation of ice will not be uniform over the sample and also because a steady buildup of ice will result in changes in the amount of ice present during the course of the data collection. The revival of the gas-stream method in the early 1950s was due to the incorporation of a means for providing a concentric sheath of warm gas around the cold gas. This outer stream can be obtained from a separate gas source, or a portion of the output of the gas generator can be warmed to room temperature. Details of the design and function of the warm outer stream are found in Section 3.2. Several other methods for preventing frost from condensing on the sample have been developed and are discussed in Appendix 9.

3.1.6. Rate of Cryogen Consumption

It is interesting to note that in most of the systems that have been described in the literature, all other factors being equal, the consumption of cryogen (in particular, liquid nitrogen) is in direct proportion to the size of the system. Even though the *lowest* attainable temperature (for a given cryogen) is primarily a function of the efficiency of the insulation on the gas generator and transfer lines, and similar temperatures have been attained for a variety of instruments, the *rate of consumption* of liquid cryogen has been found to increase with increasing volume of the system that must be cooled. The largest systems use approximately 8 liters of liquid nitrogen per hour, (including the outer stream) while the smallest systems report a consumption of less than 1 liter per hour. On the other hand, the smaller systems tend to be limited to one particular mode of operation or instrument, while the larger ones are more versatile and can be adapted for a variety of applications and X-ray instruments. However, in light of recent advances in the design of cryogenic components and the use of a nozzle heater to replace the outer stream (see Section 3.2.2), consumption rates under 2 liters per hour are becoming more common.

3.1.7. Gas-Stream Systems

An efficient *system* for cooling X-ray diffraction samples incorporates the components that have been described above, although the key to the success of this technique is the proper design and positioning of the nozzle (see Section 3.2).

There is no one system than can be used for every low-temperature X-ray diffraction application, and, in fact, it would be a waste of resources to attempt to design such a system. The designer of any system must, therefore, know what his most likely applications will be and construct the system accordingly.

Some of the questions to be considered when designing a system are:

What is the minimum operating temperature?

What is the size of the sample and sample holder?

What is the maximum length of time that the apparatus will be operated continuously?

Will a manual, semiautomatic, or fully automatic system be needed? (This refers to temperature control as well as to cryogen replenishment.)

Will the apparatus be used on one X-ray instrument only or must it be adaptable to several instruments?

How often will it be used?

What type of samples are to be studied?

Several examples follow.

a. A protein crystallographer operating at temperatures above −20°C need not concern himself with attaining temperatures of −170°C and would be advised to consider using a mechanical refrigerator (see Section 3.4.1.1) rather than a liquid-nitrogen generator (see Section 3.4.2).

b. A system used for studying liquids frozen inside capillary tubes must have provision for preparing single crystals or randomly oriented powders. These requirements are not necessary for the investigation of materials which are solid at room temperature.

c. Precise temperature control, which is necessary for the study of thermal expansion and phase transitions, may not be a critical factor in routine low-temperature crystal-structure analyses. On the other hand, crystal-structure analyses in which the primary concern is to determine the bonding electron density will require more careful temperature control (and usually lower temperatures) than those in which a knowledge of molecular conformation is the primary objective.

d. Systems employing boiling liquid hydrogen and liquid helium require more efficient insulation than those utilizing a heat exchanger immersed in a dry-ice slurry.

e. The low-temperature examination of powder samples, for relatively short periods of 1 to 24 hours, can be conducted with apparatus much simpler than that required for collecting data on an automatic single-crystal diffractometer.

As a final example we can cite the criteria used by Burbank (1973) in designing his apparatus:

1. To allow the growing of single crystals from the vapor or liquid phase enclosed in capillary tubes which may be several centimeters in length.
2. To cool a volume of space large enough to enclose the capillary tube.
3. To maintain a desired temperature for indefinite periods of time.
4. To operate automatically.
5. The consumption of LN_2 is not a limiting factor.

Items (1) and (2) will determine the size of the nozzle opening; item (3) determines the degree of sophistication of the temperature control system; item (4) affects the choice of LN_2 refill device; and item (5) reflects the economics of an organization in which consumable materials are readily available but assisting personnel are not (the reverse of the usual economics in a university laboratory). Once the set of system requirements has been properly defined, the correct choice of components is quite straightforward.

The first gas-stream system, described by Cioffi and Taylor (1922) and used in the study of the crystal structure of mercury (McKeehan and Cioffi, 1922), and the similar system described by Eastman (1924) for an investigation of crystalline benzene contained all the features found in modern gas-stream systems, with the exception of the warm outer stream. However, the gas-stream method was overshadowed by the use of the conduction and immersion techniques and fell into virtual oblivion until it was rediscovered and popularized during the early 1950s.

Cioffi and Taylor (1922) described a system in which liquefied gas from a reservoir enters a gas generator through a funnel whose lower end is below the level of the liquefied gas. A heater is used to generate cold gas, which leaves the system through a vacuum-insulated delivery tube (Figure 3–6). Although they did not incorporate it in this system, Cioffi and Taylor suggested controlling the temperature by automatic regulation of the heater. Frost is prevented from condensing on the sample by covering the sample with a thin-walled tube *or* by drawing the cold gas into a heated tube leading to a roughing vacuum pump. The use of the pump will keep the gas moving past the sample and prevent the formation of ice. Temperatures as low as 93 K are said to have been attained.

The system described by Eastman (1924) differs in that he generated the gas directly from the reservoir and surrounded the sample with two concentric celluloid tubes (total thickness: 0.2 mm) in order to prevent frost formation. He also developed the currently used method of periodically melting and refreezing the powder sample during the course of the exposure so as to insure a random orientation of the crystallites.

Gas streams were also used by Féher and Klotzer (1935) for powder samples, by Campbell and Hildebrand (1943) for a study of liquid xenon to 160 K, and by Ubbelohde and Woodward (1946) for the study of a single crystal. In Ubbelohde and Woodward's device, the goniometer head and crystal were placed inside an insulated copper tube through which the

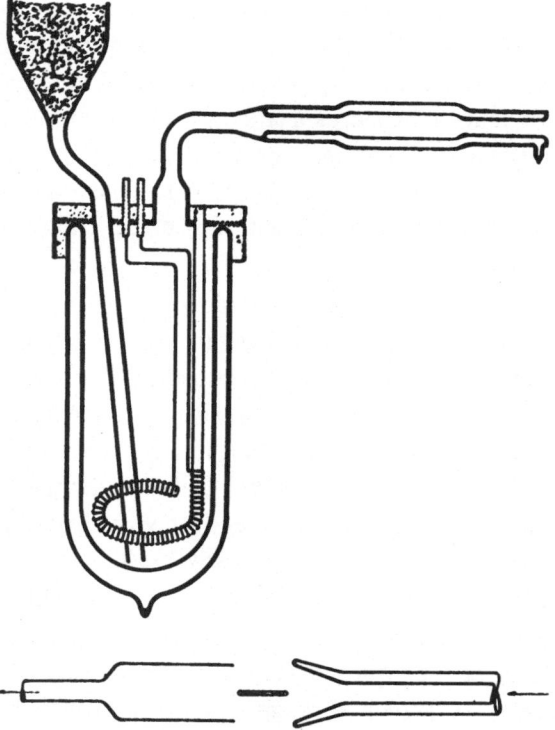

Fig. 3–6. (*Top*) Schematic diagram of the first gas-stream cooling apparatus for X-ray diffraction investigations. Liquid air is added to the dewar through a funnel stuffed with cotton (to filter out ice or other particles in the liquid air). The heater used to boil the liquid air and the insulated transfer line through which the cold gas is conducted to the sample are also shown. (*Bottom*) Diagram illustrating use of vacuum pump to draw cold gas past the sample (Cioffi and Taylor, 1922).

cold gas was moving. Cellophane-covered windows were located in the copper tube for the passage of X-rays. [This arrangement was later modified (Rhodes, 1951) so that a long strip of X-ray film could be used and 15 exposures could be taken without reloading of film.] Ubbelohde and Woodward (1946) used an outer cold stream concentric with the cold inner stream in an attempt to protect the cooling chamber. However, both these gas streams were contained within the tubes outfitted with cellophane windows (Figure 3–7).

In 1951, Post, Schwartz, and Fankuchen designed the nozzle which

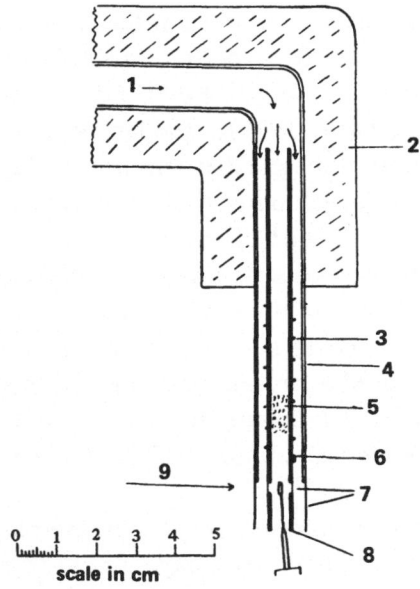

Fig. 3–7. Cold-air chamber. (1) Cold dry oxygen; (2) heat insulation; (3) heating coil; (4) glass shielding tube; (5) silver gauze to equalize gas temperature with copper chamber (6) main thermocouple junction; (7) cellophane; (8) copper chamber, walls 0.8 mm thick, internal diameter 4.5 mm; (9) X-ray beam (Ubbelohde and Woodward, 1946). (Reproduced with the permission of the Royal Society, London.)

incorporates a warm outer stream and which has been used so successfully in numerous single-crystal structure analyses. It is described in greater detail in Section 3.2.

Interestingly, the use of a flow of warm gas to prevent ice condensation had been described much earlier by James and Firth (1928). They placed a heater around the base of the sample-containing dewar so that the "up-current of warm air kept the flask free from any trace of moisture or frost."

Some of the more recently developed gas-stream devices have reverted to the earliest design: cold gas is used to cool a sample enclosed in a chamber (e.g., Sugino et al., 1973; Petsko, 1975). This has been made possible because of the development of thin, rugged window materials such as Mylar and because of the ease of using computer-calculated absorption corrections.

3.1.8. Protection of X-Ray Apparatus

Several methods have been used to prevent the cold gas stream from cooling parts of the X-ray generator and diffraction equipment that are in its path. One method is the use of electrical heating tapes, first described by Post et al. (1951). A second popular method is to blow a stream of warm air, generated by a hot-air blower, across the surface of the apparatus in the path of the cold gas stream (Hume-Rothery and Strawbridge, 1947; Clifton, 1950; Fridrichsons and Mathieson, 1958; Cole and Holmes, 1960; Kramer et al., 1963; Burbank, 1973). Another method that is used occasionally is to let water run through tubes attached to the camera (Hovi et al., 1964). The original suggestion of Cioffi and Taylor (1922) to draw the cold gas stream into the inlet of the exhaust pump (Figure 3–6, bottom) has been incorporated into several powder cameras (e.g., Campbell and Hildebrand, 1943; Dunoyer, 1952; Bouttier and Dunoyer, 1953) and at least one single-crystal system (Sugino et al., 1973). Verschoor and Keulen (1971) placed a large-diameter tube in the cold gas-stream path to direct the gas away from the instrument, while Thewlis and Davey (1955) use a hemispherical cup to break up the gas stream.

Another effective method of preventing condensation of moisture on cold parts of the apparatus is to coat them with a silicone spray.

3.2. The Cold-Stream Nozzle

The current widespread use of the gas-stream method can be traced to a series of papers that appeared in 1949–1951. In 1949, Kaufman and Fankuchen published a note in which they described the use of an evacuated, double-walled glass tube to deliver a stream of cold gas onto a sample surrounded by a cellophane container. This device was very similar to the Eastman (1924) system and was sufficient to prevent the formation of ice on the sample for temperatures as low as −50°C. In a paper appearing in 1950, Abrahams, Collin, Lipscomb, and Reed described a similar delivery tube which was operated in a dehumidified room without a cellophane cover. Finally, in 1951, Post, Schwartz and Fankuchen published what is probably the most widely quoted gas-stream reference, in which they reported the first use of a room-temperature outer gas stream concentric with the cold gas stream, which permitted frost-free operation at temperatures as low as 90 K. The use of the concentric warm outer stream resulted in a streamlined, simple system with no absorption errors caused by containers surrounding the sample (Figure 3–8).

The basic principles of single-crystal growth and alignment, *in situ,* at low temperatures, were also described in these three papers. The designers of the many gas-stream systems that have been described since then have

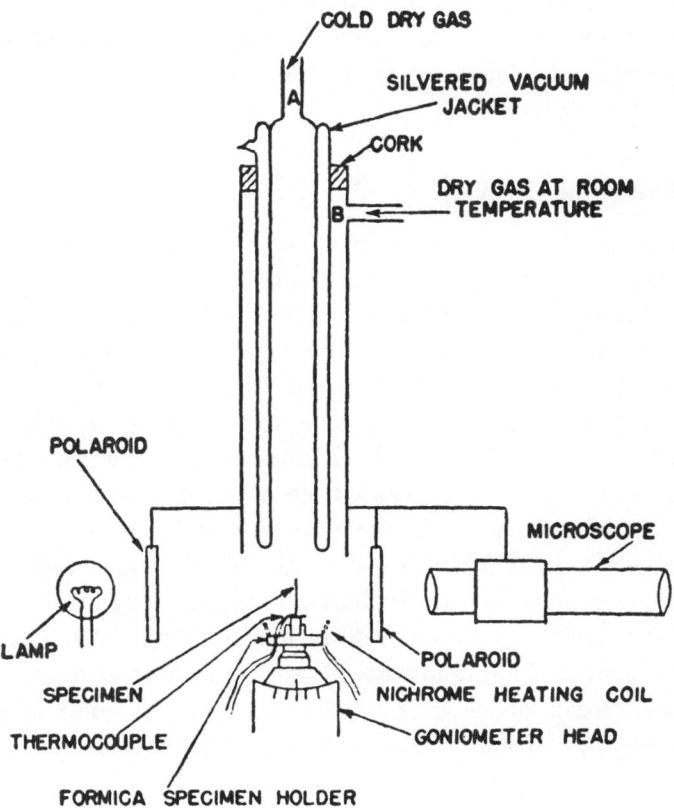

Fig. 3–8. First gas-stream cooling apparatus to employ a concentric outer stream of warm, dry air around the cold gas stream.
The warm outer stream prevents atmospheric water vapor from condensing on the sample. The use of crossed polaroids as an aid in
examining and aligning crystals grown *in situ* is also shown (Post
et al., 1951). (Reproduced with the permission of the American
Institute of Physics.)

concentrated on optimizing the nozzle shape, adapting the method to various X-ray diffraction instruments, and improving overall system efficiency
for long-term, uninterrupted, and automatic operation. More recently,
metal and foamed plastic nozzles have been described.

Several aerodynamic problems are encountered when operating with
two concentric gas streams; the possibility of turbulence at the interface of
these gas streams and the resulting intake of moisture-laden air into the
cold gas stream must be considered.

The overall efficiency of the nozzle will depend on the flow rate, the absolute and relative sizes and shapes of the concentric inner and outer gas-stream tubes, the thickness of the tube walls, and the operating temperatures of the system.

3.2.1. Size of the Opening

The inner diameter of the cold gas-stream tube (Figure 3–9) should be as small as possible for the sample being studied. Thus, if a small single crystal mounted on the end of a fiber is being cooled, the opening can be as small as 5–6 mm (i.d.), resulting in a low flow rate and economic use of cryogen. On the other hand, if the crystal is mounted inside a thin-walled capillary tube, the opening must be larger to allow for various orientations of the crystal (Figure 3–10). This is particularly important when a crystal is grown from a liquid *in situ,* and the exact location of the crystal within the tube cannot be known in advance. (However, see Section 6.2.3 for a discussion of techniques used to mount short sections of these tubes.) Similarly, a large opening will be necessary for cooling a sample for a powder diffractometer. Any system designed for use with a variety of X-ray instruments should have several interchangeable nozzles. Nozzle openings larger than 25 mm will result in excessive cryogen consumption and turbulence of the cold gas stream.

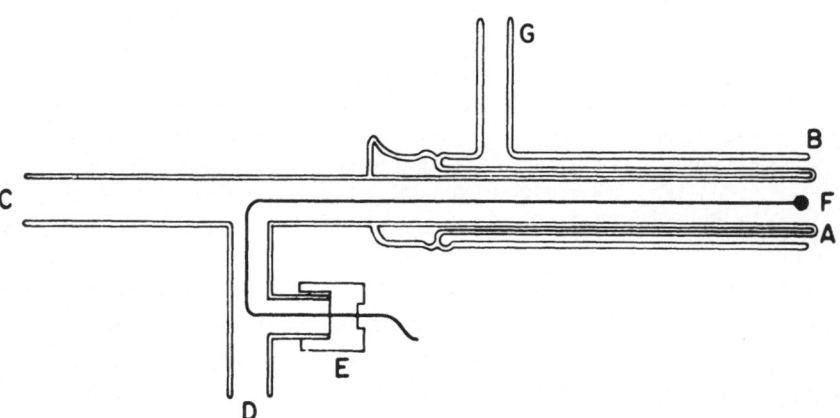

Fig. 3–9. Cooling nozzle. (A) Silvered and evacuated double-walled glass tube; (B) outer stream; (C) cold gas inlet; (D) warm gas inlet; (E) rubber serum cap; (F) thermocouple; (G) outer stream inlet (Rudman, 1967). (Reproduced with the permission of the *Journal of Chemical Education.*)

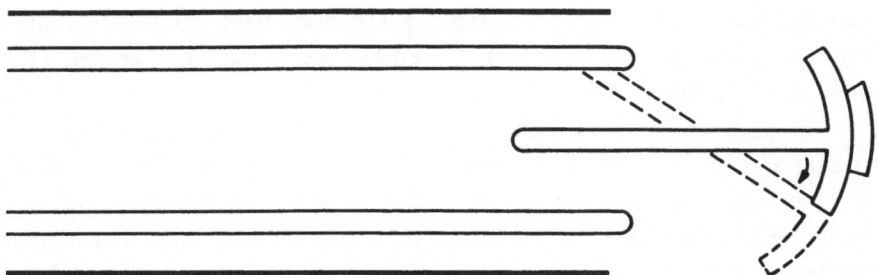

Fig. 3–10. Schematic drawing of cold-stream nozzle showing sample tube in two positions. The diameter of the nozzle opening must be wide enough to allow for various orientations of a long capillary tube containing the sample.

3.2.2. Outer Stream

The evacuated portion of the nozzle (A, Figure 3–9) should be as narrow as possible, so that the warm outer stream can sheathe the cold stream without turbulence setting in as the two streams meet. For this same reason a minimum clearance of 3–5 mm between the outside of the inner, evacuated tube (A) and the inside of the outer-stream tube (B) should be allowed.

In the diagram in their original paper, Post et al. (1951) showed the outer gas-stream tube extending past the end of the cold gas-stream tube (Figure 3–8). However, this led to turbulent mixing of the two streams and eventual formation of ice on the sample. The problem of ice formation is minimized if the outer stream is recessed between 2 and 4 mm (Burbank and Bensey, 1951, 1953; Burbank, 1973).

The aerodynamics of concentric gas streams as applied to this situation has been investigated by Young (1966). He found that the key to the success of the gas-stream technique is laminar flow in the central region of the cold gas stream. This minimizes lateral gradients, ensures dependable heat transfer from the specimen, and prevents surrounding gases from mixing far enough into the stream to reach the specimen. It is important that both the inner, cold gas stream and the outer, warm gas stream move at approximately the same velocity so as to minimize turbulence at their contact surface. In fact, an outer stream of dry gas moving at a velocity different from that of the main cold gas stream would be a detriment.

A technique for adjusting the flow rates so as to minimize interference and diffusion between the inner and outer gas streams has been described (Enraf-Nonius, 1971). A light refraction method, utilizing the

fact that the low-density cold nitrogen gas refracts light less than the surrounding room-temperature gases, is used to observe the inner stream. The
method is as follows.

A parallel beam of light is directed nearly perpendicular to the
stream and the shadow of the stream on a piece of paper or etched glass
is observed. The influence of different flow rates on the shape of the
undisturbed stream and the turbulence behavior is clearly seen. The higher
the stream velocity, the lower the diffusion process and disturbance of air
currents. In order to achieve low temperature gradients, one must adjust
for the minimum interference conditions at high flow rates commensurate
with the experimental conditions. Too high a flow rate will disturb the
sample, as well as waste cryogen. Optimum settings can result in a minimum thermal gradient near the crystal.

Several investigators have described a novel approach to the problem
of forming two concentric gas streams at different temperatures but of the
same velocity. Rather than use a separate warm outer stream, the outer
part of the cold gas stream is heated, while the inner part of the *same* gas
stream remains cold. Satisfactory results are claimed down to a temperature of 100 K.

Amoros et al. (1962) and more recently Fourme (1975) used two
heaters, one above and one below the sample, to set up an envelope of
dry air around the cold air stream. In Fourme's device the second heater
is attached to the goniometer head just below the sample and is adjusted
so that the goniometer head is at room temperature.

Previously, Renaud and Fourme (1967) had used a heated stainless
steel insert at the end of the nozzle, while Hospital (1968) and Thaxton
and Jacobson (1970) wrapped a resistance heater over the outside of the
nozzle. Hope et al. (1973) used a brass cone in which an electrical heating element is embedded in ceramic. This latter method was incorporated
in the Syntex (1974) low-temperature system (Figure 3–11). Boiko et al.
(1972a) used a novel approach by splitting the cold gas stream and heating the outer stream with a circular resistance heater (Figure 3–12).

Dietrich (1968) and, more recently, Huffman et al. (1973; Huffman,
1974) have described a different method for minimizing the interaction of
the two gas streams (and thereby minimizing icing effects). They suggest
tapering the outer stream so as to direct it toward the crystal position
(Figure 3–13). Dietrich uses a pair of plastic inserts to modify the standard nozzle. The inner cone-shaped device is constructed from a cylinder of

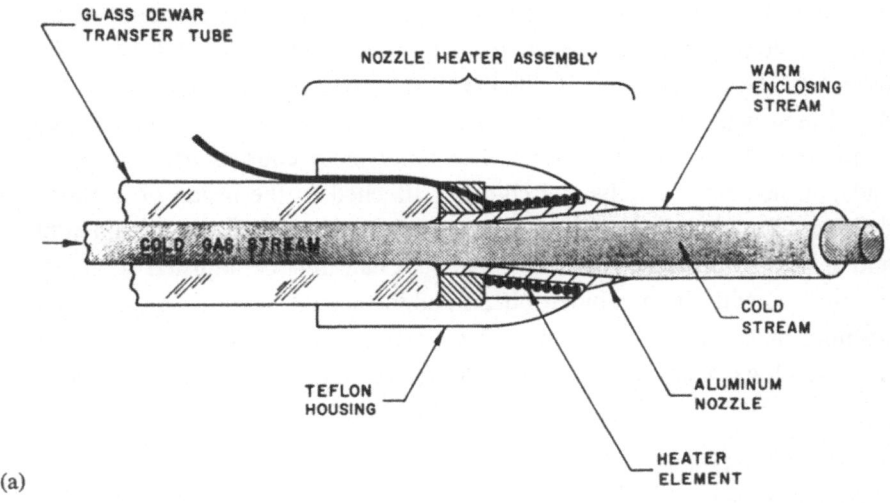

(a)

(b)

Fig. 3–11. (a) Diagram of cold gas stream with a concentric warm outer stream which is formed by heating the outer surface of the cold gas stream (Syntex, 1974, after Hope et al., 1973). (b) Photograph of low-temperature system described in (a) mounted on a full-circle diffractometer. (Courtesy of Syntex Analytical Instruments.)

thin plastic foil with an outer diameter equal to the inner diameter of the cold-gas outlet. A cone formed from thin polyester is fitted around this tube, and the empty space is filled with polystyrene insulation. The cylindrical tube is fitted into the inner tube of the nozzle, and the cone is taped to the outside of the dewar tube. The outer piece consists of a machined and polished ring of polyvinyl chloride attached to the inside of a plastic cylinder. The cylinder, in turn, fits over the outer warm gas-stream tube and is held in place with a rubber band. It can thus be adjusted so as not to interfere with the goniometer head and/or crystal mount. An inner tube opening of 6 mm with a 9-mm distance from the end of the opening to the crystal was used. The end of the outer stream adapter extends about 4

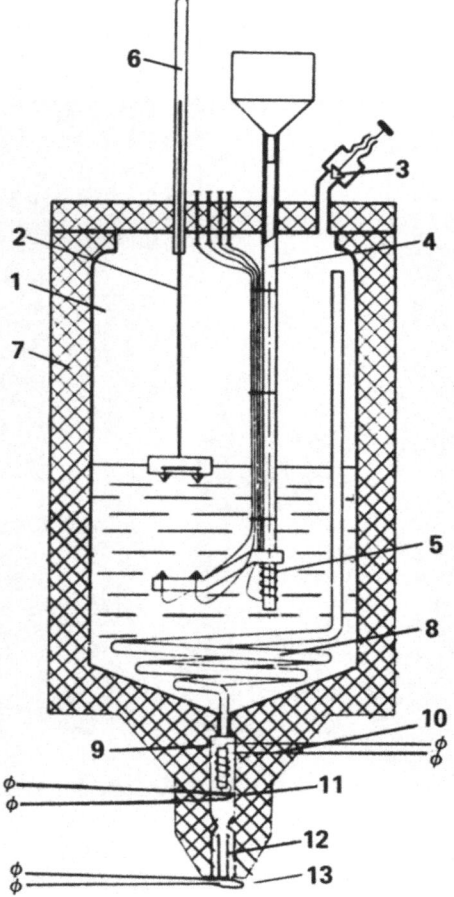

Fig. 3–12. Schematic diagram of low-temperature gas-stream apparatus (Boiko et al., 1972a). (1) Cold nitrogen gas; (2) float; (3) gas valve; (4) filler tube for liquid nitrogen; (5) boil-off heater; (6) level-indicator sight tube; (7) insulated dewar; (8) copper-coil heat exchanger; (9) chamber for controlling temperature of cold gas; (10) heater for controlling temperature of cold gas; (11) fine control of temperature of cold gas; (12) cylindrical tube; (13) heater. Note that the cold gas stream is split into two concentric streams by cylinder (12). The outer stream is warmed by the circular heater (13) to form a warm outer stream.

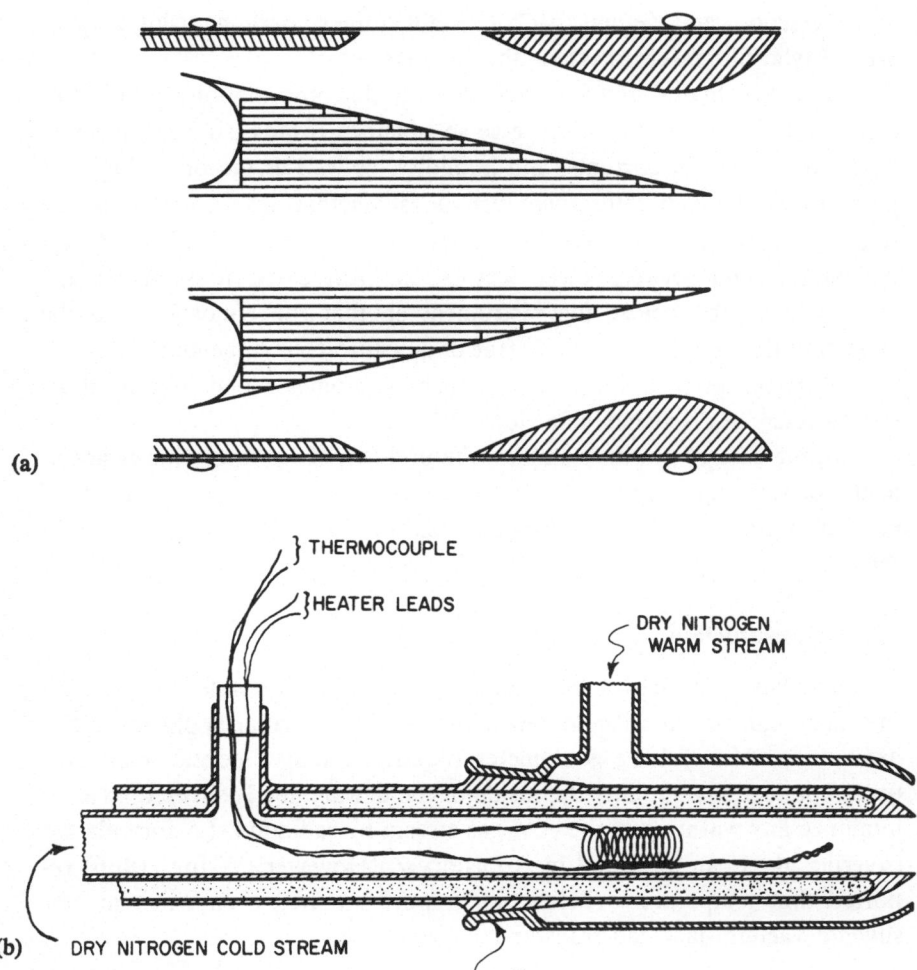

Fig. 3–13. (a) Plastic insert used to modify the flow of cold gas (after Dietrich, 1968). (b) Tapered cold stream designed by Huffman (1974).

mm past the tip of the nozzle, but turbulence is minimized since the two streams meet at the tip of the inner cone. According to the author, best results are obtained, in this case, when the velocity of the outer stream *exceeds* that of the inner stream.

Huffman et al. (1973; Huffman, 1974) described a straight inner tube with the glass tube of the outer stream curved inward, resulting in a tapered warm gas stream. This device, made entirely of glass, had an 8-mm-diameter opening, with the crystal located 5 mm from the tip.

Verschoor and Keulen (1971) modified their nozzle by using a cylindrical Mylar film insert to extend the outer stream.

It is clear that although a variety of nozzle conformations, tube diameters, and crystal positions will give satisfactory results, slight changes in flow rates, room air currents, and humidity, as well as the operating temperature of the apparatus, can turn a satisfactory instrument into one which gives unsatisfactory results. No specific formulation can be given for the interrelation between these factors. In most cases, it is easiest and most efficient to enclose the instrument so that the atmosphere in the vicinity of the sample is relatively free of moisture (see Appendix 9).

Nozzles designed for a Weissenberg goniometer may not need an outer stream, as discussed in Section 3.5.1.

In the case of powder samples, the problem of ice formation is not as acute, because the length of time the sample is cooled is usually measured in hours rather than in days. However, an outer stream is needed, even in this case.

3.2.3. Construction

The best material for fabricating the cold-stream nozzle is glass, with the exact diameter and length depending on the particular application. For example, a Weissenberg goniometer requires a longer nozzle than most other applications. Usually the inside of the dewar tube is silvered so as to minimize any warming of the cold gas stream by radiation. (A formula for silvering glass can be found in Appendix 7.) However, Young (1966) reported that temperatures below 90 K were attained even when an unsilvered vacuum-jacketed transfer tube was used.

If necessary, rubber or glass tubing wrapped with fiberglass or foamed polyurethane insulation can be used in place of the vacuum-jacketed dewar tube. In this case, Young reports that a temperature of 110 K was reached under conditions similar to those reported above. Nozzles of Teflon (Gopalakrishna and Cartz, 1972), foamed polyurethane (Marsh and Petsko, 1973), and metal (Chawdhury, 1968; Rudman and Godel, 1969; Silver and Rudman, 1971; Burbank, 1973; Darlington and Megaw, 1973; Fourme, 1975) have also been described.

Nozzles which are entirely of metal offer several advantages in design and ruggedness. A metal nozzle can be fitted with a vacuum-jacketed bayonet connector and used in conjunction with a similarly outfitted flexible metal transfer line (Silver and Rudman, 1971). The vacuum-jacketed bayonet fitting at the end of a flexible metal transfer line can also be fitted with an adapter to conduct the warm outer gas stream (Rudman and Godel, 1969).

One of the critical requirements for proper operation of the nozzle is the need for a uniform temperature gradient through the cross section of the gas stream. The inner-stream gas flow should not be so fast that it leaves the tube before it has had a chance to fill the tube. A favorite method to ensure a uniform gas flow is to put a plug of glass wool, silver gauze, or wire mesh into the inner-stream tube (e.g., Ubbelohde and Woodward, 1946; Frenz et al., 1969).

Sugino et al. (1973) reported that the temperature gradient is very sensitive to the packing of the glass wool and to the flow rate of the cold gas.

Kreuger (1955) designed an inner tube with several bulbs along its length (Figure 3–14). These expansion bulbs, which had also been described by Cioffi and Taylor (1922), serve a twofold purpose: they ease the tension caused by thermal contraction with respect to the outer tube, and they improve the uniformity of the gas stream flowing through the tube.

The outer, warm-stream tube should be recessed about 3 mm with respect to the inner vacuum-jacketed tube. An insert, such as that described by Dietrich (Section 3.2.2), can be used if found necessary, without any modifications of the nozzle itself.

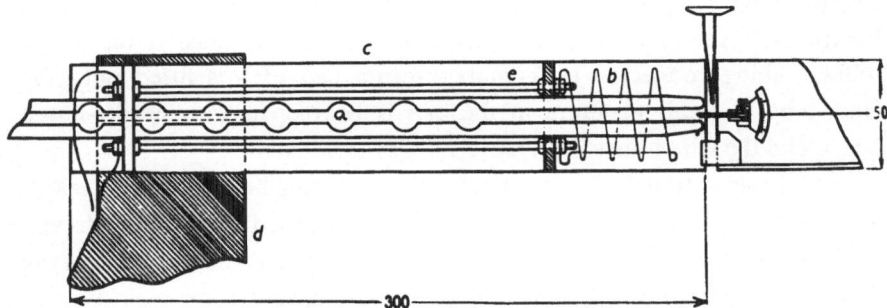

Fig. 3–14. Evacuated dewar tube with expansion bulbs (a) shown mounted on Weissenberg goniometer (Kreuger, 1955). The cold gas is heated (b) prior to leaving the compartment formed by the layer-line screen (c) which is attached to the Weissenberg goniometer at (d) and supports the dewar tube by means of two spacers (e). Dimensions are in millimeters. (Reproduced with the permission of the International Union of Crystallography.)

A difficulty often encountered in the outer stream is that the warm gas which enters the tube through a single opening tends to remain on one side of the tube and does not form a uniform sheath concentric with the cold gas stream. This phenomenon has been observed most often in short nozzles which have even shorter outer warm-stream tubes.

Here, too, the insertion of a 100-mesh wire screen (Frenz et al., 1969) or a porous cotton plug into the opening serves to spread the outer gas stream uniformly without halting it entirely. Sam LaPlaca (1965, private communication) has suggested two other methods of resolving this problem, both of which have been incorporated into several nozzles: (1) The warm gas enters through *several* inlets located around the circumference of the tube (Rudman and Godel, 1969). (2) The entrance tube for the warm stream is directed toward the *back* of the tube. This results in the breaking up of the air stream and uniform filling of the tube with cold gas (Bolhuis, 1971).

The thermocouple used to measure and/or control the temperature of the gas stream should be placed inside the nozzle. The wires can be pulled through the opening and placed as close as possible to the sample without touching it or interfering with the X-ray beam.

The temperature of the gas stream is controlled either by a heater located within the cold gas stream (which can also serve to distribute the cold gas uniformly over the inside of the tube) or by mixing a warm dry gas stream with the cold gas. In either case, provision must be made in the design of the glass or metal nozzle for the insertion of these wires and/or warm gas stream. An efficient design for a glass nozzle is shown in Figure 3–9 (the extra arms can also be formed from glass T-tubes and rubber tubing connectors). A rubber serum cap (E) is fitted onto the glass tube, a small puncture in the cap is held open with needle-nosed pliers, and the wire is inserted.

If a metal transfer line is used, the wires can be sealed with epoxy into a small tube that fits into an O-ring seal attached to the cold stream.

Experience has shown that if the apparatus is to be run for a long time at low temperatures, some ice will nearly always form regardless of the design of the nozzle. The exact conditions are difficult to predict, but if a run below 140 K and lasting more than two weeks is contemplated, the use of some auxiliary precautions, such as a cover over the apparatus (Appendix 9), is advised.

3.2.4. Calibration of Temperature Gradient

The temperature of the gas at the outlet of a typical nozzle (8 mm diameter) varies from a minimum of approximately 90 K at the center of the gas stream to room temperature (293 K) at its outer edge. Thus there is a temperature change of 200° in a distance of 4 mm. Clearly, the possibility of sharp and uneven thermal gradients near the sample must be examined. Although a properly designed nozzle will minimize these effects, the thermal gradients in the vicinity of the sample should be known.

A thermocouple constructed from very thin wires (0.08 mm) should be arranged with the thermocouple leads pointing downstream and away from the nozzle opening. This thermocouple should be attached to a mechanical stage with xyz motion. The profile of the opening at several temperatures and flow rates should then be determined. Typical curves are shown in Figure 3–15.

If a reasonably large isothermal plateau is not obtained, then the design of the nozzle should be reexamined.

Possible difficulties include nonuniform gas streams (cold or warm), obstruction of the inner tube (by the heater, thermocouple wires, or diffuser screen), and noncylindrical shape of the nozzle. The crystal position should coincide with the center of the plateau.

Plotting the temperature as a function of the distance from the tip of the cold stream is also recommended. In particular, if the nozzle is placed in a horizontal position, the cold gas stream, which is denser than the warmer surrounding air, tends to droop downward with increasing distance from the end of the nozzle (Burbank, 1973).

Special precautions must be taken when a full-circle diffractometer is used. If the nozzle follows the χ-circle, then at times the flow of cold gas is horizontal and at other times it is vertical. The temperature of the crystal should be checked at several positions of χ. Temperature variations have been found even when a fixed nozzle is used (M. Thomas, 1975, personal communication). This is due to the changing degree of gas-flow turbulence as χ changes. By carefully adjusting the direction and rate of the gas flow and by setting the end of the nozzle as close as possible to the crystal, temperature variations can be minimized.

Once the optimum position for the crystal is determined, the thermocouple inside the nozzle should be calibrated with respect to the position of the crystal (Figure 3–16). A calibration curve can be constructed relating the reading of the permanent thermocouple with the reading of the thermocouple at the crystal position. An alternative method of calibrating the thermocouple, using phase transitions of known materials, is given in Section 8.1.2.

3.3. Construction of a Simple Gas-Stream Apparatus

The devices described in this section are very simple systems that can be constructed with a minimum of effort and expense. They can be used to obtain familiarity with low-temperature systems and for preliminary or occasional low-temperature studies. These are manual systems that will

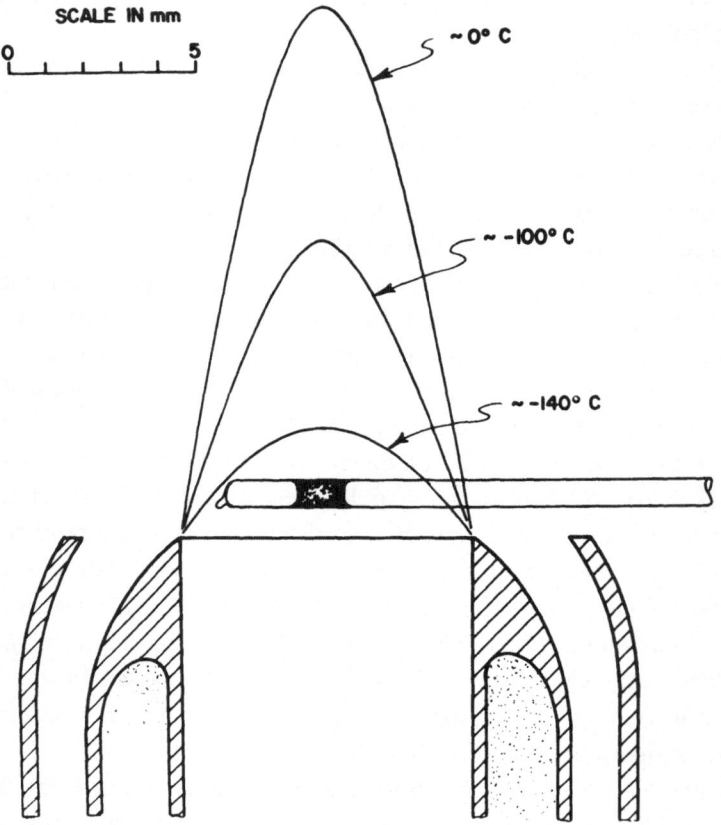

Fig. 3–15.   (a) Temperature as a function of distance from the end of the cold stream. Proper position of crystal is shown (Huffman, 1974).

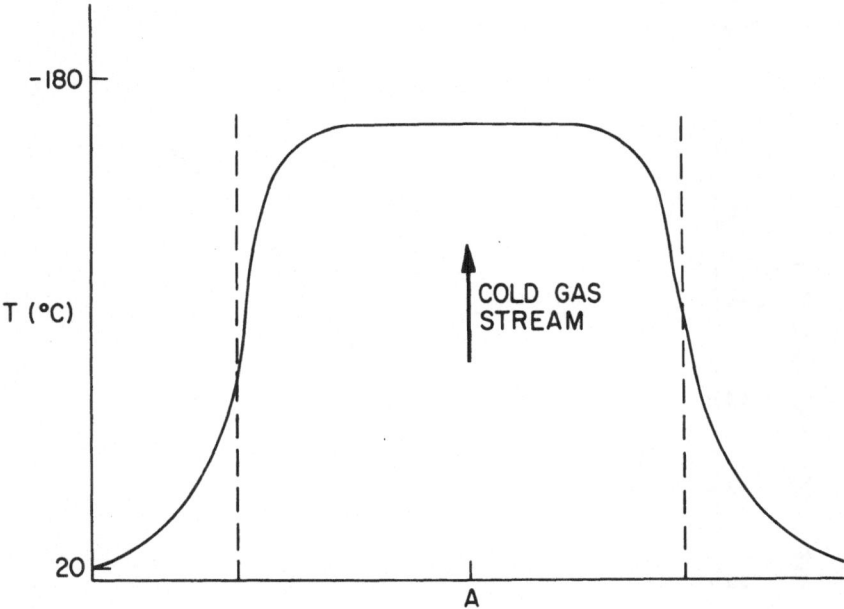

Fig. 3–15. (b) Acceptable temperature gradient at the outlet of a gas-stream nozzle measured at a distance of 6 mm from the end of the nozzle whose center is at A. The cold gas stream lies between the dashed lines, while the warm outer stream is located on either side of the two dashed lines.

operate unattended for short periods of time. In some cases they can operate for long periods of time if sufficient care is taken.

These systems can be used with single-crystal or powder instruments and can be adapted easily for other uses such as low-temperature optical microscopy or nmr and ir spectroscopy.

3.3.1. The Nozzle

The nozzle can be constructed by an experienced glassblower as shown in Figure 3–9. This is a useful piece of equipment to have on hand and is a recommended device for all X-ray laboratories. If one is not available, a simpler model can be made by forming a simple vacuum-jacketed tube and attaching an outer tube for the warm gas stream by means of a cork ring at the upper end, as shown in Figure 3–8. Glass T-tubes attached by rubber tubing can be used for the inlets for mixer stream (or heater wires) and for the thermocouple wires. If a plastic tent is used in

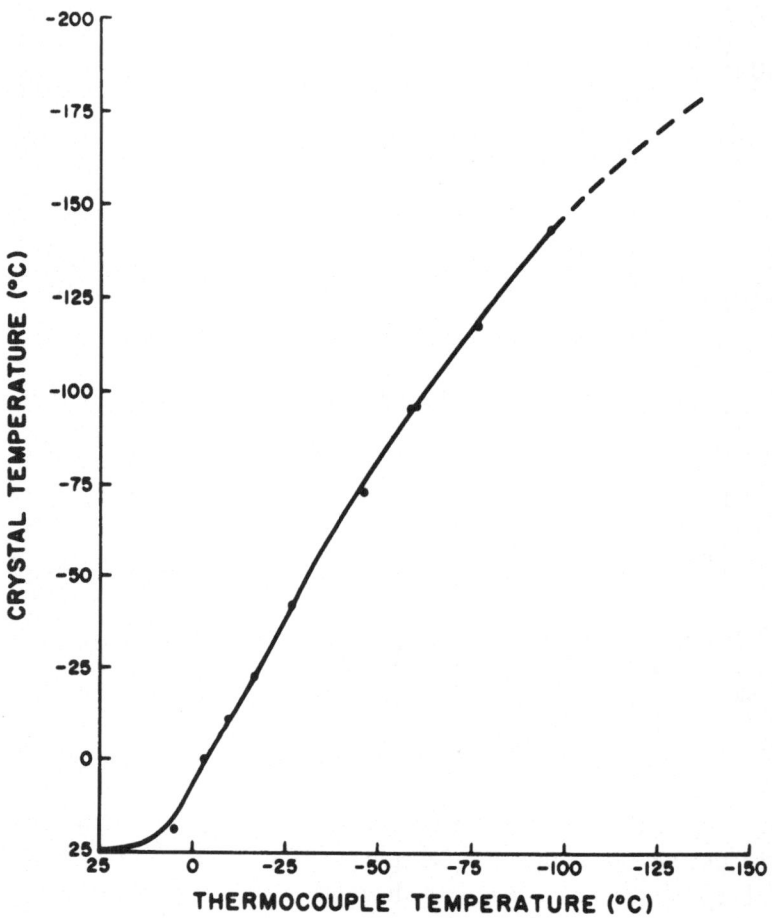

Fig. 3–16. Temperature calibration curve. The abscissa corresponds to the temperature recorded by the thermocouple sensor while the ordinate is the actual specimen temperature (Abowitz and Ladell, 1968). (Reproduced with the permission of the Institute of Physics.)

place of an outer stream, the cold stream can be fabricated from a single piece of Armaflex tubing insulation (Marsh and Petsko, 1973).

3.3.2. The Heat Exchanger

The heat exchanger is formed by placing a coil of several turns of copper refrigeration tubing (¼ in. i.d.) inside a large dewar flask (a 4-liter size works well) or a stainless steel or glass beaker placed inside a styrofoam-insulated box. The copper coils should first be cleaned with dilute nitric acid and dried by baking in an oven at 110°C.

Two heat exchangers should be prepared if liquid nitrogen is used, with the first acting as a precooler. If an automatic liquid-nitrogen refilling system is available (along with an adequate supply of liquid nitrogen), it may be easier to use liquid nitrogen in both containers. If not, a dry-ice slurry should be used in the first container. If only moderately low temperatures, such as −50°C, are required, only the dry-ice slurry is required.

3.3.3. Dry Gas

Very fine air dryers are available commercially, but may be too expensive to purchase for occasional use (Appendix 10). Air can also be dried by passing it through a drying tube containing molecular sieves. There should be two of these tubes in parallel, so that the drying agent in one can be replaced or dried while the other is in use. The length of time a tube will dry the air depends, of course, on the moisture content of the air, the rate of gas flow, and the type of desiccant being used. Most compressed gas is very moist and oily, and if the compressor is not equipped with a water and oil filter, it will be necessary to install a filter to protect the drying tubes. It is also possible to use a small oilless reciprocating compressor, which can usually be obtained through a general laboratory supply house. (Also see Underwood and Chapman, 1976.)

A better alternative is to use a cylinder of dry, compressed nitrogen gas. In this case, two inlets should be attached to the system so that one tank can be operating while the other is being changed. The proper operating pressure is a function of the diameter and length of the gas lines and heat exchanger, and is generally between 5 and 15 psig (3–10 N/cm²).

3.3.4. Temperature Control System

The temperature control system consists of a set of three valves. One line leads to the heat exchanger and nozzle; a second goes to the mixer; and the third controls the warm outer gas stream. A standard brass X-connector with four openings, one to the compressed gas line and three others for the valves, is very useful.

The heat exchanger is connected to the nozzle with rubber tubing. The tubing should fit tightly but should not be clamped or tied. Thus, if the coils become clogged, the unclamped tubing will act as a safety valve. Standard insulation tubing (e.g., Armaflex) or fiberglass insulation should be wrapped around the tubing connecting the nozzle with the heat exchanger. The dewar (containing the heat exchanger) should be placed on the X-ray unit near the sample so that the heat-exchanger-to-nozzle distance is as short as possible. The nozzle is held in place with ordinary laboratory clamps. A schematic diagram of this simple system is shown in Figure 3–17.

Descriptions of simple systems are found in the following references:

1. General purpose: Abrahams et al. (1950); Post et al. (1951) (Figure 3–8)
2. Single crystal camera: Reed and Lipscomb (1953)
3. Powder diffractometer: Miksic et al. (1959)
4. Debye–Scherrer powder camera: Rudman (1966a,b, 1967)

Wallwork and Harding (1954) describe a very simple apparatus that can be used for a short exposure with a vertically mounted oscillation–rotation camera. Cold nitrogen gas is generated by a heater placed in a small dewar vessel and directed at the crystal via an insulated glass tube fitted through a hole in the cover of the camera.

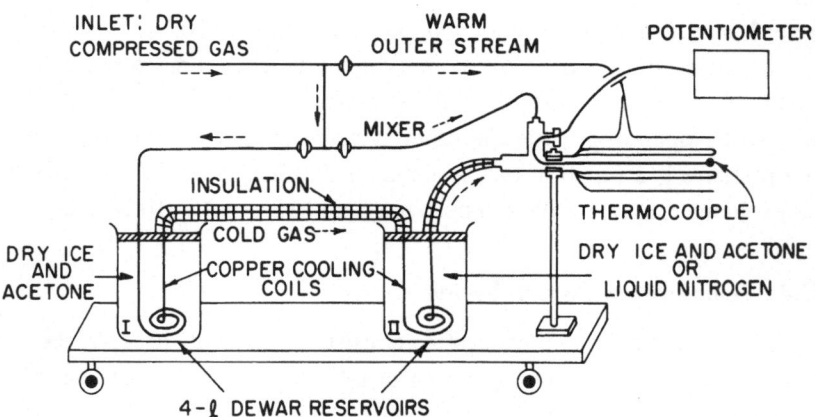

Fig. 3–17. Schematic diagram of simple low-temperature gas-stream apparatus. The device shown is mounted on wheels for ease in positioning it near a Debye–Scherrer camera. For other applications, the low–temperature apparatus is placed on the table (Rudman, 1967). (Reproduced with the permission of the *Journal of Chemical Education*.)

Chawdhury (1968) generated dry nitrogen gas by boiling liquid nitrogen in a sealed dewar with only an inlet and outlet. A funnel, with its lower end immersed in the LN_2, is placed in the inlet and is used for replenishing the LN_2. A copper coil is immersed in the LN_2 with one end of it inside the dewar above the LN_2 level and the other end passing through the outlet. As nitrogen gas forms from the boiling LN_2, it is recooled as it passes through the coil to the nozzle. A dry box is placed over the camera.

Another simple gas generator was described by Mehl and Barrett (1930). They bubbled nitrogen gas into LN_2 inside a closed dewar and allowed the cold nitrogen gas to flow over the sample. This is similar to the system described above, but does away with the copper-coil heat exchanger. A similar approach has been used by Wheeler (1968).

3.4. More Sophisticated Low-Temperature Gas-Stream Systems

The negligible absorption of the incident and scattered X-ray beams is responsible for the popularity of the gas-stream method. However, the requirement that the end of the nozzle be located near the crystal, and, therefore, near the diffraction apparatus, often results in certain restrictions on the number of reflections that can be measured. The resulting variations in instrument design, coupled with the different methods of generating cold gas, are described in the literature. Several of these systems are now available commercially.

Automatic LTXRD apparatus, that is, a cooling device that can keep a sample cooled to a constant temperature for many hundreds of hours without ice formation and with a minimum of operator attention, has its principal application in the field of single-crystal crystal-structure determinations. The development of the four-circle automatic diffractometer, with its many modes of operation and large range of goniometer head positions, has resulted in novel designs for the transfer lines and nozzles. However, with increasing complexity of design, the sophistication and cost of the instrument also increase. Electronically controlled feedback systems, evacuated flexible metal transfer lines, and specialized automatic filling devices are all expensive and must be compatible with the rest of the system.

In the following descriptions of low-temperature apparatus, only special features are discussed. Readers interested in constructing any one of these systems are referred to the original article for further details. These systems do not all operate at the same efficiency and the reader is warned that any duplication of these instruments will not necessarily reproduce all their reported properties. The two key features for which reproducibility is not always guaranteed are the reported rate of cryogen consumption and the lowest temperature attainable. This is due to minor variations in the construction of the apparatus, in local laboratory conditions (e.g., air current, humidity, etc.), and in the use of the instrument (e.g., different gas-stream velocities), as well as a possibly excessive zeal in reporting these factors on the part of the original authors.

3.4.1. Gas Stream from External Source

The passage of dry compressed gas through a heat exchanger immersed in an unpressurized cold bath eliminates the need for using a liquid-nitrogen gas generator and eliminates any fluctuations in the sample temperature during replenishment of the coolant. The recent development of mechanical refrigerating units capable of cooling gas streams to 155 K (Rudman, 1972; Lippman and Rudman, 1976) extends the use of this method to temperatures below that obtainable with dry ice alone (Figure 3–18).

One problem that is occasionally encountered is the clogging of the heat exchanger by moisture condensed from insufficiently dried air. Several investigators have suggested enlarging the first section of the heat exchanger that is immersed in the cold bath, so that any moisture left in the air after passage through the dryers will be deposited on the walls of this chamber without restricting the passage of air. Owen and Williams (1954) utilize three heat-exchange coils in three separate dewars, with metal boxes at the bottoms of the first two. Harding (1956) (Figure 3–19) and Olovsson (1960) report using a large copper cylinder (3 cm diameter) in the precooler dewar. This method is particularly recommended if drying is accomplished only by means of a drying tube and not by means of an automatically regenerating gas dryer. Dry gas can also be obtained from a compressed-nitrogen cylinder or by using the gas outlet on a large (110–175 liter) LN_2 dewar (see Appendix 6).

3.4.1.1. Mechanical Refrigeration

The current availability of large-capacity mechanical refrigerators capable of operating at cryogenic temperatures allows the use of mechani-

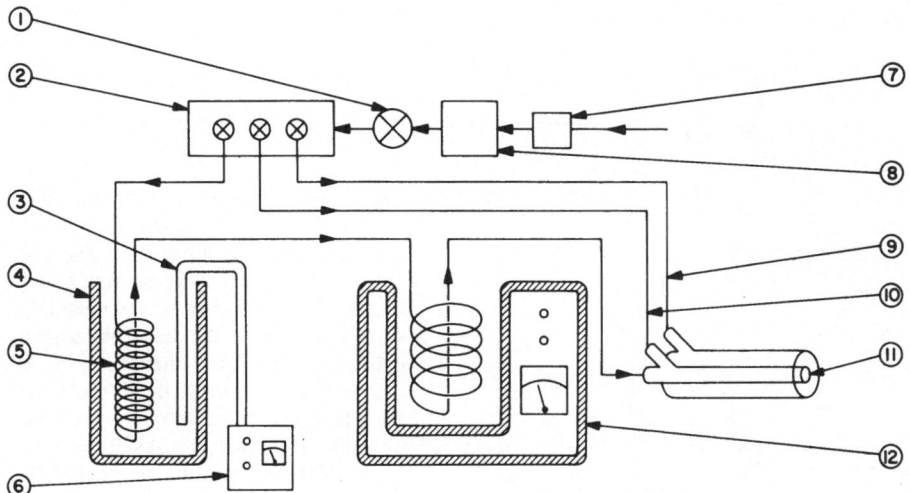

Fig. 3–18. Schematic diagram of gas-stream cooling system employing closed-cycle refrigeration. Dry air passes through a precooler (cooled by a mechanically refrigerated probe) and is cooled to 155 K inside a heat exchanger immersed in a refrigerated dewar maintained at 135 K (Lippman and Rudman, 1976). (1) Pressure regulator; (2) manifold; (3) flexible refrigerated probe (180 K); (4) precooler; (5) copper coil heat exchanger; (6) closed-cycle refrigerator; (7) oil and water filters for compressed air; (8) self-regenerative molecular sieve air dryer; (9) dry, warm outer air stream; (10) dry, warm mixer air stream; (11) cold gas stream; (12) refrigerated dewar (to 135 K).

cally refrigerated probes (Rudman, 1972; Marsh and Petsko, 1973) and/or mechanically refrigerated dewars (Lippman and Rudman, 1976). The advantages of these systems are obvious. They operate virtually indefinitely (in the absence of power failures), use no consumable cryogens, require no attention on the part of the user, and, for long-term studies, are more economical than LN_2 or dry ice. The major disadvantage is that the lowest temperatures currently attainable are limited. The refrigerator can cool the bath to between 145 K and 135 K. When air flows through the system, the exchange of heat between the incoming warm gas and the cold bath results in a sample temperature of about 155 K (Figure 3–18).

A single refrigerating unit (see Appendix 3) can cool a sample to approximately $-80°C$ in a properly insulated system. In the temperature range $-60°$ to $-120°C$, compressed gas passed through an automatically regenerating molecular sieve air dryer can be used for the gas stream. Sat-

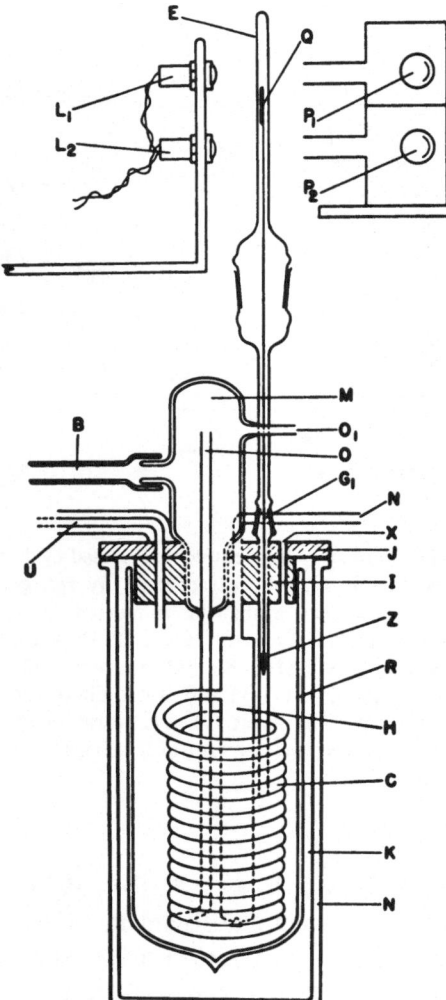

Fig. 3–19. Cooling system showing heat exchanger immersed in a cold bath. Any residual moisture in the gas entering the heat exchanger (upper N) is condensed in the large cylinder (H) without clogging the heat exchanger coils (C). The liquid-nitrogen level is automatically controlled by the float (Z) which raises and lowers an opaque cylinder (Q) that breaks the light beam from L_1 and L_2 to photocells P_1 and P_2. The temperature of the cold gas that blows over the sample (B) is controlled by mixing it with warm gas that enters at O_1. Other parts of the apparatus include: (E) level-indicator sight tube; (G) ground-glass joint; (I) insulation; (J) cover; (K) felt insulation; (M) chamber for mixing cold and warm gas; (N) (lower) copper container; (O) outlet of cold gas into mixing chamber; (R) dewar; (U) liquid-nitrogen refill port; (X) exhaust vent for nitrogen gas (Harding, 1956). (Reproduced with the permission of the National Research Council of Canada.)

isfactory operation of the apparatus, with no icing problems, has been obtained over several months of continuous operation.

In an adaptation of this method for protein studies in the range of $-10°C$, a refrigerated probe is in contact with a fin tube inside a cylindrical plexiglass container. Air entering this container is cooled as it passes the fin tube and is then directed over the sample (Tropp, 1975).

3.4.1.2. Heat Exchanger Immersed in Cold Bath

Dry-ice slurries have been used successfully (Kaufman and Fan-kuchen, 1949; Abrahams et al., 1950; Rudman, 1966a,b), but they are not satisfactory for many applications since the lowest sample temperature that can be reached is only $-70°C$.

Liquid-nitrogen baths are often used in such systems (Post et al., 1951) (Figure 3–19). The major disadvantage of using a cold bath is its inefficiency due to the loss of boiling LN_2 to the atmosphere. Furthermore, an external gas source must be supplied. Since compressed air cannot be used with liquid nitrogen as coolant, gaseous nitrogen must be obtained from cylinders of the compressed gas or by boiling a separate container of liquid nitrogen. A large cylinder of compressed gas will last, at the most, for 12 hours; most applications will require more frequent changes. This is a time-consuming and difficult task if it must be done regularly over a period of several weeks. The advantage is that it is a very easy system to set up, does not require any special apparatus, and is recommended for investigations of short duration.

A more efficient method is to obtain dry nitrogen gas from boiling LN_2, as described in Appendix 6. This is essentially the simple system described in Section 3.3, which is based on the papers by Cioffi and Taylor (1922), Eastman (1924), Kaufman and Fankuchen (1949), Abrahams et al. (1950), Post et al. (1951), and Rudman (1966a,b). A more complex version was described by Clifton (1950).

If boiling liquid nitrogen is used as the source of nitrogen gas, care must be taken not to disturb the gas pressure unduly during refill cycles or dewar changes. Since such disturbances lead to difficulties similar to those encountered when the boiling gas is used without an intermediate cooling bath (Section 3.4.2), there seems to be no particular reason to use this method, except that with a 160-liter dewar of liquid nitrogen as the gas source, the gas fluctuations only occur occasionally.

A closed system in which the dry, cold air is recirculated by a rotary pump was constructed by Owen and Williams (1954) for a back-reflection camera. The pump draws recirculated air into the camera through a heat exchanger immersed in a cold bath.

Harding (1956) built a similar system in which he used flowmeters on all the gas lines. The temperature is controlled by mixing warm *air* with cold gaseous *nitrogen* in a relatively large, vacuum-insulated glass chamber. The liquid-nitrogen level is controlled by a float, the top of which intercepts beams of light aimed at photoelectric cells (Figure 3–19).

A gas-stream system in which the sample is surrounded by a chamber constructed from a double layer of 0.25-mm-thick lucite windows was designed by Attard and Azaroff (1960). An outer stream is not used, but the freedom of access to the sample, which is characteristic of the gas-stream method, is lost.

Prakash (1966) developed a system in which nitrogen gas is passed through a heat exchanger immersed in liquid oxygen. The apparatus is designed so that the cold oxygen gas formed by the boiling liquid cryogen mixes with the cooled nitrogen gas.

The system constructed by Davies et al. (1970) is of interest primarily because of the way in which it is attached to a single-crystal diffractometer; it is described in Section 3.5.3.

Gobrecht et al. (1971) designed a system expressly for the precession camera, in which nitrogen gas, obtained from boiling liquid nitrogen, is passed through a heat exchanger immersed in liquid air. However, the design of the instrument in no way restricts the source of nitrogen to the boiling liquid; compressed nitrogen gas could be used just as easily (Section 3.5.2).

The system designed for the Syntex single-crystal diffractometer also operates on this principle. An interesting feature of this system is the use of short lengths of silvered, vacuum-insulated tubing connected by Teflon joints. Recent improvements in the design of these joints resulted in a moveable joint which does not frost over while in use (Figure 3–20). Several investigators have used a mechanically refrigerated probe to cool a cold bath of ethanol in the Syntex system. For operation above $-75°C$ this is an economical, stable system. The nozzle is mounted on the χ-circle opposite the crystal and directs the stream of cold gas along the ϕ-axis regardless of the orientation of the crystal (Section 3.5.3). This system has been modified for use with other full-circle diffractometers.

3.4.2. Gas Stream from Boiling Liquid Nitrogen

Systems employing a gas stream generated from boiling liquid nitrogen require a generator in which liquid nitrogen is converted to cold gaseous nitrogen. Methods of overcoming the difficulties encountered in refilling a pressurized dewar have been described in Section 3.1.1.

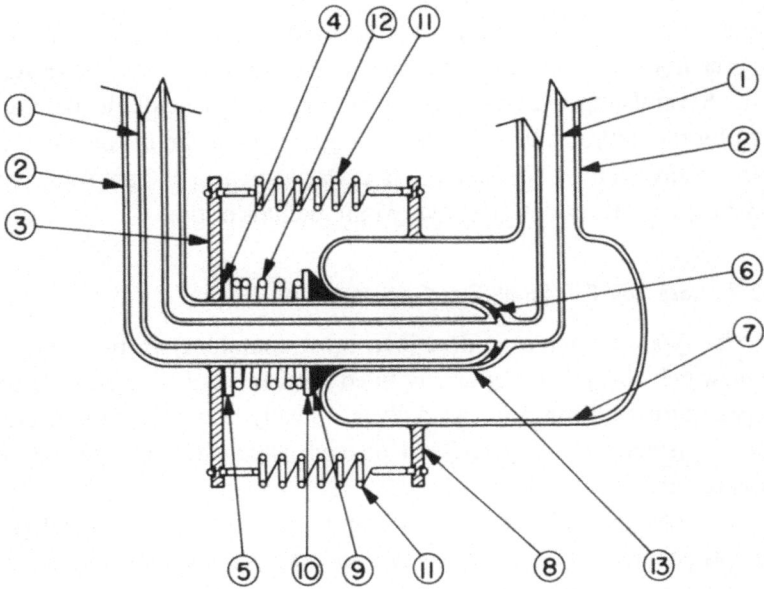

Fig. 3–20. Cross section of joint between two evacuated tubes that does not frost over while in operation: (1) Pyrex tubing 7 mm o.d. × 5 mm i.d. (2) Pyrex tubing 17 mm o.d. × 14.6 mm i.d.; (3) Al spring support; (4) Teflon bushing; (5) Al ring; (6) Teflon seal bushing; (7) Pyrex tubing 51 mm o.d. × 47 mm i.d.; (8) Al spring ring; (9) Silastic 732 RTV; (10) Teflon ring; (11) tension spring—two for each tube; (12) compression spring; (13) Pyrex tubing 22 mm o.d. × 19 mm i.d.; (Strouse, 1975).

3.4.2.1. The Reservoir as Gas Generator

An early version of this type of device was reported by Burbank and Bensey (1951). In this system, a large (100 liter) dewar is used as the gas generator. No special precautions to minimize the pressure variation during the filling cycle are reported.

Frenz et al. (1969) report a device for obtaining a constant flow of cold nitrogen gas at constant temperature, directly from a commercially obtained 110- or 160-liter self-pressurized dewar. A styrofoam-filled control box with connections for two large dewars is used for manually switching from the nearly empty to the full container. A constant temperature within the control box is maintained by utilizing a thermostatically controlled cryogenic valve. If the temperature is too high, the valve opens and the warm gas is vented and is replaced by colder gas from the larger dewar. Other features include the use of room-temperature, dry, compressed air as the source of the outer stream, a consumption rate of 2.7 liters per hour of liquid nitrogen, and a minimum sample temperature of 140 K. Several improvements over their basic, homemade device are suggested by the authors. Its major advantage seems to be its very low price, but the many manual adjustments necessary for proper operation militate against the use of this device for very long periods of time.

3.4.2.2. Reservoir Fills Small Gas Generator

The type of apparatus described most commonly in the literature is one in which the gas generator is filled from a liquid-nitrogen reservoir. The generator itself, in order to deliver a steady flow of cold gas, must be a closed system with only one exit through which the cold gas passes on its way toward the sample.

This category of instruments can be divided into two subgroups: small gas generator (2 liters or less) and large gas generator (greater than 2 liters).

Altona (1964a,b) described a device designed for a horizontal Weissenberg goniometer but which could be easily adapted for any instrument (Figure 3–21). The gas generator is a 170-ml-capacity dewar and the cold stream is a double-walled vacuum-insulated sidearm (3.5 mm i.d.) tube attached directly to the dewar. Since it was intended for use on a

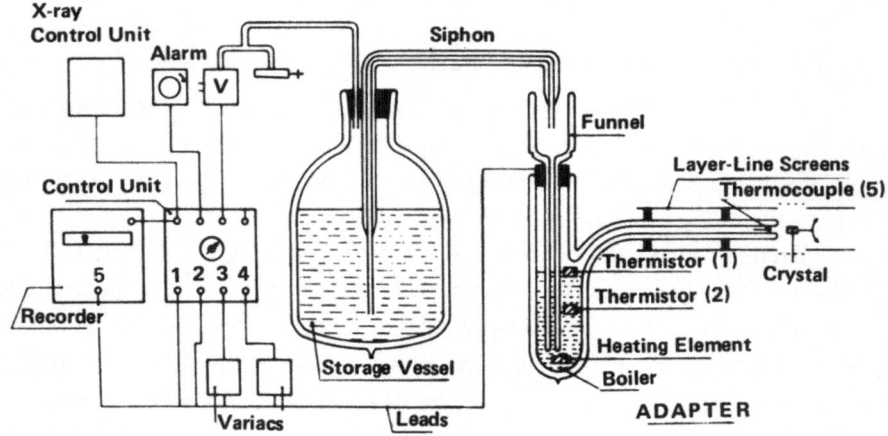

Fig. 3–21. Schematic diagram of Altona (1964a,b) device described in text. (Reproduced with the permission of the International Union of Crystallography.)

Weissenberg goniometer, no outer stream is employed (Section 3.5.1). Cold gas is generated by a heater placed in the bottom of the generator, and the liquid level is controlled by a miniaturized thermistor-controlled refill system. The temperature is varied from 85 K to 295 K by means of a small heating coil placed in the cold gas stream.

One thermistor is used to maintain the liquid-nitrogen level within a narrow limit of ±3 mm in height; a second thermistor, situated in the lower half of the generator, is used as a safety control. If the liquid reaches this level, an alarm sounds and power is cut off to the X-ray unit and Weissenberg goniometer.

The consumption rate, inclusive of losses from the storage vessel and from siphoning, is reported as 5.5 liters in 24 hours. The liquid nitrogen is forced into the generator by a compressed-nitrogen source which is connected to the generator by a thermistor-controlled solenoid valve. No provision is shown for automatic filling of the 10-liter reservoir.

The construction of a small sidearm dewar with provision for a warm outer stream is described by Dickens (1966) (Figure 3–30).

A small gas generator has been marketed since 1969 by Enraf-Nonius. Called the "Universal Low Temperature (ULT) Device for X-Ray Diffraction Cameras and Diffractometers," it is a modification of a system designed by Bolhuis (1971). It differs from Altona's instrument in that (1) the gas generator is a 2-liter dewar to which several different types of nozzles can be attached, (2) there is provision for generating a warm outer stream directly from the gas generator, and (3) the liquid-nitrogen reservoir can be of any size from 10 liters to 40 liters. The consumption rate, when the outer stream is also used, is about 2 liters per hour at a minimum temperature of 100 K.

Hospital (1968) has also described a system in which the liquid-nitrogen level is maintained to within 5 mm. Two solenoids (one between a tank of compressed nitrogen gas and the reservoir and one between the reservoir and the gas generator) are opened during the filling cycle. An interesting feature of this apparatus is that four resistance heaters are used. The first is used to boil the liquid nitrogen; the second is located in the cold gaseous nitrogen which collects at the top of the gas generator and is used to adjust the temperature of the gas in the generator; the third heater, which is placed in the nozzle, is used for fine control of the gas-stream temperature; and the fourth is attached to the end of the nozzle to prevent ice formation. The system is steady to within 0.2° from 100 K to

230 K and to within 0.5° from 230 K to ambient. Since it is used with a Weissenberg goniometer, no outer stream is used.

A further variation of this approach is the use of an intermediate reservoir between the main reservoir and the gas generator (Thaxton and Jacobson, 1970). In this device, a 700-ml gas generator (very similar to Altona's) is located directly below a 1.8-liter dewar outfitted with a magnetically controlled valve at the bottom. As the liquid level in the generator drops, the switch is opened and liquid nitrogen is gravity-fed into the generator. The level of the intermediate dewar is maintained by a thermistor-controlled solenoid valve connected to a pressurized reservoir of liquid nitrogen. The advantage of this system is that the intermediate reservoir does not have to be pressurized, inasmuch as the liquid nitrogen is gravity-fed to the gas generator. Therefore, the intermediate reservoir can be vented to the atmosphere and any pressure changes that take place during the refilling of this dewar do not affect conditions in the gas generator (Figure 3–22). On the other hand, automatic refilling of the Altona and Enraf-Nonius–Bolhuis reservoir dewars without affecting the pressure inside the gas generator can only be accomplished with some difficulty [use of cryogenic pressure relief valves and solenoid valves; see Danielsson et al. (1976) for details]. However, since these reservoirs need only be filled once a day, this is not a critical problem.

Another variation of this approach is found in the apparatus described by Renaud and Fourme (1967). Automatic refilling from a 50-liter reservoir maintains the level of a 2-liter gas generator to within 10 mm by means of a photoelectric eye activated by the opaque top of a float (this method was also used by Harding, 1956). A glass nozzle is attached to the styrofoam-insulated stainless steel gas generator. The temperature of the gas is controlled by two heaters placed in the cold stream: the first acts as a coarse heater, and the second offers fine temperature control and is regulated by feedback from the thermocouple. No outer stream is used, but a heated stainless steel insert is placed at the end of the cold stream. This prevents the formation of ice at the end of the cold stream (see Section 3.2.2) by warming the outside of the cold gas stream. A modification of the goniometer head helps deflect the cold gas in such a way as to prevent frost formation. Consumption rates of 0.7 to 0.8 liters per hour at 110 K and 1.4 liters per hour at 90 K, with a temperature stability of less than 0.25°, are reported. In a recent modification of this device (Fourme, 1975), a metal transfer line and nozzle were used in conjunction with a modified goniometer head.

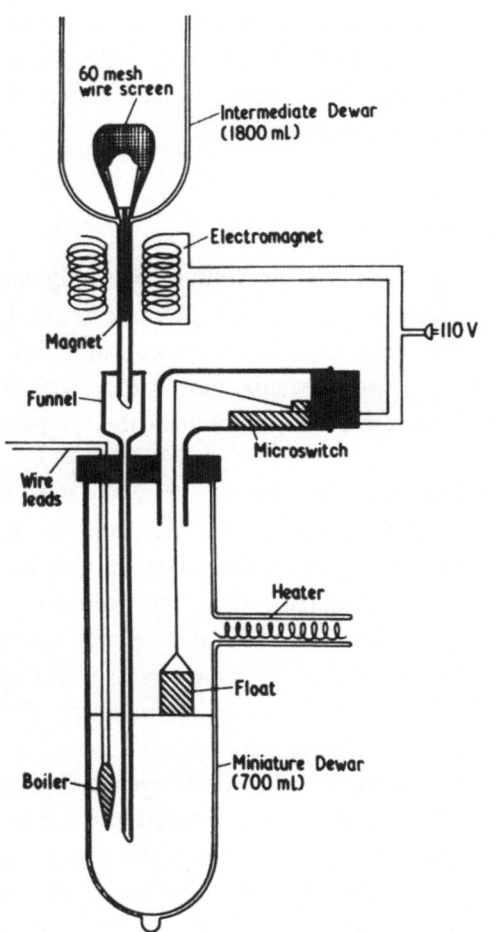

Fig. 3–22. Schematic diagram of miniature and intermediate dewars of Thaxton and Jacobson (1970), as described in the text. (Reproduced with the permission of the Institute of Physics.)

Stoe Instruments offers a 2-liter gas generator that can be mounted on any of several instruments. A small rotary pump gas compressor is used to remove the evaporated LN_2 from the top of the generator. This gas is then compressed and pumped through a heat exchanger immersed in the LN_2. From there it passes through an insulated transfer line and the nozzle to cool the sample. In this manner the cold gas flow is maintained at a constant flow rate resulting in a stable temperature at the sample. [This is similar to the system used by Silver and Rudman (1971), as described in the following section.] The gas generator is filled by a LN_2 pump located in the reservoir which is controlled by a sensor in the generator.

An interesting device, based on the conduction-cooling apparatus of

Ubbelohde and Woodward (1947), has been reported by Kay and Vousden (1949) and by Cimino et al. (1959). A copper tube, with plastic-
film-covered windows for passage of the X-ray beam, is placed over the
crystal and cold gas is allowed to flow over the sample.

3.4.2.3. Reservoir Fills Large Gas Generator

Large gas generators, generally 20–40 liters in volume, naturally con-
sume larger amounts of liquid nitrogen than do smaller generators. As the
volume of the system is reduced, the rate of liquid-nitrogen consumption is
also reduced. Large dewars are often used because they are compatible
with the most readily available components used to control the temperature and pressure of the cold gas.

Since low temperatures are involved, the initial tendency has been to
use specialized (and expensive) cryogenic equipment, particularly cryogenic valves. Silver and Rudman (1971) designed a system that uses
standard, noncryogenic components. At the same time, the problems of
regulating gas flow and controlling pressure variations during refill cycles
were also solved. The key to the success of this instrument is to warm the
cold gas generated by boiling liquid nitrogen to room temperature. This
warm gas is regulated by a standard, noncryogenic (and economical)
pressure reducer. It is then passed back into the liquid-nitrogen bath (15-
liter dewar) via a heat exchanger (Figure 3–23). The cooled gas is deliv-
ered to the crystal through a flexible metal transfer line fitted at the end
with a metal nozzle. A 2.5-cm-diameter pop-off relief valve maintains the
pressure in the gas generator, thereby reducing the load on the pressure
regulator and permitting short refill times. The warm gas stream is split
into three separately controlled components: one to be recooled in the
heat exchanger, one for the outer warm stream, and one to mix with the
cold gas stream for temperature control.

The specific heat of nitrogen gas is approximately 0.25 cal/g, while
the heat of vaporization of liquid nitrogen is 47.6 cal/g. Thus the energy
released by cooling 1 g of nitrogen gas from room temperature to 77 K is
more than enough to vaporize an equivalent amount of liquid nitrogen.
However, the system is not self-sufficient in the generation of nitrogen gas
because some of the gas is diverted for use in the outer and mixer gas
streams. A small auxiliary heater is set to operate continuously, although
an improvement would be to connect it to a relay activated by a pressure

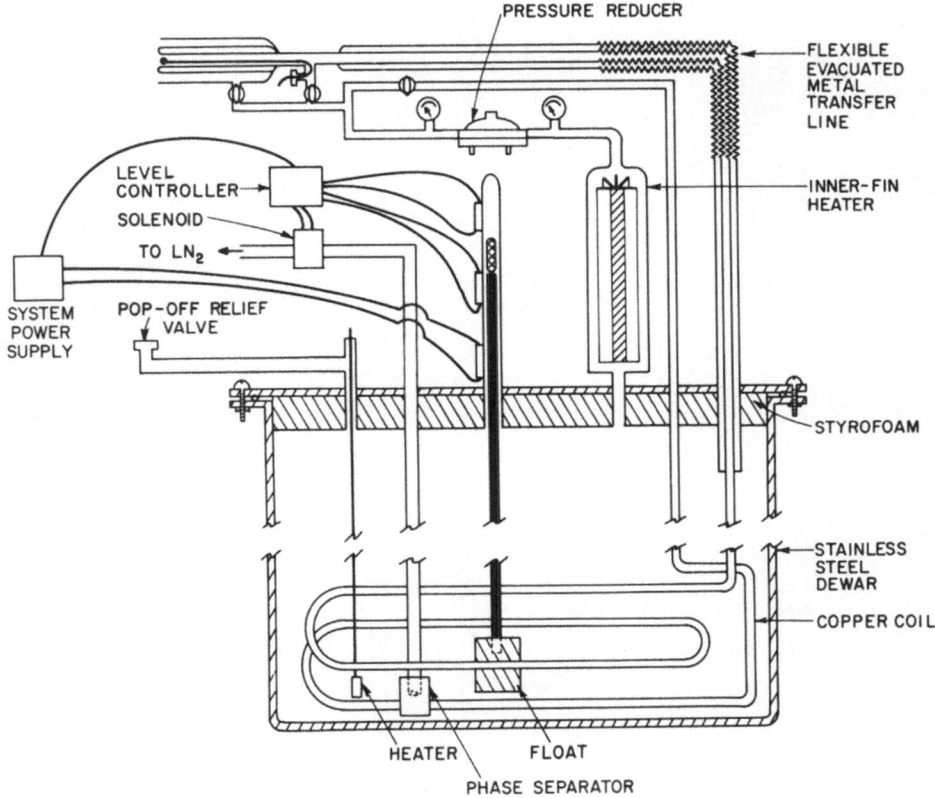

Fig. 3–23. Schematic diagram of Silver and Rudman (1971) apparatus described in text. The nozzle is attached to the system through a flexible transfer line and plastic tubing so that it can be positioned in any convenient manner. The top of the 15-liter dewar is covered with styrofoam insulation. (Reproduced with the permission of the American Institute of Physics.)

switch. The liquid-nitrogen level is maintained by a magnet attached to the top of a float which activates a reed switch which in turn controls the solenoid valve to the reservoir (Figure 3–24). The system operated nearly continuously for 1200 hours with a liquid-nitrogen consumption rate (including the outer stream and losses due to venting and refill) of approximately 3 liters per hour. The temperature at the outlet remained stable within 0.2° at all operating temperatures between 220 and 100 K. The basic design is very economical, since all components are standard items; the use of a small dewar for the gas generator and a shorter transfer line should further reduce the liquid-nitrogen consumption rate. This

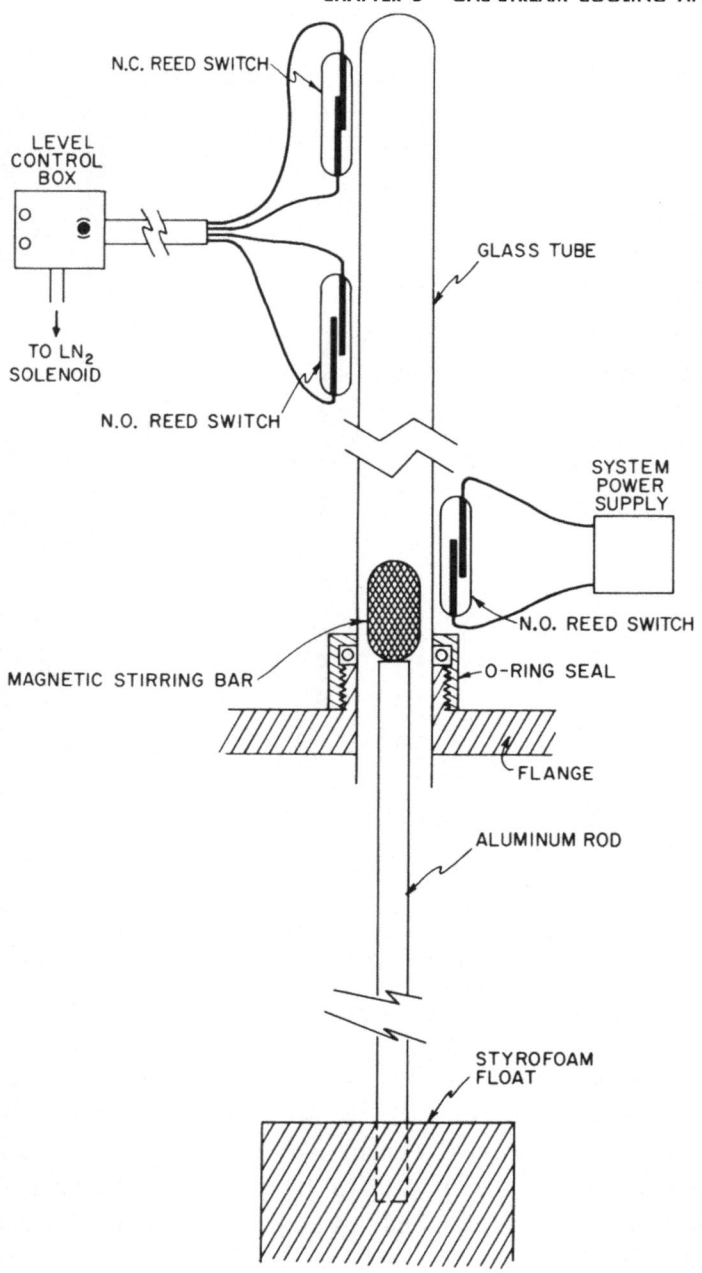

Fig. 3–24. Detail of use of floating magnet and reed switches to control liquid-nitrogen level controller (Silver and Rudman, 1971). (Reproduced with the permission of the American Institute of Physics.)

system is similar to those in which an external source of warm nitrogen gas is used, except that the "external source" is the generator itself.

A floating magnet and reed switches to control the LN_2 level (Figure 3–24) have also been used more recently by Hori and Matsuno (1972) and LaCour (1974). Systems in which the LN_2 level-control sensing unit is placed in a pressurized dewar cannot use the common gas-filled sensor (which is pressure sensitive). Therefore, these systems usually use an electronic sensing device (thermistor, resistor, etc.) or a special float-activated sensing device (e.g., reed switch, photoelectric cell).

Abowitz and Ladell (1968) designed a sophisticated system in which an electropneumatic proportional control system maintains the temperature by continuously adjusting both the flow rate of gaseous nitrogen through a pneumatic cryogenic valve and the boiling rate of liquid nitrogen. The periodic transfer of liquid nitrogen from a 160-liter reservoir to the 45-liter stainless steel double-walled vacuum-insulated gas generator utilizes automatic venting of the transfer line and a pop-off relief valve (see Section 3.1.1). A modified device of this type was constructed at Brookhaven National Laboratory (Rudman and Godel, 1969) and has been used extensively on the High-Flux Neutron Beam Reactor for low-temperature neutron diffraction studies (Figure 3–25). The complexity, high cost, and relatively high consumption rate of liquid nitrogen make this device impractical for all but the largest laboratories. At the time of its construction it was one of the most stable, completely automatic, and versatile low-temperature instruments, but recent improvements in low-temperature apparatus (e.g., Silver and Rudman, 1971) allow nearly equivalent operational characteristics at lower construction and operational costs.

Another method of controlling pressure surges during refill cycles has been utilized in several instruments (Hori and Matsuno, 1972; Burbank, 1973). In Burbank's device (Figure 3–26) a 40-liter stainless steel dewar is divided into two chambers by a stainless steel cylinder. The bottom of this cylinder is covered with a flat plate in the center of which is a 3-mm-diameter aperture. The passage of liquid nitrogen into the outer chamber from a self-pressurized reservoir is aided by a large vapor vent connecting the outer chamber with the atmosphere. The limitation of liquid-nitrogen transfer to the inner chamber through the small aperture reduces temperature variations during refill cycles to a rise of $3°$. Several other novel features also found in this device have already been discussed: the flow of cold gas onto the X-ray instrument was prevented by directing a stream of

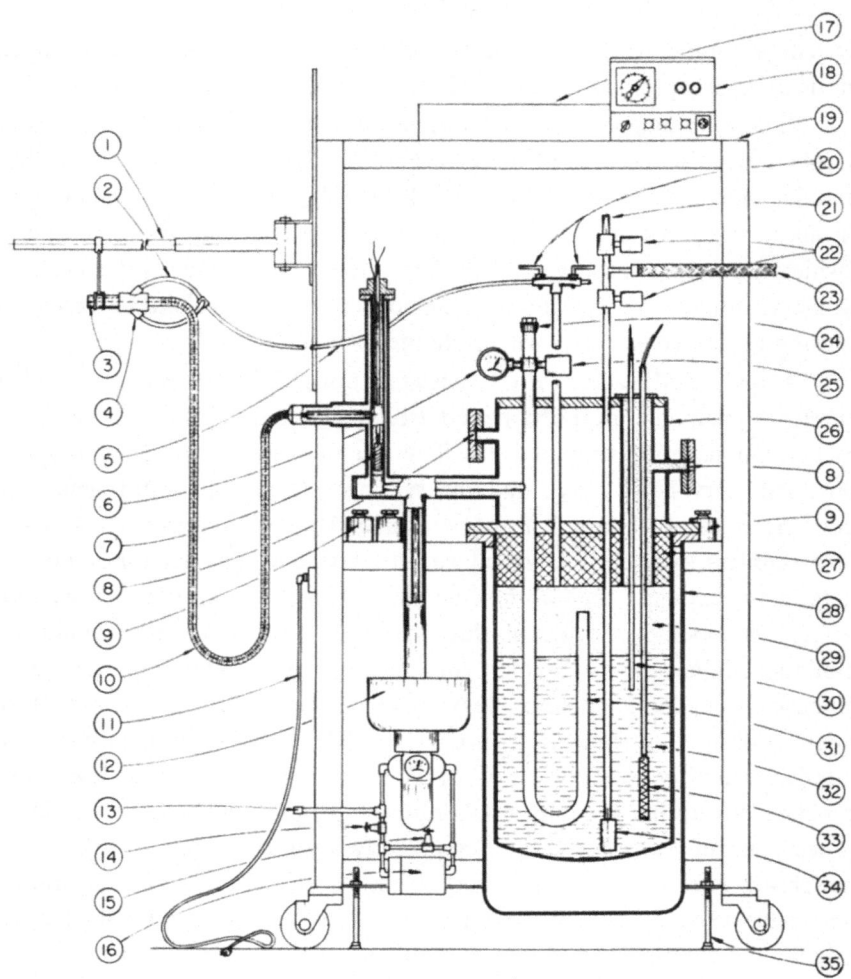

Fig. 3–25. Schematic diagram of automatic low-temperature apparatus (ALTA) (Rudman and Godel, 1969; after Abowitz and Ladell, 1968). (1) Adjustable boom; (2) rubber tubing for warm outer stream; (3) thermocouple in cold nitrogen gas stream; (4) adapter for outer stream; (5) copper tubing for warming cold nitrogen gas to room temperature; (6) pressure gauge; (7) heater to control temperature of nitrogen gas; (8) Mylar burst disc; (9) Variac; (10) flexible metal transfer line; (11) electric cord; (12) control valve for cold nitrogen gas; (13) compressed-air inlet; (14) reducing valve; (15) bypass valve; (16) electropneumatic converter; (17) precision set point control system, Leeds and Northrup; (18) liquid-nitrogen refill controls; (19) metal frame, 60 in. high, 32 in. wide, 23 in. deep; (20) dewar vent valves; (21) fill line vent; (22) fill line solenoid valves; (23) liquid-nitrogen fill line; (24) adjustable pop-up relief valve; (25) adjustable pressure switch; (26) vacuum insulated dewar cover; (27) styrofoam insulation; (28) dewar; (29) cold nitrogen gas; (30) liquid-nitrogen level sensor; (31) outlet for cold nitrogen gas; (32) liquid nitrogen; (33) heater to boil liquid nitrogen; (34) cryogenic phase separator; (35) stabilizer. (Reproduced with the permission of the International Union of Crystallography.)

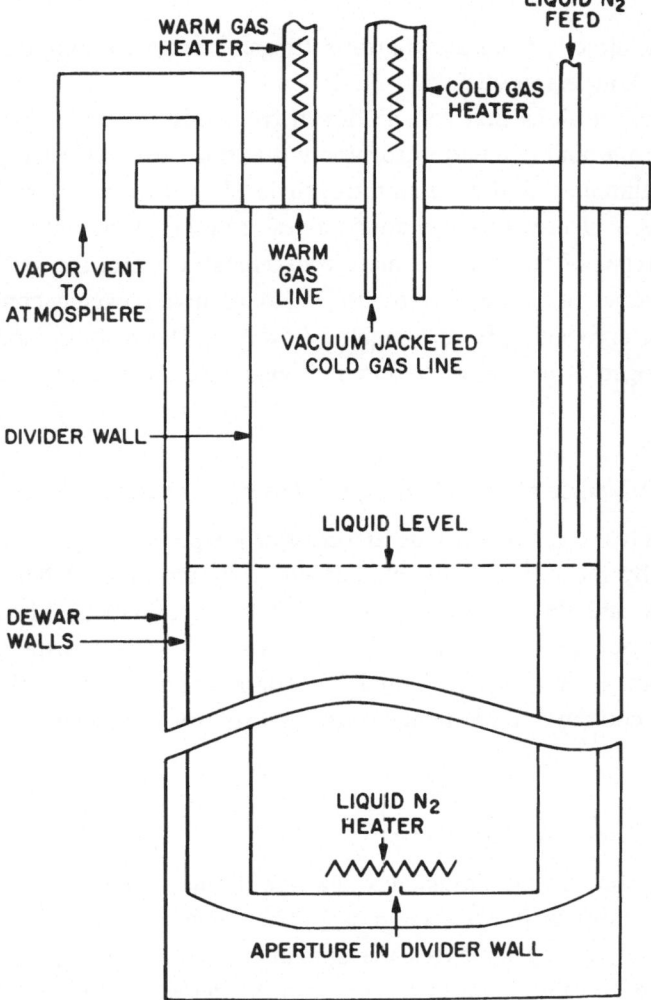

Fig. 3–26. Schematic diagram of gas generator divided into two compartments with a small aperture connecting them. Pressure fluctuations in the outer compartment (e.g., during LN_2 refill) are minimized in the inner compartment, resulting in a stable temperature output (Burbank, 1973). (Reproduced with the permission of the International Union of Crystallography.)

warm air from a heat gun through a knife-edge aperture to form a warm air stream normal to the cold gas stream; the temperature at the crystal position was determined for a number of control settings; and no thermocouple was used during the X-ray investigation, with temperatures being determined directly from the control settings. A large consumption rate of 6 liters per hour is reported.

The conclusions that can be drawn from this survey of gas streams obtained from boiling liquid nitrogen are that the most stable temperatures can be maintained if the liquid-nitrogen level is maintained within very close limits or if pressure-regulating valves are used in the outlet line. For the latter types of devices, it is more economical to warm the gas to room temperature so as to be able to use standard noncryogenic components than to use cryogenic pressure-regulating valves. The consumption rate of liquid nitrogen is lowest when small dewars and short transfer lines are used.

3.4.3. Gas Stream from Boiling Liquid Hydrogen or Liquid Helium

The advantages of working at very low temperatures (Section 8.1.1) are offset by the difficulty of maintaining a stream of cold hydrogen or helium gas and the increased problem of frost formation. Most of the work at temperatures below 77 K is carried out with cryostat systems. However, in a few instances, suitable apparatus has been developed for gas-stream cooling at these temperatures, mostly for investigations of short duration.

3.4.3.1. Weissenberg Goniometer

Robertson (1960) designed a dewar with an evacuated delivery tube leading out through its base (Figure 3–27). Hydrogen gas, generated by boiling liquid hydrogen, leaves the system by flowing through a cooling coil immersed in the liquid hydrogen, into the delivery tube, and onto the sample which is placed directly below it. The 3-liter dewar lasted about 18 hours (0.16 liters per hour), and a minimum temperature of 23 K was obtained. A separate outer stream of dried hydrogen gas was used to keep ice off the crystal, but icing problems restricted use to about 48 hours of operation. Further details on the use of this apparatus are reported in a later paper (Robertson, 1965). The original instrument was used for a vertical Weissenberg goniometer, but could be adapted easily for a precession camera or a horizontal Weissenberg goniometer.

Sugino et al. (1973) described a Weissenberg camera in which the specimen can be cooled to 20 K and heated to 700 K, using one of several tubes that fit into the layer-line screen opening. Cold helium or nitro-

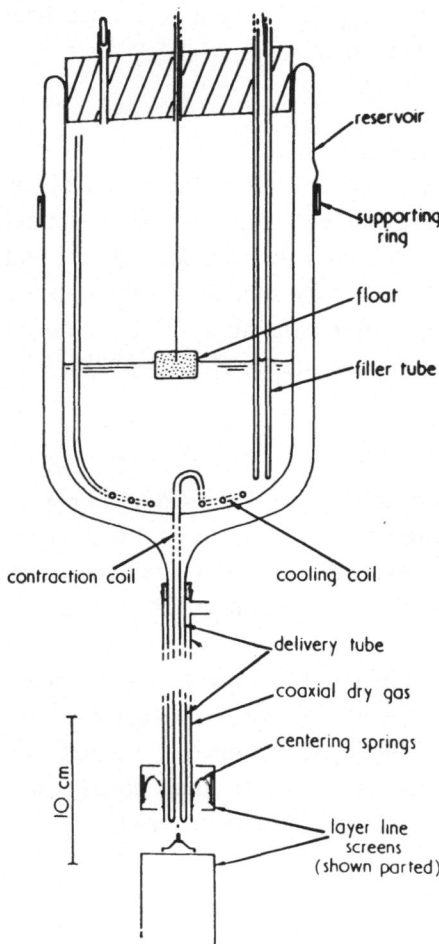

reservoir

supporting ring

float

filler tube

contraction coil

cooling coil

delivery tube

coaxial dry gas

centering springs

layer line screens (shown parted)

10 cm

Fig. 3–27. Schematic diagram showing design of a gas-stream apparatus for use with a vertical Weissenberg goniometer. The dewar is filled with liquid hydrogen (Robertson, 1960). (Reproduced with the permission of the Institute of Physics.)

gen gas is blown over the sample, and the temperature is controlled by regulating the gas flow and heater current so as to keep the temperature differential between two strategically placed thermocouples less than 0.1°. Mylar is used to cover the layer-line screen opening and the collimator, with the spent gas exiting coaxially with the incoming cold gas. Gas-tight seals around the layer-line screen and collimator are made with O-rings, so that the helium gas can be recovered. A suction pump is used to draw the cold gas through the system.

Altona (1964a,b) suggests the extension of his device (see Section 3.4.2.2) to very low temperatures by filling it with liquid helium or hydrogen and immersing the gas generator in liquid nitrogen to reduce heat leaks. No experimental results have been reported.

3.4.3.2. Single-Crystal Diffractometer

Dernier (1972) designed a baffle constructed from lucite and Mylar film which was connected directly to the cradle of a quarter-circle manual diffractometer. Cold helium gas is directed onto the sample from the end of an adapted transfer line; the gas flows over the sample and through the baffle. Several hours of relatively trouble-free running at 15 K could be obtained, at a nominal investment of time and money (Figure 3–28).

3.4.3.3. Rotation Camera

A standard goniometer head is placed in an enclosed chamber in which a gas stream generated by boiling liquid helium can cool the crystal to as low as 15 K. The X-ray beam passes through Mylar windows and the film can be loaded from outside the cold chamber (Beckman and Knox, 1961).

In another rotation camera, a miniaturized hydrogen gas Joule–Thomson liquefier, only 2 cm in diameter, fits inside a 3-cm-diameter film cassette. The crystal is attached to the liquefier, which is "rotated" or oscillated to obtain the diffraction pattern (up to a 300° range in 2θ), and is cooled by cold gas from the liquefied hydrogen (Forrestor, 1961).

3.4.3.4. Powder Camera

Pinot et al. (1965) describe a modified Debye–Scherrer camera in which the sample is centered inside a vacuum-insulated transfer line through which cold helium gas is flowing. This line passes along the sample axis from the back of the camera and out through a modified cover (Figure 3–29). Mylar windows incorporated in the transfer line allow passage of the X-rays and a heater permits the use of any temperature between 10 K and 300 K.

Solente (1965) blows cold helium gas (12–77 K) over a sample inside a chamber with beryllium windows. A pump is used to remove excess gas. A similar system was described by Gränicher et al. (1959).

Höhne et al. (1968) describe the use of the gas-stream technique as applied to single-crystal, Debye–Scherrer, and Guinier cameras at temper-

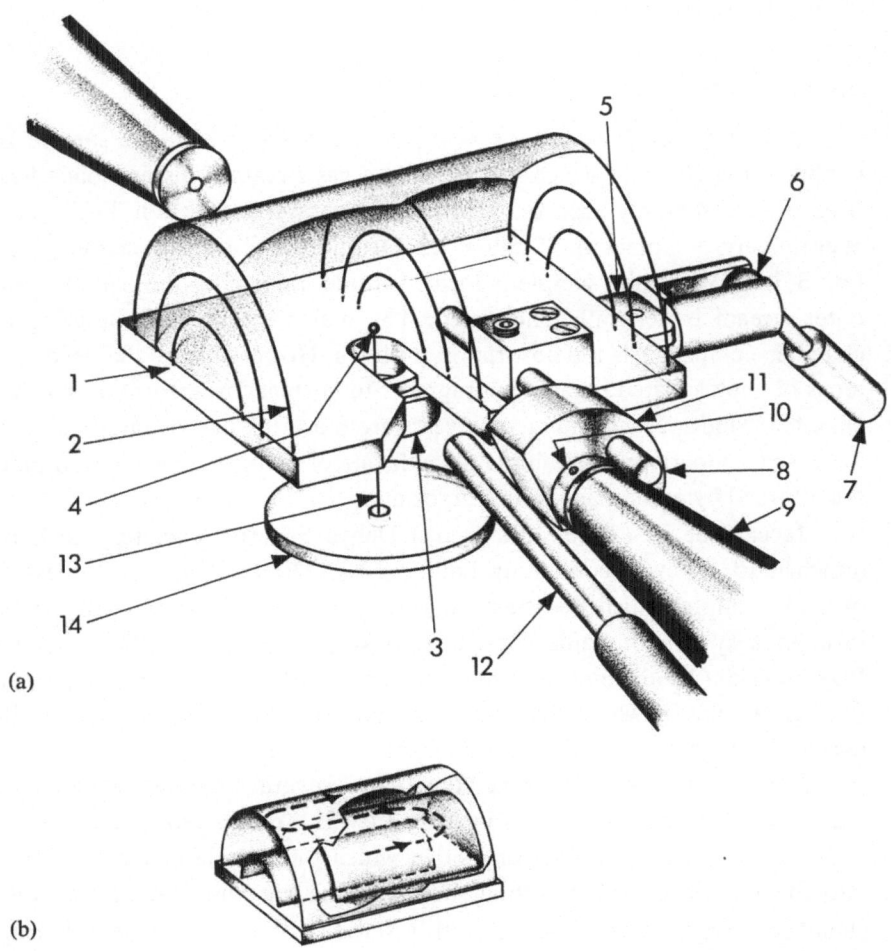

(a)

(b)

Fig. 3–28. (a) Schematic diagram of a baffle designed for directing a stream of cold helium gas over a single crystal mounted on a diffractometer. The baffle is supported by collimator (9) and brace (5, 6, 7). Helium enters through transfer line (12). (1) Lucite base (4.5 × 4.0 × 0.3 cm); (2) wire hoops embedded in base (1); (3) nylon bushing; (4) sample; (5) stainless steel tab; (6) socket joint; (7) socket stem; (8) tie rod assembly; (9) X-ray collimator; (10) brass rings (two) to hold central ring in place; (11) central ring attached rigidly to tie rod (8) but rotates on collimator (9); (12) helium gas transfer line; (13) glass fiber sample support; (14) lucite disk (3.0 cm diameter) attached to goniometer head. (b) View of the baffle covered with aluminized Mylar film. The arrow shows the direction of flow of the exiting helium gas. (Dernier, 1972). (Reproduced with the permission of the American Institute of Physics.)

atures between 4.2 K and ambient. Details for constructing the various chambers are given in the article.

3.4.4. Chambers Cooled by Gas Streams

A number of devices have been constructed in which the sample is confined in a *chamber* which is cooled by a gas stream. This approach has been most commonly used for very low temperatures (Section 3.4.3), for Weissenberg goniometers (Section 3.5.1), and for powder cameras (Section 3.5.4). When the sample is located inside a chamber the need for an outer stream is generally eliminated. The major reason for not using a chamber (cryostat) is the absorption problem. However, with the development of sophisticated computer programs to correct for absorption by the chamber windows, a number of investigators are taking advantage of the increased temperature stability and small rate of cryogen consumption that are afforded by gas-stream-cooled cryostats.

Jaccard et al. (1953) designed a Debye–Scherrer camera, which is attached to a dewar containing boiling LN_2. The cold gas rises, passes over the sample, which is suspended vertically above the dewar, and exits through a system of baffles designed to keep moist air off the sample. Przedmojski (1966) built a similar camera in which the cold gas passes through a tube in which the sample is rotated. The gas is warmed as it leaves the camera in order to prevent frost formation.

Boiko and co-workers have constructed several instruments employing gas-cooled chambers. Boiko et al. (1973a) developed a gas-cooled cryostat for a powder diffractometer in which the sample can be inserted into the sample chamber without breaking the vacuum seal around the chamber. Temperatures to 4.2 K with a stability of 0.05° can be attained.

Boiko et al. (1973b) developed a chamber for a precession camera in which the specimen is cooled by a stream of cold nitrogen gas. A gas-stream-cooled chamber for divergent-beam photography was described by Boiko and Shmytko (1974) and by Aknazarov et al. (1974).

A very simple device for use on a full-circle diffractometer has been described by Petsko (1975) for use in protein structure studies. The cryostat, which was designed by G. A. Petsko, M. Pickford, and D. C. Phillips, consists of a double-walled cylinder of polystyrene polished to a high transparency. The space between the walls is evacuated. The crystal can be mounted in the inner cylinder either in a capillary tube or in a flow cell.

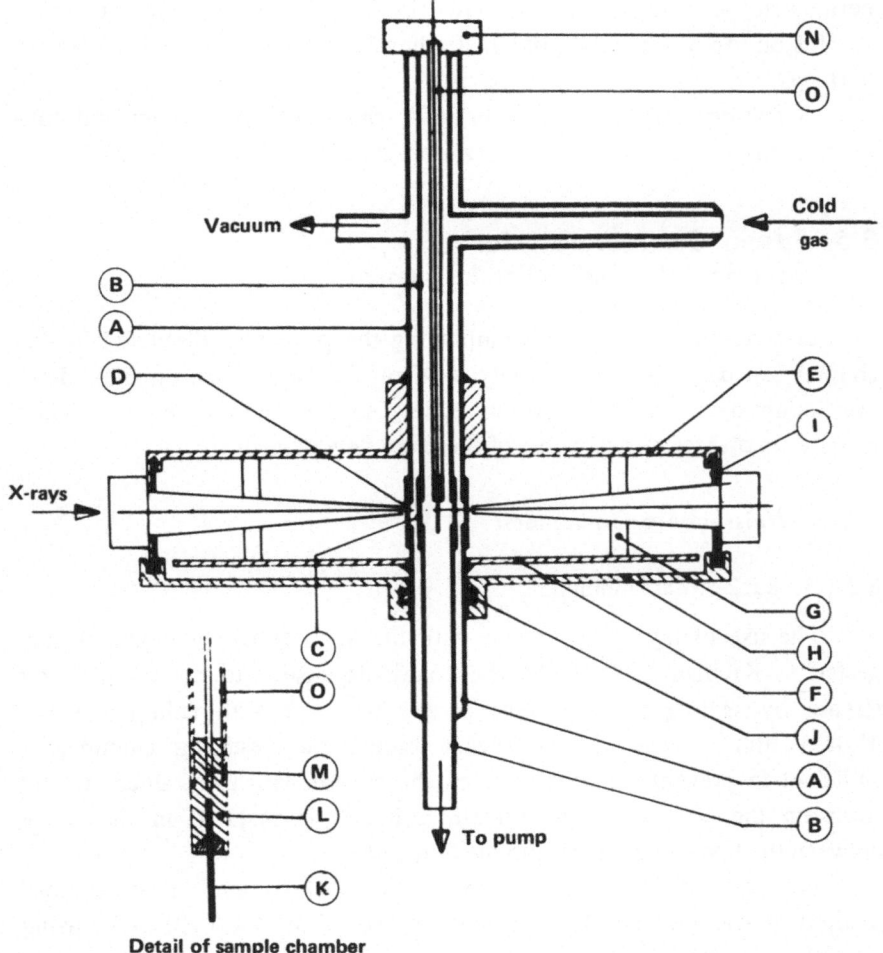

Fig. 3–29. Schematic diagram of low-temperature Debye–Scherrer camera for use between 10 K and 300 K, as seen from above. (A) Outer tube; (B) inner tube; (C) Mylar tube; (D) Mylar tube; (E) camera chamber; (F) platform supported by G; (G) three supporting columns; (H) removable cover; (I) X-ray film; (J) joint; (K) sample; (L) copper cylinder; (M) thermocouple; (N) device for centering the sample; (O) stainless steel tube (Pinot et al., 1965). (Reproduced with the permission of the *Journal de Physique.*)

Cold dry gas is forced into the top of the cryostat and can escape out of one of the ports; only a slight positive pressure (equivalent to the consumption of 0.1 liter of LN_2 per hour) is required to cool the crystal. The connection at the top of the cryostat is via a ball-and-socket joint and permits full ϕ rotation and essentially the full 2θ range as well. The cryostat fits on top of a regular goniometer head and has a temperature stability of 0.3°C.

A few cryostats have been built in which both gas-stream and conduction cooling are used (e.g., Gervais et al., 1966; Thomas, 1972).

3.5. Special Considerations for Specific X-Ray Diffraction Instruments

Low-temperature systems employing the principles discussed in this chapter can be used with a variety of X-ray diffraction instruments. However, adapting a device to a specific type of instrument may require special techniques or accessories. Some of these are now described.

3.5.1. Weissenberg Goniometer

3.5.1.1. Back-Stream Method

The use of layer-line screens with the Weissenberg goniometer suggested to Kreuger (1955) that he could dispense with the warm outer stream by sealing the layer-line screen openings with cellophane (or Mylar) film. The cold gas flowing through the resulting chamber is sufficient to prevent ice formation on the crystal. However, since ice can form on the cold layer-line screen, a heater is inserted in the space between the layer-line screen and the film cassette.

This method has become known as the "back-stream" or "back-flow" method. It was modified by Fridrichsons and Mathieson (1958) by using a window of 0.005-in. beryllium foil (in place of the Mylar) and blowing warm air across the outside of the layer-line screen (in place of the heater).

In a variation of this method, Brown and Wallwork (1962) blocked off the *ends* of the layer-line screen with foam rubber washers and allowed the gas to escape from the layer-line screen opening.

Singh and Ramaseshan (1964) used an approach reminiscent of the early work of Cioffi and Taylor (1922). They blocked off both the ends and the opening of the layer-line screen and drew the cold air out through a cup surrounding the goniometer head by means of an exhaust pump.

Ichikawa (1974) eliminated the problem of ice formation by putting the entire Weissenberg goniometer into a box filled with dry nitrogen gas and blowing cold gas over the sample. Trotter (1959), on the other hand, used a warm outer stream and claimed that ice formation is prevented, because of better circulation of cold gas, if the layer-line screen tube has a large opening cut out of the top.

Dickens (1966) also used an outer stream on his device, but covered the entire barrel (and layer-line screen) with a cylindrical 1-mil-thick Mylar tube (Figure 3–30). The layer-line screen could be slid out from under the Mylar tube for adjusting the crystal (through a small hole in the Mylar tube) and for oscillation and rotation photographs. The layer-line screen was permanently located on the instrument and was supported over the nozzle.

3.5.1.2. Layer-Line Screens

Olovsson (1960) describes a telescoping layer-line screen in which

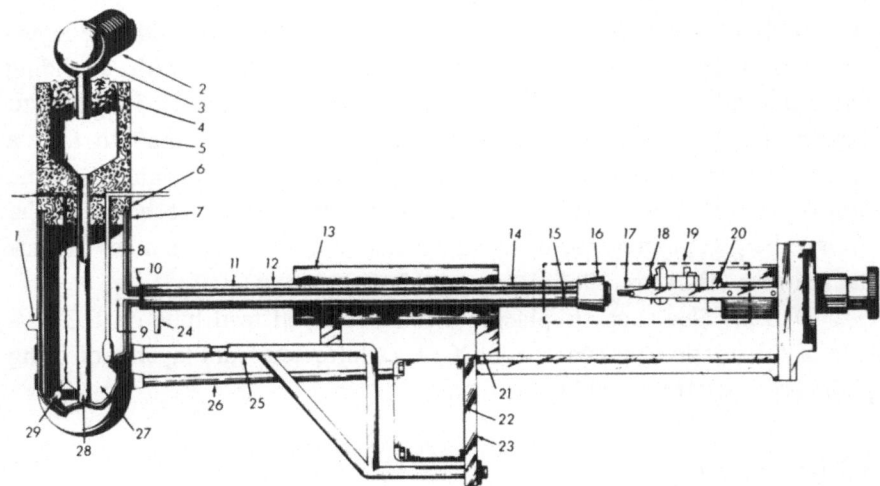

Fig. 3–30. Low-temperature dewar in position on the Weissenberg camera (Dickens, 1966). (1) Degassing attachment; (2) transfer tube; (3) solenoid valve; (4) paper tissue; (5) funnel head; (6) stopper; (7) 1-cm gap between dewar walls; (8) depth probe; (9) 3 mm i.d. × 37-cm-long Pyrex tubing; (10) brass shim; (11) 11 mm i.d. × 35.7-cm-long Pyrex tubing; (12) 17 mm i.d. × 35-cm-long Pyrex tubing; (13) layer-line screen; (14) arm of low-temperature dewar; (15) spacer; (16) funnel-shaped lead foil; (17) capillary containing crystal; (18) aluminum foil skirt; (19) Mylar screen; (20) oscillation beam stop; (21) removable support for layer-line screen; (22) Weissenberg motor; (23) Weissenberg motor plate; (24) inlet for dry air stream; (25) primary support for low-temperature dewar; (26) secondary support; (27) dewar reservoir; (28) funnel tube; (29) 10-watt heater.

the left-hand side (through which the nozzle passes) is divided into two cylinders, with the section closest to the crystal able to slide over the fixed part (which also supports the nozzle). Collin and Lipscomb (1951) describe a three-piece layer-line screen that can be telescoped over the spindle of the Weissenberg goniometer.

Another approach which can be used if a warm outer stream is employed entails the construction of a layer-line screen split horizontally for ease of removal during crystal growth and alignment procedures. Layer-line screens in the form of two semicylinders hinged along one edge and with catches on the other edge have been described by Abrahams and Lipscomb (1952) and Clark and Baxter (1966).

3.5.1.3. Nozzle Support

If removable layer-line screens are used, the nozzle should be supported by a bracket fixed to the end of the Weissenberg goniometer (Figure 3–30). Otherwise, it must lie on a support that fits inside the layer-line screen (Figure 3–14), which must then be fitted with a sliding panel that allows access to the goniometer head adjustments. It is best to support it firmly from the end of the instrument rather than to take a chance on the nozzle's moving during upper-level adjustments. The bracket is attached by drilling and tapping new mounting holes in the Weissenberg goniometer or by using existing motor-mounting holes or similar features (Dickens, 1966; Rudman, 1966a).

Dickens also suggests placing a funnel-shaped lead tube on the outside of the nozzle to prevent stray X-rays (diffracted from layers not being photographed) from striking the film.

3.5.1.4. Film Cassette

The standard cylindrical film cassette will generally not fit over the nozzle [although if it has a continuous slot along the back, this slot can be enlarged to permit it to slip over the layer-screen (Collin and Lipscomb, 1951; Dickens, 1966)]. Steinfink et al. (1953) designed a special split film cassette that is now available from most Weissenberg goniometer manufacturers. This cassette is split horizontally (parallel to the cylinder axis); the upper half is easily removed while the nozzle is in place. If a narrow-diameter nozzle (less than 35 mm o.d.) is used, the lower half of the cassette can usually be removed if the carriage of the Weissenberg goniometer is moved to the extreme right-hand position (beneath the barrel) and the cassette is slid off to the left.

Removal of the cassette is not generally required once the crystal is aligned and the various layers are being photographed. However, it is very useful if the crystal is being grown *in situ* and must be examined between crossed polaroids.

A split film cassette requires the use of a good film envelope that can be loaded in the darkroom and then placed in the cassette. A description of such an envelope is given by Dickens (1966). They key to the success of this envelope is to prepare cardboard spacers (1–2 mm thick, depending on the number of films in a film pack) by soaking them in hot water (in a dewar) for about one hour. They are then bent around a 5.7-cm-diameter paper roll and held in position with wide rubber bands until dry. A piece of light-tight black paper, 16 cm wide and 35 cm long (enough to form both sides and the flap of the envelope) is then folded to fit and pasted onto the spacers. The envelope should be checked to be sure that it is light-tight.

A number of commercially sold split film cassettes require the use of two pieces of film, one for each half. However, the use of the film envelope can allow a single sheet of film to be placed on the cassette. A slight modification, i.e., the removal of two extra restraining clips, one from each half of the cassette, is all that is necessary.

Gopalakrishna and Cartz (1972) described a novel modification to the Stoe low-temperature system in which a tapered Teflon nosepiece was machined to fit into the slot of a standard film cassette. The flow of gas lies nearly parallel to the collimator and requires *no* modification of existing apparatus. However, the apparatus was run at relatively high temperatures. How it would react at lower temperatures (below $-50°C$), *vis à vis* the formation of ice, is not known. This approach is certainly worth further investigation, although its use for growing crystals *in situ* is not recommended.

Viswamitra (1962) also bypassed the problem of using a split film cassette by designing a new arrangement for the spur and bevel gears that couple the worm gear to the shaft which rotates the specimen. The goniometer head support and the entire driving system are mounted on a right-angled vertical aluminum post shaped so that it is possible to load or remove a cylindrical film cassette from the side opposite the nozzle. He also suggests using a dummy camera to maintain the proper environment for the sample whenever the film cassette is removed for processing the exposed film.

However, Singh and Ramaseshan (1963) claim that Viswamitra's camera has a large amount of backlash. Instead, they use a standard Weissenberg goniometer with a film cassette that slides over the cold-stream support so that it can be removed without disturbing the system.

3.5.1.5. Flat-Plate Adapter

Collin and Lipscomb (1951) used a flat-plate film cassette attached to the base of the Weissenberg goniometer to align crystals grown at low temperatures.

The use of a Polaroid film cassette, which allows the rapid examination and orientation of crystals, is especially helpful in low-temperature work. A device to replace the removable film cassette carriage on a Nonius–Weissenberg goniometer with a Polaroid film cassette support was described by Rudman (1968). This adaptor fits over the rails of the instrument and can be interchanged with the cylindrical film cassette carriage while the cold stream is in place (Figure 3–31).

A procedure for aligning the crystal with the Polaroid flat-plate cassette and also suggestions on how to use it for obtaining flat-plate powder photographs and for checking the quality of single crystals are given in Section 8.3.2. The logical extension of this device, to hook the base of the flat-plate cassette holder onto the worm gear and obtain a flat-plate Weissenberg photograph, is described by Hope (1969) along with a nomograph for the interpretation of the resulting photograph.

3.5.1.6. Upper-Level Photographs

For full use of the Weissenberg equiinclination capabilities, the low-temperature apparatus must allow photography of the upper levels. The problem of allowing this freedom of motion has been solved in several ways. Most common is the use of a small dewar that rides with the camera and is not rigidly attached to the reservoir (see Section 3.4.2.2). It is also possible to place the gas generator on a small dolly or a piece of Teflon sheet and to move it as the equiinclination angle is set. All gas lines on the warm, inlet side of the gas generator should be flexible. A third suggestion is to use a flexible, insulated transfer line between the generator and the cold stream so that the generator does not have to be moved (Olovsson and Templeton, 1959).

It is important to remember that the Weissenberg goniometer is normally set at a take-off angle of 4–6°. As the equiinclination angle is adjusted, a dewar attached to the camera will change its height relative to the table top. Sufficient clearance must be allowed for all settings of the camera. If the X-ray tube is mounted so that its ports are close to the table top and the cable is suspended from above, the Weissenberg goniometer will be slanted downward toward the tube. Under these conditions, the clearance must be checked at the *highest* equiinclination angle. In any case, smooth adjustment of the instrument must be possible.

3.5.2. Precession and Oscillation–Rotation Cameras

The precession camera is an open instrument, and so special precautions must be taken with respect to frost prevention. This entails the construction of a low-humidity chamber around the camera (see Appendix 9) or the use of the outer stream.

What is the proper position for the nozzle? Several investigators prefer to place the nozzle at an angle of about 60° to the direction of the collimator, parallel to the top of the X-ray generator (Burbank and Bensey, 1951; and others). If the precession arc is then placed in a horizontal position and a suitable precession angle is set, the cold gas stream is directed along the spindle axis (Figure 3–32). This is very useful for crystal growing and centering procedures, since the thermal gradients in the vicinity of the crystal are minimized. If the crystal is solid at room temperature, there are those who prefer to direct the cold gas stream down onto the crystal from in back of the X-ray tube, so that there is minimum interference with the precession camera.

Some precession cameras must be swung away from the port in order to center the crystal. In such a case, a movable gas generator (as described in the previous section) must be used. However, it can be eliminated by using a telescope mounted perpendicular to the spindle axis (Burbank, 1973) (Figure 3–32). This location for the telescope has also been used when the 60°–horizontal position is used (Renaud and Fourme, 1967; Lippman and Rudman, 1976).

A cryostat, with a temperature range of −175°C to +450°C, has been described in which the crystal is enclosed within a quartz chamber

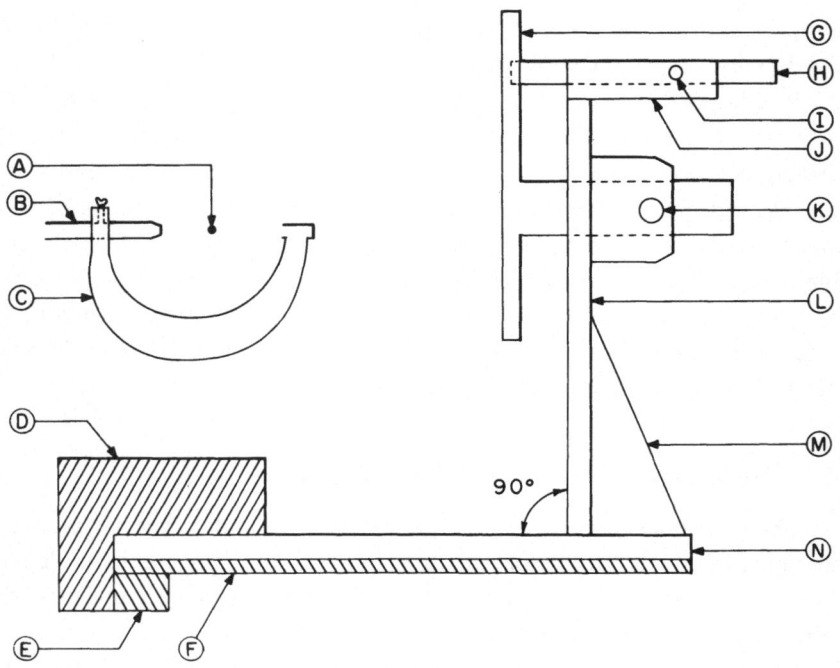

Fig. 3–31. Schematic drawing of adapter: (A) Crystal; (B) collimator; (C) beam-stop; (D) lead counterweight; (E, F) lucite (to prevent damage to support rods on goniometer); (G) Polaroid XR-7 Precession Mount #57–5; (H) sliding rod to set crystal-to-film distance; (I) setscrew; (J) bracket; (K) setscrew; (L) aluminum plate (10 cm × 6 cm × 6 mm); (M) brace; (N) aluminum plate (13.5 cm × 10 cm × 6 mm) (Rudman, 1968). (Reproduced with the permission of the International Union of Crystallography.)

outfitted with Mylar film windows (Figure 3–33). After the crystal is centered, the cooling chamber is attached to the goniometer head base and supported, at the other end, by the heat exchanger, which rests on the telescope mounting dovetails (Gobrecht et al., 1971). Another precession cryostat utilizing a cold gas stream has been described by Boiko et al. (1973b). The cryostat is placed above the crystal and the cold gas is warmed as it leaves the cryostat in order to eliminate frost formation.

There are times, e.g., when growing a single crystal of a disordered phase *in situ,* when it is more convenient to interpret an oscillation or rotation photograph than a precession photograph. An adapter which converts a precession camera to an oscillation–rotation camera has been constructed (Lippman and Rudman, 1976). A reversible, high-torque, low-speed motor with adjustable limit switches is mounted with the shaft of

Fig. 3–32. Low-temperature apparatus for precession camera (Burbank, 1973). (Reproduced with the permission of the International Union of Crystallography.)

the motor colinear with the spindle axis of the precession camera (when $\bar{\mu}$ = 0°). This device can be mounted and removed without disturbing the camera or crystal alignment. It can be used to obtain flat-plate rotation or oscillation photographs with regular or Polaroid film. The oscillation range is completely adjustable from 5° to 300°. It is also possible to use this adapter for aligning crystals according to the method described in Section 8.3.3.

Oscillation–rotation cameras are becoming quite popular with macromolecular crystallographers. Many reflections can be recorded rather quickly and their intensities measured on automated microdensitometers. These cameras are similar to the precession camera in that they include a flat-plate cassette and readily accessible sample. Therefore, a simple support or bracket can be constructed to hold a cold-stream nozzle near the crystal for low-temperature studies.

3.5.3. Single-Crystal Diffractometer

A fundamental problem encountered in adapting low-temperature apparatus for use in single-crystal diffractometry is the same as that mentioned for the precession camera: proper positioning of the nozzle so as to avoid interference with the operation of the diffractometer. The standard approach is to locate the nozzle as nearly parallel to the collimator as possible. Gopalakrishna and Cartz (1972) have carried this to its logical conclusion by surrounding the collimator with a Teflon cone through which

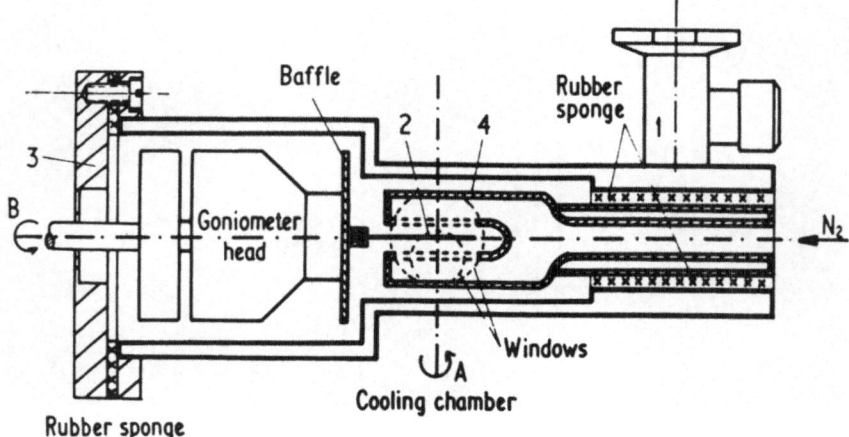

Fig. 3–33. Heat exchanger and cooling chamber for precession camera. (1) Valve for evacuated chamber; (2) tube for release of nitrogen gas; (3) aluminum disc;

the cold gas is blown (Figure 3–34). Since they are operating at fairly high temperatures, as mentioned in the discussion of their Weissenberg apparatus, insulation requirements are minimal and no outer stream is required. Earlier attempts to use this approach with lower temperatures met with failure for several technical and practical reasons: the short nozzle resulted in lack of laminar flow; moisture formed on the inside of the collimator; and there were sharp thermal gradients at the crystal due to the presence of the collimator in the center of the cold gas stream. Recently, however, Sharon (1975, personal communication) has designed a collimator through which the cold gas stream passes along with the X-ray beam. This is used in conjunction with a dry box around the diffractometer.

A second approach has been to place the nozzle on the χ-circle diametrically opposite the crystal, so that the cold gas stream blows along the φ-circle axis. This has been used by Cucka et al. (1970) and is also

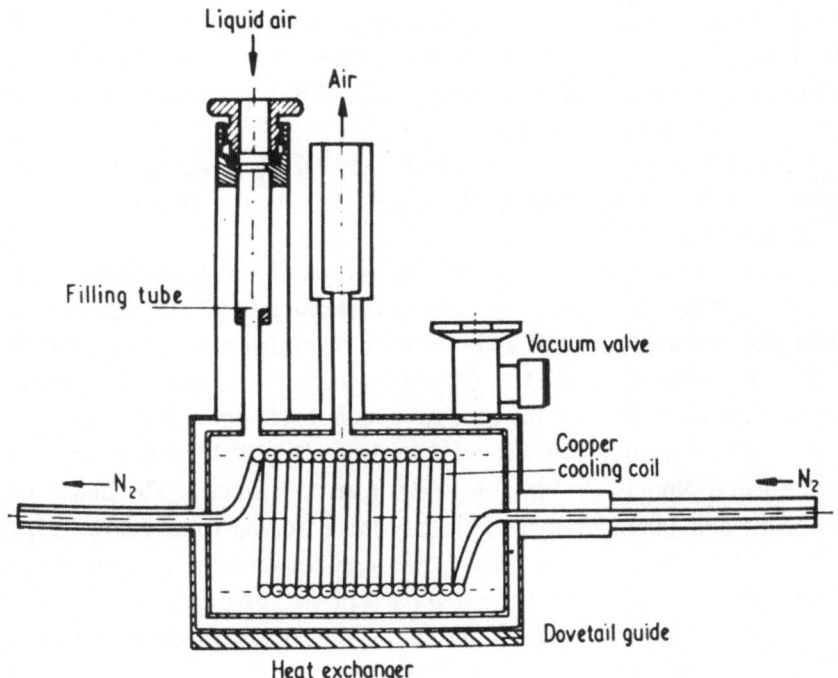

(4) quartz cooling chamber; (A) axis of precession camera; (B) axis of gonimeter head (Gobrecht et al., 1971). (Reproduced with the permission of the Institute of Physics.)

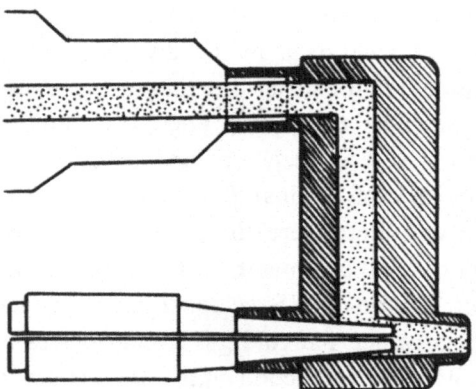

Fig. 3–34. Low-temperature Teflon nozzle for diffractometer collimator (Gopalakrishna and Cartz, 1972).

incorporated in the Syntex system. Sturdy mounting and flexible lines, as well as diffractometer settings that will not foul the lines, are necessary. Cucka et al. operated at relatively high temperatures ($-13°C$) so that flexibility of the transfer lines was no problem. The Syntex system uses a series of short dewar tubes connected with Teflon joints, which are themselves covered with rubber to prevent icing.

Strouse (1975) improved on this design by devising glass-to-glass joints that do not ice over as the nozzle follows the crystal around the χ-circle (Figure 3–20).

Davies et al. (1970) designed a brass sphere, made in segments, that mounts within the χ-circle of the diffractometer. These segments are arranged so that the χ- and φ-circles have no restrictions, while 2θ is limited to the range of $2.5–57.0°$. The cold nitrogen gas is directed so that the crystal mounting fiber and the cold gas stream are at about a $45°$ angle.

A different approach was tried by Bolhuis (1971) and incorporated in the Enraf-Nonius Universal Low-Temperature Device. The dewar tube is curved and mounted so that the crystal is cooled from below (Figure 3–35). The cold gas-stream flow rate is very low and the cold gas spreads downward so as not to cool the metal parts of the instrument.

Fourme (1975) places the nozzle so that data collection is restricted to the lower hemisphere in χ. A heater is mounted on the modified xyz-goniometer head, with two sliding contacts on the base of the goniometer head used to deliver electric current to the heater at all positions of ϕ.

A novel nozzle design has been described by Colapietro (1975). A

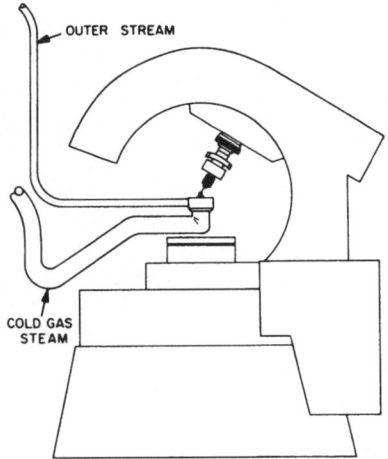

Fig. 3–35. Cooling of crystal mounted on diffractometer from below (Enraf-Nonius, 1971).

dewar is constructed with a rigid sidearm extended downward from its base at approximately a 45° angle. This arm serves as the nozzle. Cold gas, from above the LN_2 level in the dewar, passes through a copper-coil heat exchanger immersed in the LN_2 into a German silver tube extending into the sidearm nozzle. This metal insert is cooled by LN_2 and reaches to the end of the nozzle (Figure 3–36). In this manner, the nitrogen gas is cooled by LN_2 to within a few centimeters of the crystal. A heater is placed at the end of the nozzle to prevent ice formation and a reed switch system is used to control the LN_2 refill cycle.

3.5.4. Powder Cameras and Diffractometer

The cooling of powder specimens, whether on a flat-plate camera, Debye–Scherrer camera, Guinier camera, or powder diffractometer, is very similar to the cooling of single-crystal specimens. Although the nature of diffractometer specimens suggests the use of conduction methods as the preferred cooling method, a cold gas stream can be used effectively and is recommended when low-temperature studies are performed only occasionally. A distinct advantage of using gas-stream cooling on a powder diffractometer is that instrument misalignment is minimal and very easily corrected.

The earliest application of this technique consisted of surrounding the specimen, in a capillary tube, with a glass or cellophane tube so as to prevent ice formation (McKeehan and Cioffi, 1922; Eastman, 1924). After

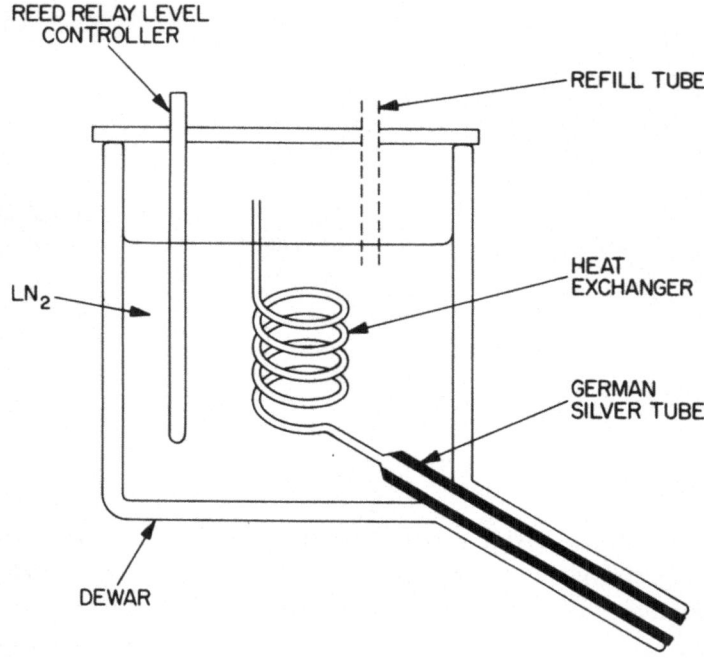

REED RELAY LEVEL
CONTROLLER

REFILL TUBE

HEAT
EXCHANGER

LN_2

GERMAN
SILVER TUBE

DEWAR

Fig. 3–36.   Schematic diagram of the system constructed by Colapietro (1975). Gas passes through the heat exchanger and exits through the sidearm of the dewar. Liquid nitrogen cools the heat exchanger up to the end of the sidearm which is located within a few centimeters of the crystal.

the "outer-stream" method was developed, this precaution was no longer necessary.

McKeehan and Cioffi (1922) mounted the sample tube in a metal bearing that was outfitted at one end with a light screw *propeller*. The cooling air stream thus supplied the "motive power" for rotating the sample.

In the case of the "open systems" (flat-plate cameras, Debye–Scherrer cameras in which the sample is not enclosed in the film cassette, and horizontal diffractometers with the sample accessible from above), most of the systems described in Sections 3.3 and 3.4 can be used.

Thewlis and Davey (1955) warn that metal specimens which are in contact with the body of the camera or diffractometer can be warmer than the cold gas stream and should be insulated from this source of heat.

Several cooling devices have been designed specifically for the

Debye–Scherrer camera. Hume-Rothery and Strawbridge (1947) placed the sample in a chamber equipped with cellophane windows and cooled the outer portion of the chamber. The film is placed outside of the sample chamber, while the outside of the camera is bathed in warm gas to prevent ice formation. Temperatures as low as $-100°C$ were achieved.

Debye–Scherrer cameras in which the sample axis is vertical have been adapted for low-temperature use in several ways. Dunoyer (1952; Bouttier and Dunoyer, 1953) supported the sample from the *top* of the camera chamber and drew the cold gas up a tube extended from the gas generator, which was below the camera, by means of an exhaust pump.

On the other hand, Campbell and Hildebrand (1943) supported the sample from the bottom and used an exhaust pump to force the cold gas down over the sample. They adjusted the opening of the tube by inserting or removing thin glass rods.

Pearson (1954) also preferred to support the sample from the bottom of the camera and to blow cold gas or let liquid nitrogen drip onto the specimen. The camera was placed in a larger chamber which had warm helium gas flowing through it.

Francombe (1957) divided the inside of the camera chamber into two parts separated by a window made of film base. The sample was cooled but the film and camera body remained near room temperature. The cold gas exited through a tube coaxial with the cold stream.

Taylor (1960) designed a vertically split film cassette that can be loaded and removed from the camera without disturbing either the cold gas stream or the sample.

Hovi et al. (1964) were interested in obtaining a built-in calibration curve. Rather than mixing the sample with a known material, they constructed an apparatus with a collimator which divides the X-ray beam. The sample is attached to the bottom of the camera with the cold gas stream blowing onto it from above. The sample tube is slipped through a hole in an iron disc which is located at the center of the split X-ray beam. Two samples are placed in the sample tube, one of them being a material whose low-temperature behavior is known, and two powder patterns are obtained simultaneously. The working parts of the camera are kept warm by a brass-cone shield placed over the sample tube and by flowing water that runs through the camera body.

A Debye–Scherrer camera with a diameter of 160 mm for use with liquid helium was described by Lévy (1969). The LHe dewar is protected by a LN_2 outer bath; cold helium gas passes through the sample chamber on its way out of the camera.

Several investigators have designed special covers that can be used to modify the standard Philips Debye–Scherrer camera (with a horizontal sample axis) for LTXRD studies. A permanently modified cover can be made by drilling two holes: one just above the sample to which the cold stream is connected and the second for the cold gas to escape (Kramer et al., 1963). The interior of the camera is then flushed constantly with dry cold gas and no outer stream is required. In order to prevent light from leaking in, the inlet and outlet tubes are formed into short coils, while warm air is used to keep the pulley from freezing. A similar cover was designed by Crandall (1969). In this device, the camera chamber is divided into two parts; cold gas cools the sample, but warm gas keeps the film warm and ice-free.

An earlier device (Wood, 1953) used a single tube passing through the cover of the chamber at two points. The cold gas flowed into and out of the camera chamber through this tube and cooled the sample, which fitted into a slot cut in the tube (Figure 3–37).

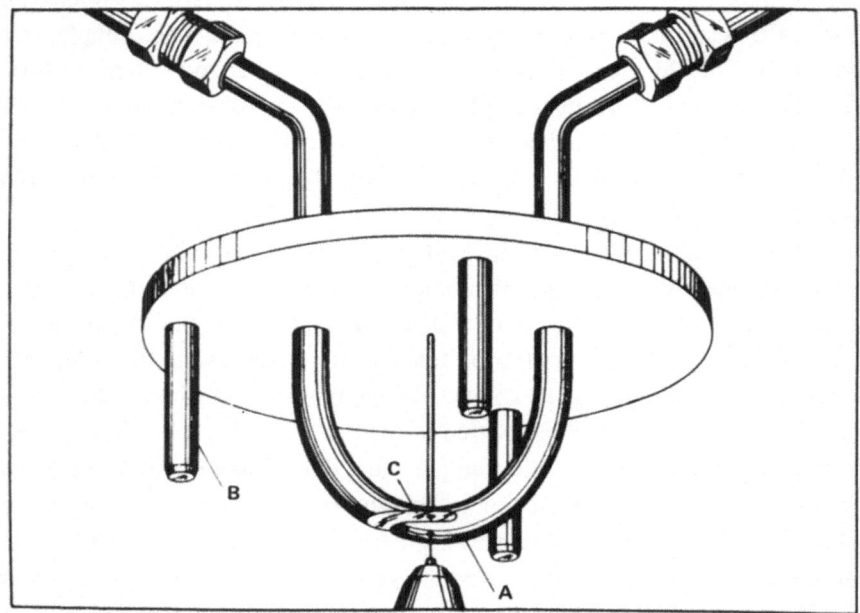

Fig. 3–37. Modified cover for Debye–Scherrer camera. Cold gas flows through tube A and cools specimen. X-ray beam passes through slot C (Wood, 1953). (Reproduced with the permission of the American Institute of Physics.)

Wheeler (1968) also replaced the cover plate of a Philips Debye–Scherrer camera, but at the same time he modified the sample rotation mechanism to allow the sample tube to move perpendicularly to the beam while rotating. This additional sample motion improved the quality of the powder photographs.

An alternative method, good for occasional research needs (Rudman, 1966a,b) or for student use (Rudman, 1967) consists of preparing a light-tight film envelope for the Philips camera so that it can be used without the standard cover. Rubber washers cut from black rubber tubing are fitted over the collimator and beam-stop and pushed up against the film envelope to prevent light from entering the envelope. A simple system is shown in Figure 3–17. The system can be placed on a rolling platform made by clamping a strip of ¾-in.- thick plywood to the top of the X-ray generator and nailing wooden guiderails in place. A wooden "stop" is located so as to permit reproducibility in rolling the device up to the camera after the sample has been centered and the camera loaded with film.

A vertical diffractometer can be modified by cutting a hole in the safety shield surrounding the sample; in some models, the central section of this shield is removable. The cold stream is then brought in close to the sample (Figure 3–38), with the axis of the nozzle lying along the axis of the $\theta/2\theta$-circle (Miksic et al., 1959; Sakurai and Suzuki, 1959). It may be necessary to use small sample holders or to use a large nozzle opening so as to ensure uniform cooling of the specimen.

Kellett and Steward (1962) designed a specimen mount for the Philips vertical diffractometer in which a stream of cold gas is directed through a flared opening at a 45° angle to the sample. The diffractometer is covered with a plastic tent or bag to prevent ice formation.

Cole and Holmes (1960) and Bonfiglioli and Testard (1964) describe methods of cooling samples mounted on horizontal diffractometers. The latter blow cold gas over the sample, which is contained in a cryostatlike chamber equipped with beryllium windows. The gas leaves through the outer of two concentric tubes. A simpler version employing Mylar windows is also described. Boiko et al. (1973a) have also constructed a cryostat for liquid-helium temperatures (see Section 3.4.4).

A description of low-temperature apparatus for a small-angle scattering diffractometer is given by Naudon and Jaulin (1968).

Simon (1971) described a moving film camera based on a modified

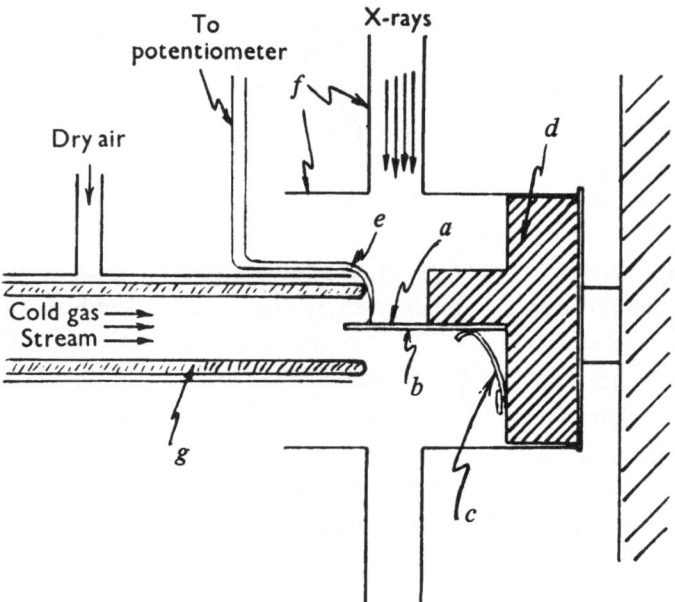

Fig. 3–38. Gas-stream cooling of sample on vertical powder diffractometer (Miksic et al., 1959). The sample holder (b) is held firmly to the goniometer (d) by a metal spring (c). The specimen (a) is cooled by a cold gas stream (g) whose temperature is measured by a thermocouple (e). The standard radiation shield (f) is used. (Reproduced with the permission of the International Union of Crystallography.)

Guinier design in which the temperature can be maintained to within 0.1° from −190° to +400°C. This camera is used to obtain the powder pattern of a sample as a function of temperature. A modification of this camera, operating in the range of −180° to 1100°C, is sold by Enraf-Nonius. The low-temperature operation consists of a stream of cold gas blown over the sample, using a modified "Universal Low-Temperature" Device (see Section 3.4.2.2).

It is also possible to approximate the Guinier–Lenné or Simon device by using a standard Weissenberg goniometer adapted for low-temperature work. Place the film cassette on the goniometer and take a series of powder photographs on different sections of the film through a half-inch layer-line screen opening at different temperatures. If carefully done, it is possible to observe the change in lattice constants as a function of temperature. Any interesting regions can be studied at shorter temperature intervals in a second photograph.

A standard Guinier camera has been used for low-temperature studies by modifying the sample holder. The sample is placed in several thin-walled capillary tubes which are placed side by side in the modified sample holder and then cooled (Veith, 1975). This method is good for handling liquid or reactive samples.

3.6. Summary of Gas-Stream Methods

A. Properties
   1. Ready access to crystal
   2. Crystal visible
   3. 0.2° stability at best
   4. 2–5° stability usual
   5. No absorption problem
   6. Some frost problem
   7. 1–8 liter/h consumption rate
   8. Usual T: 100 K
   9. Easily adapted to other instruments

B. Gas-generating methods
   1. Heat exchanger with mechanical cooling
   2. Heat exchanger with cold bath
      a. Ice–salt–water
      b. Dry-ice slurry
      c. Liquid nitrogen
   3. Boiling liquid nitrogen
   4. Boiling liquid hydrogen or helium

C. Maintaining temperature stability during refilling of generator
   1. Boil gas inside inverted bell
   2. Boil gas in second chamber connected to first by small opening
   3. Frequent refilling with little change in level of cryogen
   4. Use intermediate dewar
   5. Use pressure-regulating valve
      a. Special cryogenic valve
      b. Warm gas to room temperature with standard valve and then recool

D. Liquid-nitrogen level controller

   1. Standard level controller (if system not pressurized)

   2. Thermistor sensor

   3. Resistance sensor

   4. Float with magnet to sensitize reed switch

   5. Float with opaque top to control photoelectric cell

E. Nozzle—materials of construction

   1. Glass

   2. Metal

   3. Teflon

   4. Foamed polyurethane

F. Nozzle—proper location (single-crystal diffractometer)

   1. Nearly parallel to collimator

   2. Cocentric with collimator

   3. On χ-circle diametrically opposite crystal

   4. Mounted below sample with cold gas directed upwards

G. Methods for frost prevention

   1. Enclose sample in dry chamber

   2. Enclose X-ray camera or diffractometer in dry chamber

   3. Use warm outer stream

      a. Separate warm and cold gas streams

      b. Heat outer portion of cold gas stream

   4. Draw cold air past sample with an exhaust pump

H. Temperature Control

   1. Mix warm and cold gas streams

   2. Heater in nozzle controlled manually

   3. Heater controlled by feedback from thermocouple

   4. Two heaters—for coarse and fine control

   5. Use flowmeters on all lines for reproducible gas flow and temperature

CHAPTER 4

Conduction-Cooling Apparatus

4.1. General Principles

4.1.1. Introduction

Conduction-cooling techniques have been developed for a variety of applications in the field of X-ray diffraction. Systems have been described for attaining minimum temperatures of 1.3 K or 250 K, for studying thermal diffuse scattering or thermal expansion, for collecting single-crystal intensity data or powder data, and for growing crystals from gases or cooling metallic specimens. Clearly, no one system will do a satisfactory job in all these cases. Most systems have been developed for a specific purpose and for use on a specific instrument. The universal adaptability of gas-stream cooling is not common to the conduction method of cooling.

The basic cryostat consists of a cold reservoir which cools a sample chamber or mounting block which, in turn, cools the sample by conduction (Figure 4–1). The sample is usually placed inside a chamber so that

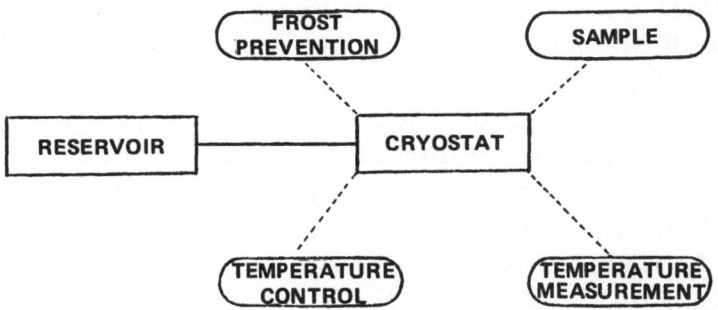

Fig. 4–1. Schematic outline of basic cryostat.

it will not gain heat by thermal radiation or by contact with the atmosphere. This chamber is either evacuated or filled with a dry gas. Since the cold reservoir is not normally in direct contact with the specimen, the problem of automatic refill of cryogen, which plagues the gas-stream method, is not evident in the design of conduction-cooling apparatus. However, the problem of absorption of the X-ray beam by the sample-chamber windows, which is virtually nonexistent in the gas-stream method, is of paramount importance in the design of a cryostat.

Several methods have been used for cooling the cold sample holder. These include the use of consumable cryogenic materials such as liquefied gases (He, H_2, N_2, O , Ar, air) or dry-ice and water–ice slurries, self-contained sealed refrigeration units (operating from an electric outlet), Joule–Thomson expansion coolers (operated by compressed nitrogen or helium gas), and thermoelectric devices (operated by electricity). The basic principles of these devices are described in Appendices 3–5. The properties of liquefied gases are discussed in Appendix 2, while Appendices 6–8 are devoted to dewars and transfer lines.

Very often, the use of a conduction cryostat for low-temperature X-ray diffraction studies requires extensive modifications of the X-ray equipment itself and/or an expensive cryostat (whether commercially obtainable or constructed by the user). In particular, the design and construction of the sample chamber and windows requires an expertise not generally developed by the X-ray crystallographer.

Normal laboratory workshops do not have the necessary machinery to fabricate an efficient, well-constructed cryostat. Therefore, the finished product is often larger and heavier than necessary, with the results that it takes longer to evacuate the cryostat for each experiment, and the equipment consumes more cryogen than is necessary. For this reason, specific construction details are not included in this chapter. If any of these instruments are to be constructed by the user, he is referred to the original literature, or, if possible, to the original designer. On the other hand, a number of simple powder-diffractometer adapters have been described in the literature; more details are given for these devices.

Also included in this chapter are some instruments designed for use with neutron diffractometers, either because they have been adapted already for use with X-ray diffraction instruments or because the authors suggest such modifications as are necessary to make them useful for X-ray diffraction apparatus. These changes generally entail a reduction in size and/or the substitution of Mylar or beryllium windows for aluminum or quartz windows.

4.1.2. Advantages of Conduction-Cooling Techniques

As was already mentioned, the sample chamber is usually separated from the cold source. Thus the *refill* problem, so critical in the design of gas-stream devices, is negligible. *Frost* formation is a negligible problem in any properly designed cryostat because the crystal is in a chamber that is either evacuated or protected by a vacuum-insulated cover.

Exceptional *temperature stability* is attained by the use of thermocouple-controlled heaters acting in conjunction with a constant-temperature cooling block. Temperature variations of as little as 0.01° have been reported (Losee and Simmons, 1968; Girt et al., 1974). The *lowest temperature* reached during an X-ray diffraction investigation has been reported as 30 mK (Simmons, 1976). Temperatures in the 10 to 20 K range can be easily attained for both powder and single-crystal studies. The temperature is easily *controlled* by a heater attached to the sample holder or by using different cooling baths.

Finally, *thermal equilibrium* is readily attained in a cryostat, with the only heat input resulting from small heat leaks into the system. In contrast, when the gas-stream method is used, warm gas is constantly entering the heat exchanger and must be cooled down, while the liquefied gas is being boiled off at a rather rapid rate. Consequently, the *consumption* rate of liquefied gases is considerably lower in a cryostat than in a gas-stream generator. Consumption rates of 0.1 to 0.2 liters of liquid helium per hour and 0.4 to 0.5 liters of liquid nitrogen per hour are easily attained.

4.1.3. Disadvantages of Conduction-Cooling Techniques

Major deterrents to using conduction techniques are those of design, complexity, and initial cost. As the basic design is modified to solve a particular problem, the complexity of the instrument (both in construction and in use) increases and so does the cost. Each instrument, therefore, is based on a delicate balance of research needs, available funds, and designer and operator patience. Several other disadvantages common to many cryostats are discussed in the following paragraphs.

4.1.3.1. Limitations on Data Collection

While many cryostats have been designed for use with powder diffractometers (e.g., Barrett, 1956; King and Preece, 1967; Himmler et al., 1969) or for the collection of zero-level single-crystal data (e.g., Chopra, 1962; Streib and Lipscomb, 1962; Morosin, 1966), only a few have been designed to collect full three-dimensional single-crystal data (e.g., Smith and Lipscomb, 1965; Smith, 1966; Coppens et al., 1974; Air Products, 1973). This is due, primarily, to the *difficulty in orienting a crystal* located inside the cryostat. The following methods have been used to solve this problem: (1) moving the entire cryostat together with the crystal (Abrahams, 1960; Heaton et al., 1970; Air Products, 1973; Coppens et al., 1974); (2) the use of complicated gearing so that the crystal inside the cryostat can be rotated about one or more axes (Atoji, 1965); and (3) the utilization of Weissenberg geometry, in which the crystal has no χ motion but the detector, which is located outside the cryostat, moves (Smith and Lipscomb, 1965; Smith, 1966).

Other factors restricting the collection of data in a cryostat include the presence of *blind spots* due to support rods or to the window size and limitations caused by the *physical construction* of the apparatus. If the cryostat contains liquefied gases, then it cannot be tilted past a certain point or the liquid will spill. Even in the case when liquid is not used (see Sections 4.5, 4.6, and 4.7), restrictions in the mode of data collection are caused by cables between the "cooling-energy" source (gas cylinder or electric outlet) and the cryostat. In most cases, these restrictions are not as severe as might be expected. All the necessary data can usually be collected with one mounting of a sample, except in the case of single-crystal data from a triclinic crystal (which may sometimes require mounting along a second axis).

In addition, a cryostat attached to the sample holder (e.g., diffractometer sample-mounting block or goniometer head) requires a rigid support. The design requirements in such a case often add to the complexity of the instrument and restrict its mobility and, therefore, the data that can be collected with it (e.g., Coppens et al., 1967).

4.1.3.2. Absorption of X-Ray Beam

A second major disadvantage of conduction cooling is the absorption of the incident and diffracted X-ray beams by the cryostat windows. This problem has been overcome in three ways. The most common method is to use very thin Mylar or beryllium windows. These windows are quite fragile and must be securely supported (occasionally these supports cause blind spots), especially if the sample chamber is evacuated. A less common approach is to use more penetrating radiation than would normally be used (e.g., Mo K_α, McDonald et al., 1966; Ag K_α, Lytle, 1964). Finally, and this approach is often used in conjunction with one of the others, absorption corrections are applied to the observed data (see Section 8.4.1).

4.1.3.3. Thermal Gradients

A third problem is that of temperature gradients within the sample. The sample must be in good thermal contact with the cooling block. The problem is minimized with large, flat, thin powder diffractometer samples by mounting them on a copper plate. Single crystals are usually mounted on a metallic pin (which should be kept out of the beam or, at the very least, accounted for in the observed pattern). McDonald et al. (1966) pointed out that the conduction method is good for metallic specimens. However, in the case of materials with low thermal conductivity, such as AgI, the exact specimen temperature is not, in general, known. These thermal gradients are not usually very large and may offer difficulty only if the temperature need be known to within 0.2°.

An excellent systematic study of the best method for preparing conduction-cooled powder samples was recently published (Etourneau et al., 1975). The investigation was carried out using a liquid-helium cryostat with provision for recovering the helium gas. Three methods of sample preparation were used.

T_1: A suspension of the sample in alcohol or acetone was deposited in a thin layer on a copper sample holder. After the carrier evaporated, the sample was used without further treatment.

T_2: The sample was mixed with a thermally conducting grease or varnish and was spread onto the sample holder.

T_3: A special sample cell was constructed in which the sample was in contact with cold helium gas. The cell was cooled by contact with the cooling block of the cryostat. The sample, which was deposited on the sample holder without any adhesive, was cooled by conduction with the cold carrier gas.

The temperature variation of the unit-cell dimensions of known materials was used to calibrate the temperature recorded by a thermocouple embedded in the cold block (the usual position for a thermocouple in a conduction cryostat) vs. the actual temperature of the sample.

The results of this study showed that when the thermocouple read 4.2 K, the T_1 samples registered unit-cell dimensions corresponding to 160 K while the T_2 and T_3 samples indicated 4.2 K. However, the peak width at half-height varied inversely with the temperature for the T_2 samples, but remained constant (and narrow) for the T_3 samples. This is correctly explained as being due to the different rates of thermal contraction exhibited by the sample and sample binder, resulting in increased mechanical strains in T_2 samples as the temperature is lowered.

One can conclude from this that the most efficient method for cooling a powder specimen using conduction techniques is to use a sample cell in contact with the cold block and filled with a carrier gas. If constant peak width is not crucial, thermal gradients in the sample can be eliminated by using a conducting grease or varnish to bind the sample to the sample holder.

Although not the result of a systematic study, similar conclusions were reached by Cocks et al. (1966) and Eubig and Tomizuka (1972). The latter observed that, in a vacuum, the temperature of the sample is sensitive to thermal contact resistance between the sample and sample holder. In order to minimize thermal gradients inside the cryostat and to ensure temperature uniformity, the sample chamber was filled with helium gas cooled to liquid-nitrogen temperature. Critical parts of the cryostat were gold plated to reduce thermal contact resistance.

Cocks et al. (1966) observed line-broadening effects in colloidal gold powder mounted with a collodion–amyl acetate binder and cooled to 77 K. An improved method for mounting powder samples that will eliminate strain effects to 100 K is described in Section 6.5.

Burton and Oliver (1936) reduced the thermal gradient to negligible proportions by placing one thermocouple and one heater on each side of the sample. They then adjusted the two heaters till both thermocouples read the same temperature.

4.1.3.4. Sample Preparation *in Situ*

A special problem occurs when samples that are gaseous or liquid at room temperature are studied. It is generally difficult to grow a single crystal *in situ* in a cryostat unless observation ports and fine control of the specimen temperature are provided for in the original design. It is also difficult to obtain randomly oriented powder samples from these materials.

The use of fiber optics to observe the sample in a very-low-temperature cryostat has been reported (Williams and Packard, 1974). The possibility of incorporating fiber optics into an X-ray diffraction cryostat, so as to be able to observe the sample, should be carefully investigated.

4.2. Cryostat Design Requirements

The final choice of cryostat will depend on the answers to the following questions:

- a. Which X-ray instrument will be used?
- b. What type of sample will be studied?
- c. What is the lowest temperature desired?
- d. What degree of temperature control is needed?
- e. How is the temperature to be monitored?
- f. What materials will be placed in the path of the incident and diffracted beams?

Moreover, all conduction-cooling devices are guided by the following basic requirements:

1. Minimal interference with the X-ray beam. The thermal shielding and vacuum shrouds should not reduce the X-ray beam intensity by an appreciable amount. The cryostat size and configuration should not interfere with the operation of the X-ray instrument and should allow maximum data collection.
2. Good temperature control of the sample. There should be a uniform cooling of the sample with good thermal conductivity between cooling source and sample. Provision should be made for accurately measuring and controlling the temperature.
3. Rigid and vibration-free mounting of cryostat and sample.
4. Efficient use of cryogen.
5. Ease of operation.

The ability to observe the sample for crystal growth and alignment purposes is helpful, but not critical, in most applications.

In general, the cryogenic dewar is designed so that the specimen is thermally connected to a liquid coolant but is not actually immersed in it. This arrangement eliminates the absorption and scatter of the X-ray beam by the liquid and by the vessel containing it and allows the variation of specimen temperature above the boiling point of the liquid. However, this design also requires the specimen to be loaded into the specimen chamber at temperatures above the boiling point of the liquid. Such a procedure is normally acceptable, but in certain cases (e.g., if metastable phases are to be retained) it is undesirable (McDonald et al., 1966).

Atoji (1965) has published an excellent review of the basic principles of cryostat construction for neutron diffraction instruments. Most of these principles are equally applicable to X-ray diffraction instruments. Temperatures down to 4.2 K and both powder and single-crystal instruments are discussed.

Many commercially available cryostats can be adapted for use with X-ray instrumentation if the above requirements are met. In this area of low-temperature apparatus, more so than in other areas discussed in this book, the X-ray investigator must normally depend on individuals or companies with experience in cryogenic design. A list of those manufacturers known to have experience in X-ray applications of cryogenic instrumentation is given in Appendix 12.

4.2.1. Frost

The condensation of water vapor on the outside of the cryostat does not usually occur if an evacuated sample chamber is used. Brady and Reuth (1966) report that a vacuum of 1 torr is sufficient to provide a frost-free surface. However, in several instruments in which moisture tends to condense on the surface of the dewar or on the windows, frost formation has been prevented by blowing dry, warm gas over the windows and outer surfaces of the cryostat (Claus, 1931; Williams, 1933; Jetter et al., 1957; Nikolaenko and Karpukhin, 1967).

4.2.2. Beryllium Windows

Beryllium windows (of the order of 0.25 mm in thickness) have been constructed in several shapes and sizes. Circular discs are easiest to obtain and handle but restrict the available data. Semicylindrical, cylindrical (Smith and Lipscomb, 1965; Sampson, 1970), and hemispherical windows (Lynch and Morosin, 1971; Renton and Baun, 1963) have also been used. Two precautions are in order. The first deals with safety: beryllium dust is poisonous when inhaled. Machining of beryllium should be carried out only by experienced personnel under proper conditions. This is best done by dealers in beryllium (e.g., Beryl Co. of America, Brush Beryllium Co.). The second warning deals with the data: the normally used hot-pressed beryllium is polycrystalline and can give a diffraction pattern that can interfere with the low-temperature data (Sampson, 1970; Air Products, 1973). A blank should be run to determine any possible interference effects (Figure 4–2). If a counter is used, the background may be affected unevenly over the scan range. When a slit system is used, this effect may not be easily recognized (Coppens et al., 1974). Other window materials may also interfere. For example, Mylar has a large, broad peak at a d-spacing of 3.1 Å (Abell and King, 1970). This should be avoided by changing windows [e.g., use Teflon as suggested by

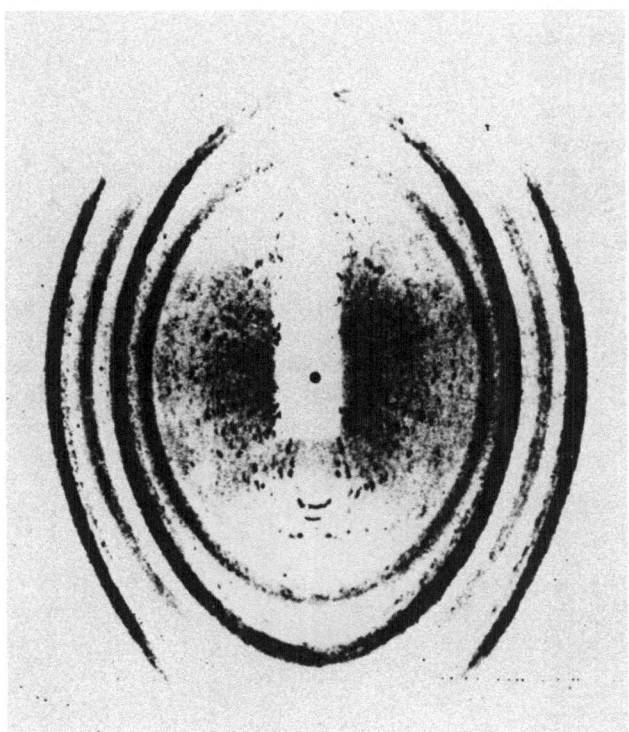

Fig. 4–2. Powder pattern obtained from a beryllium window (Air Products and Chemicals, 1973).

Barrett et al., (1967)] or by carefully treating (or neglecting) data collected in this range.

4.2.3. Temperature Gradients

See comments in Section 4.1.3.

4.2.4. Continuous-Flow Cryostat

A variation of the standard conduction-cooling method is to cool the sample holder with a continuous *flow* of liquid cryogen or cold gas. The reservoir is not rigidly fixed to the sample chamber but is connected by a flexible transfer line (Figure 4–3). In modern devices utilizing liquid

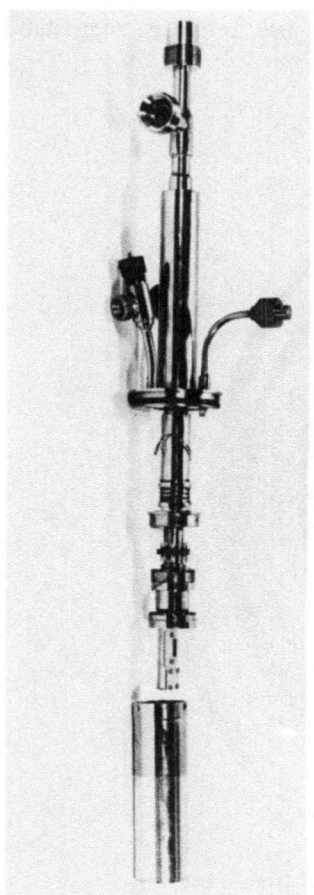

Fig. 4–3. Continuous-flow cryostat for powder samples, shown with radiation shield removed to reveal specimen holder. (Courtesy of Oxford Instrument Co.)

helium, the returning flow of cold gas is used to cool a radiation shield surrounding the specimen chamber and transfer line. It is vented through an opening a considerable distance away from the cryostat. As a result, an outer liquid-nitrogen dewar is not required and no ice forms on the cryostat.

Two major advantages result: greater flexibility is allowed in the orientation of the sample, and the low-temperature sample chamber that is affixed to the X-ray diffraction apparatus can be reduced in size.

The earliest use of this method was the Debye–Scherrer camera described by Simon and Simson (1924a,b), in which they passed cold gas through a narrow tube in the center of the camera. A gaseous sample was introduced to the previously evacuated camera; it condensed on the cold tube and was then photographed (see Section 4.4.3). A stream of continuously circulating cold acetone was used to cool a specimen holder in a flat-plate camera (Hanson and Halverson, 1948); this is similar to the cameras of Barnes (1929), Cox (1932), Barnes and Hampton (1935), and Zubenko and Umanskii (1957).

This method has been adapted recently for use on a powder diffractometer (Penfold, 1971; King, 1975, personal communication; Oxford Instrument Co.; Air Products and Chemicals, Inc.); a full-circle single-crystal diffractometer (Coppens et al., 1974); and a high-pressure, low-temperature cell (Mills and Schuch, 1974). It is a very useful method and will become more popular in the future. Further details are given in Section 4.4. A similar concept is also used in the Joule–Thomson expansion devices (see Section 4.7).

4.3. Special Considerations for Individual X-Ray Instruments

The majority of cryostats utilizing liquefied gases have been designed to be used in a vertical position with the axis of the cylindrical dewar parallel to the sample axis. This shortens the path length to the sample chamber and simplifies design. However, there are also a number of cryostats in which a right-angle bend in the transfer line allows a vertical dewar to be used with a sample whose rotation axis is horizontal or close to horizontal. The recent use of the continuous-flow cryostat eliminates this problem to some extent.

Thermoelectric and Joule–Thomson expansion devices can be operated in any orientation and so do not have to be designed for a specific mode of operation.

Relatively few cryostats have been constructed for use with *Weissenberg* and *precession* instruments. A vertical Weissenberg goniometer was constructed to be used with a specially designed cryostat (Keeling et al., 1953; Pavlovic, 1956). The only chambers built for a precession camera used gas-stream cooling and are discussed in Section 3.5.2 (Gobrecht et al., 1971; Boiko et al., 1973b). However, a Mylar-covered chamber for relatively high temperatures has been reported for a thermoelectric cooler used on a precession camera (Agron et al., 1972). [Also, Morosin and Schirber (1974) describe a high-pressure cell for the precession camera; see Section 5.2.]

A serious limitation of many *single-crystal diffractometer* cryostats is the degree of tilt allowed the dewar. This leads to restrictions on angular setting of ω, 2θ, χ, and ϕ. A ϕ motion of 360° is usually attainable, provided the diffractometer is programmed not to go above 360° but to control the ϕ motion so as to wind and unwind the tubing and wires connecting the dewar to the reservoir.

Powder cameras can be classified into two groups: those with the sample located at the outer edge of the instrument (e.g., symmetric back-reflection camera) and those with the sample placed in the center of the camera (e.g., Debye–Scherrer camera). In the first case, the sample can be cooled by placing it on a cold block without interfering with the operation of the camera in any way (Weigle and Saini, 1936; Mascarenhas and Mascarenhas, 1967). (A powder diffractometer can be considered a special case of this arrangement, but the size of the cryostat window will place some restrictions on the quantity of data that can be collected.) The second type of camera, which is more common, can have the film placed either inside the cold chamber (McFarlan, 1936), in which case there is no restriction on data but changing the film is a major undertaking, or outside the chamber, in which case the quantity of recordable data depends on the size of the window.

A popular method of cooling condensed gaseous samples in a Debye–Scherrer camera is to condense the sample onto a narrow-diameter tube or wire which is cooled either by conduction or by a cold liquid flowing through it. This method has been described as early as 1922 and as recently as 1969 (see Section 4.4.3).

The type of cooling apparatus used for *powder diffractometry* depends on the minimum temperature that is required. If liquid-nitrogen temperatures are used, then a simple cold-finger type of sample-holder adapter can be employed (e.g., Abrahams and Kalnajs, 1954), while if lower temperatures are needed, it may be necessary to replace the entire sample-holder assembly with a specially designed cryostat.

The choice between using a horizontal or vertical powder diffracto-meter also depends on several factors. The horizontal diffractometer is better suited for mounting a cryostat, since the weight of the cryostat can be balanced by aligning its axis with the axis of the diffractometer. This is a very important feature and is commonly used. On a vertical diffractome-ter, it is very easy to clamp a specially designed sample-holder adapter in place, and it is not much more difficult to replace the shaft of the instru-ment with a new shaft to which the cryostat is attached. However, in the latter case, the weight of the cryostat, which may cause misalignment of the instrument, must be considered when designing the apparatus.

If the sample is liquid at room temperature, the use of a vertical dif-fractometer with a horizontally placed sample holder should be considered. However, a sample support with a Mylar cover for use with a horizontal diffractometer has been successfully used (Abell and King, 1970). It is also possible to condense the sample on the cold sample support by spray-ing it from a vaporizer (Section 6.2.2).

Cryostats designed for use in *X-ray topography* must cool a relatively large surface uniformly and efficiently. Otherwise, inhomogeneous sample temperatures can cause deterioration of the resolution. Cryostats for this purpose have been described by Batterman and Barrett (1966), Ando (1973), Boiko and Shmytko (1974), Aknazarov et al. (1974), Coulon et al. (1974), and Mathiot and Petroff (1975).

Finally, special mention is made of the apparatus constructed by Goldak (1964) in which a complete X-ray powder diffractometer is oper-ated in a high-vacuum chamber to eliminate X-ray windows, air scatter, and the usual cryostat-imposed restrictions on data collection (Figure 4–4). Data between 77 K and 600 K can be collected automatically by computer-controlled step scanning and data recording. The major difficulty in using an X-ray tube in a vacuum was arcing when the high-voltage connection was in the evacuated chamber. This problem was solved by making the connection in air and utilizing a vacuum feed-through connec-tion for the horizontal X-ray tube (Figure 4–5). The preamplifier tube on the scintillation counter failed in about three hours because of inadequate cooling. Therefore, the detector assembly was mounted in a vacuum-tight box through which compressed air was blown to provide the necessary

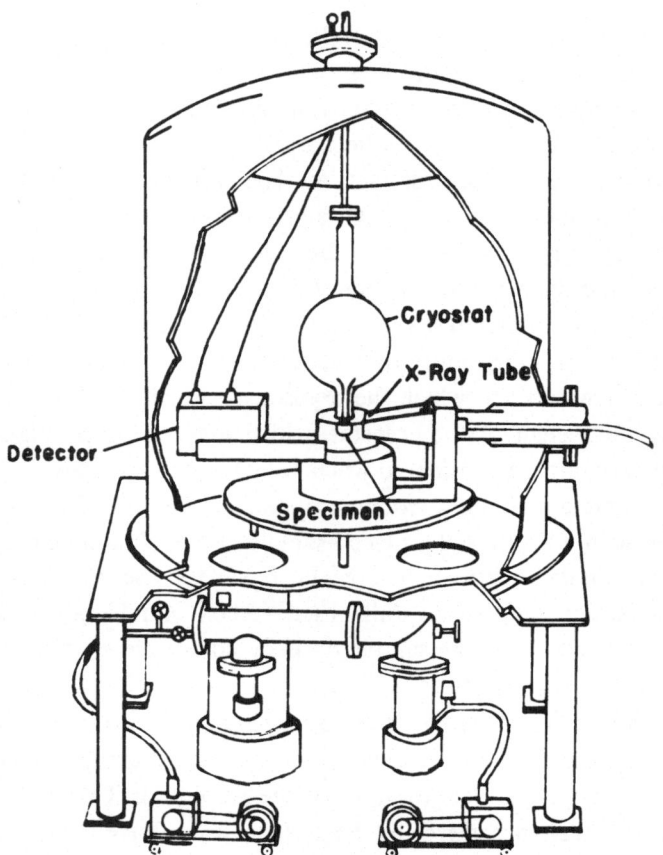

Fig. 4–4.   Schematic diagram of a powder diffractometer and cryostat which were placed inside a vacuum chamber (Goldak, 1964). (Reproduced with the permission of the Institute of Physics.)

cooling. Modern solid-state devices may not require this degree of cooling.

A summary of the general properties of conduction-cooling systems is presented in Table 4–1.

Table 4–1.   Summary of Conduction-Cooling System Parameters

Poor access to crystal	Crystal not usually visible
0.005° stability at best	0.1° stability common
Absorption corrections necessary	Frost no problem
0.1–1 liter/h consumption rate	4 K or lower attainable
Not easily adapted to other instruments	

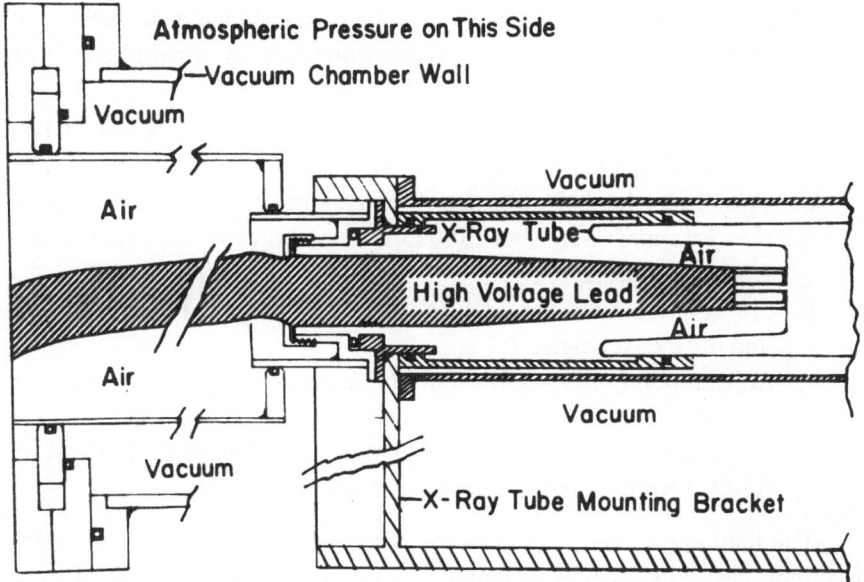

Fig. 4–5. High-voltage connection to X-ray tube operating in vacuum chamber (Goldak, 1964). (Reproduced with the permission of the Institute of Physics.)

4.4. Conduction Cooling via Liquefied Gases

4.4.1. Single-Crystal Cryostats for Use with Liquid Helium

This discussion is limited to apparatus used for collecting data suitable for crystal-structure determinations. The cryostats that are used to cool single crystals during thermal expansion and phase transition investigations are discussed in Section 4.4.3 since they are generally limited to rotation only about the $\theta/2\theta$ axis.

Three types of X-ray instruments will be discussed: single-crystal diffractometers utilizing Eulerian geometry (three- or four-circle); single-crystal diffractometers based on Weissenberg geometry; and Weissenberg goniometers for film use. Recent representative examples of each type will be given.

A cryostat based on original specifications by P. Coppens, W.C. Hamilton, J. Godel, and R. Leonard was designed and constructed to permit unrestricted collection of three-dimensional data on an automatic full-circle diffractometer at liquid-helium temperatures (Coppens et al., 1974). The original requirements can be summarized as follows:

a. The instrument should be operable at liquid-helium as well as at intermediate temperatures; temperature stability should be 0.1° or better.

b. There should be no blind angles limiting data collection to less than a full sphere in reciprocal space.

c. To prevent loss of accuracy, the crystal should not change position by more than about 0.025 mm on rotation around the diffractometer axes.

d. No condensation should be formed in the path of the incident and diffracted beams.

e. Consumption of cryogen should be minimal to ensure economy of operation.

f. To enable routine use of the equipment, conditions a–e should be fulfilled without excessive experimental complexity.

The final apparatus (Figure 4–6), based on the continuous-flow principle, consists of a cryostat mounted on the ϕ-circle and supported by a movable bracket attached to the χ-circle at a point diametrically opposite the ϕ-circle. The base of the cryostat is a modified xyz-goniometer head, which is used to center the crystal. No arc corrections are used; all angular and counter settings are computed from an orientation matrix calculated from a few observed reflections. The cryogen reaches the copper block on which the crystal is mounted through a flexible vacuum-jacketed transfer line. The used gas is returned through the transfer line and vented near the supply dewar to eliminate condensation at the cryostat. There is no permanent cryogen reservoir at the cryostat and the need for a liquid-nitrogen jacket is eliminated by the use of superinsulation. The absence of these reservoirs allows the cryostat to be tilted to any angle permissible within the limits of the transfer line. A rotating seal around the fixed tip of the transfer line allows unrestricted motion of the ϕ-circle, although the computer software was modified to eliminate motion through the $\phi = 180°$ position so as not to tangle the electrical connections.

The crystal is surrounded by three thin-walled beryllium radiation shields (only two are necessary for operation above 30 K). Temperature control is achieved through feedback from the resistance thermometer to a heater in the copper block. At 30 K, the liquid-helium consumption varied between 0.3 and 0.5 liter per hour; at 78 K, 0.1–0.2 liter of liquid nitrogen was consumed per hour.

The outer beryllium wall has a thickness of 0.25 mm, while each of

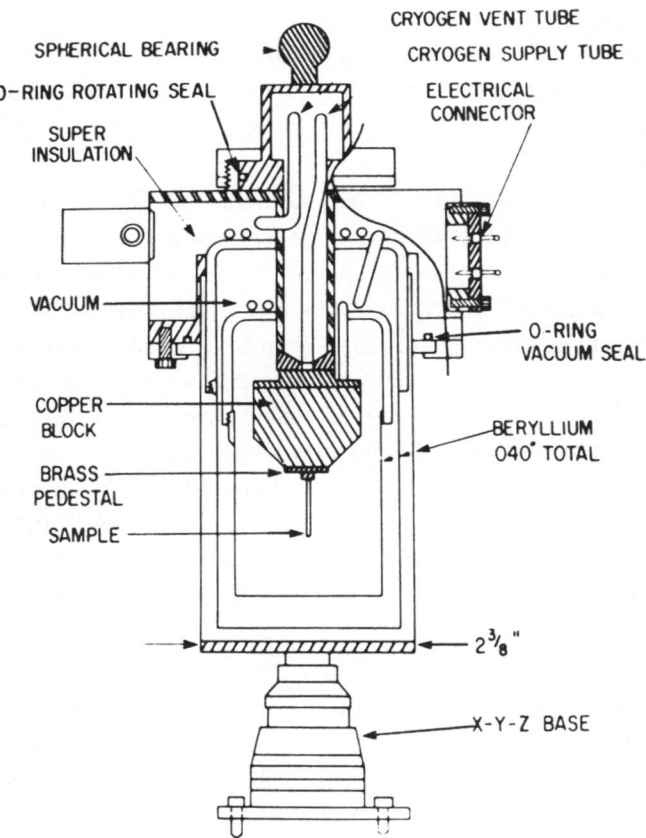

Fig. 4–6. Schematic diagram of liquid-helium single-crystal cryostat that can be tilted to any angle (Coppens et al., 1974). (Reproduced with the permission of the International Union of Crystallography.)

the two inner walls is 0.125 mm thick. Absorption is quite small, but, because the containers are cylindrical and not spherical, absorption is a function of χ and 2θ. Maximum corrections, which are easily calculated, are 2.5% for two shields used with Mo K_α radiation up to $2\theta = 90°$. Corrections must be applied; otherwise, these systematic errors would affect the calculated thermal parameters.

The authors state that no beryllium powder lines are observed in a scan of the empty cryostat. This is attributed to the location of the beryllium far from the center of the diffractometer, the use of a finely collimated beam, and the use of a receiving aperture in front of the counter.

The temperature of the copper block is controlled to within 0.1°. The possibility of a temperature differential existing between the copper block and the specimen has been investigated using a known phase transition in the region of 32 K. On this basis, the temperature difference is estimated at $0.5 \pm 0.5°$.

Ross and Williams (1974) and Lenhert and Takagi (1974) discuss some other aspects of this cryostat. The temperature stability was found to have some long-range variation. Therefore, the temperature at which each reflection is measured is recorded using a physical-interrupt meter tied to a proportional temperature controller. When LN_2 is used, it is found that the temperature depends on the delivery pressure, which in turn depends on the level of LN_2 in the reservoir. Collision protection is afforded by the use of both computer control and safety switches at critical contact points.

Single crystals can be grown from liquids by using an external cold gas stream, maintaining the sample by cooling the cold block, covering with the radiation shields, and rapidly evacuating the chamber. This cryostat is a modification of a previously constructed neutron diffraction cryostat.

Heaton et al. (1970) described a neutron diffraction cryostat which they suggested could be modified for X-ray studies. This cryoorienter consists of a double-dewar cryostat with full 360° rotation of ϕ and a $\pm 50°$ limit on the χ motion. The temperature stability is 0.02° between liquid-helium and room temperatures, and the apparatus is capable of collecting three-dimensional data over three-fourths of all reciprocal space. An interesting feature of this device is the use of spiroid gears for all drives. These gears offer the advantage that 20 to 30 teeth are engaged at once, and, consequently, the deviations of the gear teeth are averaged. Various mechanical techniques are used to minimize backlash, and final positioning of each angle with a positive movement of the angle is used to eliminate backlash.

The sample is inside a tube extending to the center of the cryostat. This tube is filled with a dry heat-exchange gas and is not evacuated. The sample can be changed without affecting the vacuum insulation.

The use of Weissenberg geometry eliminates the necessity of tilting the cryostat. In a Weissenberg diffractometer, the crystal can be kept stationary if peak height measurements are used, or it need only be rotated about its axis if integrated intensities are to be obtained. A cryostat based on this principle has been described by Smith and Lipscomb (1965;

Smith, 1966) and is based on an earlier low-temperature camera (Streib and Lipscomb, 1962). This cryostat is designed for use with the normal-beam Weissenberg technique and so need not be moved for upper-level photographs. It can operate as low as 20 K and has provision for growing single crystals *in situ*.

The apparatus consists of a vertical dewar and a Weissenberg diffractometer with a vertical crystal-rotation axis. The sample chamber is evacuated, with the sample mounted on a fiber placed in a thin-walled capillary tube, and is surrounded by a 0.3-mm-thick beryllium cylinder. The entire dewar can be rotated for intensity-data collection. A film cassette is used for the initial study and alignment of the crystal, after which it is replaced by a counter detector. The single crystal is centered, but is not aligned by any arc corrections. All determinations of unit-cell parameters, crystal orientation, and counter detector position are calculated.

The third example is that of a cryostat for a vertical-axis Weissenberg goniometer that can be operated at any temperature between 7 K and room temperature (Thomas, 1972). The cooling is achieved by enclosing the crystal, which is glued to a copper conducting mount, in an inert atmosphere of cooled helium gas. The specimen is thus cooled both by conduction along its mounting pin and by conduction/convection through helium gas at a pressure of a few millimeters of mercury (Figure 4–7). Thermal contact between specimen chamber and liquid refrigerant, and hence the lowest temperature and rate of cryogen consumption attainable, are controlled by the number of copper wires connecting the reservoir to the sample chamber.

A 2.5° temperature differential was measured between the specimen and specimen chamber. The vacuum-insulated chamber is separated from the sample chamber and from the outside of the cryostat by three symmetrically placed windows of 0.1-mm-thick Melinex supported on mounting struts. These three struts block a total of 90° around the circumference of the cryostat. A tilt of up to 3° from the vertical is allowed.

Several other devices possessing somewhat limited data collection capabilities have been reported. Abrahams (1960) constructed a dewar which is mounted on a large-scale version of the standard goniometer head. This cryostat is used on a neutron diffractometer and is limited to a maximum angular setting of $\pm 20°$ on each arc.

Chopra (1962) and Morosin (1966) described devices for single-crystal diffractometers that are capable of measuring only zero-level data.

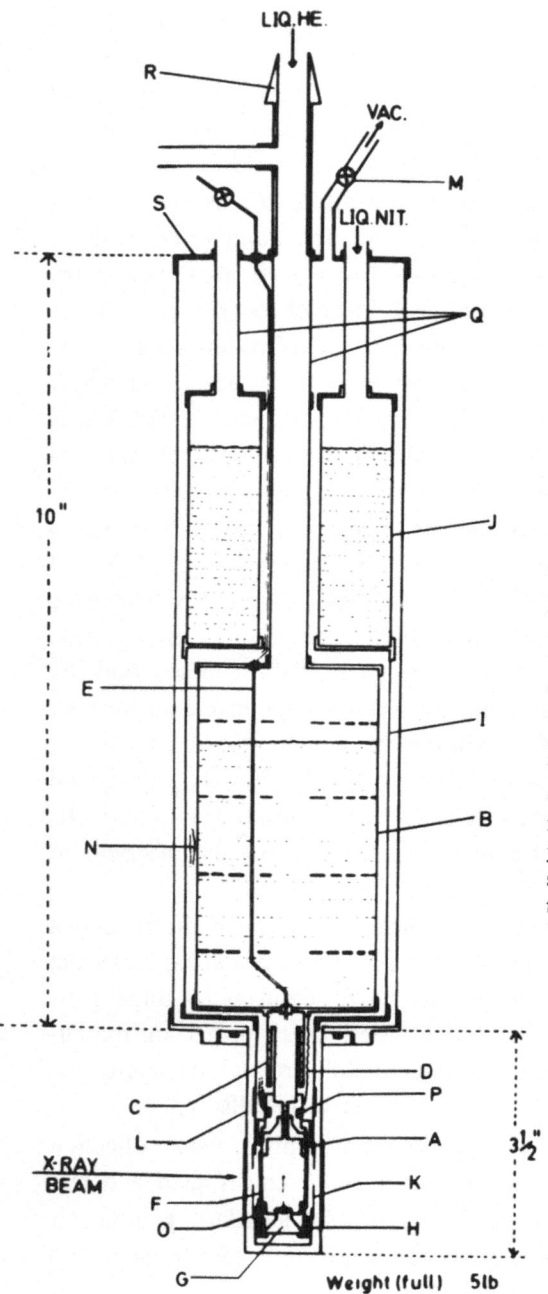

Fig. 4–7. (a) A cryostat for single-crystal X-ray diffraction studies down to liquid-helium temperatures. (A) Specimen chamber; (B) helium can; (C) "trombone" tube; (D) thermal link; (E) German silver capillary; (F) Melinex windows; (G) crystal mount; (H) locking collar; (I) radiation shield; (J) nitrogen can; (K) aluminum windows; (L) outer tail tube; (M) evacuation valve; (N) thermocouple reference junction; (O) measuring junction; (P) electric heater; (Q) support tubes; (R) cone fitting; (S) top plate.

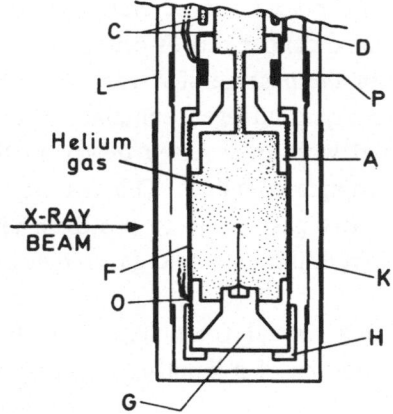

Helium
gas

X-RAY
BEAM

Fig. 4-7. (b) An enlarged view of the tail of the cryostat (Thomas, 1972). (Reproduced with the permission of the International Union of Crystallography.)

Upper levels can be measured in the Chopra device by changing the orientation of the X-ray tube. Finally, Woodley et al. (1971) report the construction of a cryostat which replaces the Eulerian cradle of a standard four-circle diffractometer, resulting in a three-circle diffractometer with fixed χ geometry. The sample, which is mounted on a rotatable shaft (ϕ-circle axis) that is inserted in a conduction-cooled chamber, is also bathed by a stream of cold helium gas that passes through a heat exchanger immersed in the liquid-helium dewar.

4.4.2. Single-Crystal Cryostats for Use with Liquid Nitrogen

An early version of a low-temperature diffractometer was described by Williams (1933). The crystal was mounted on a pin in contact with a reservoir and placed in a chamber having celluloid windows. Data were collected with an ionization chamber during a study of diffuse scattering.

Keesom and Taconis (1935) reported the construction of an interesting device in which a copper rod in contact with a dewar was attached to a dish of mercury into which a thin-walled (0.006 mm) glass tube (1 mm diameter) was placed. By adjusting the pressure, the mercury level was brought up to the height of the X-ray beam. Liquid air was then introduced into the cryostat and cooled the copper rod which froze the mercury. A gaseous sample, introduced into the sample tube, froze on top of the mercury at the height of the X-ray beam. The sample temperature was adjusted by means of a heater wound around the mercury, and a single crystal was grown. Rotation and primitive Weissenberg (film translation, but no layer-line screen) photographs could be taken.

Ubbelohde and Woodward (1947) placed the crystal in a small cavity in the bottom of a conduction-cooled copper rod. A cellophane-covered hole was drilled through the rod for the passage of X-rays. An interesting feature of this device is that the liquid-nitrogen container, located on the top of the cryostat, was covered with an inverted dewar so that boiling cold gas flowed down over the sample container. This feature was a forerunner of the use of a concentric dry gas stream to protect against ice formation (see Section 3.2), except that in this case cold gas, rather than room temperature gas, was used.

Keeling et al. (1953) constructed a special goniometer head from coin silver and silver-soldered it to the bottom of a liquid-nitrogen dewar. The dewar was placed on the axis of a vertical Weissenberg goniometer and both dewar and crystal were rotated during data collection. Both the crystal and the liquid-nitrogen dewar were enclosed in a single chamber which was sealed at the lower end, with a 10-mm-diameter polystyrene cylindrical cover over the crystal. This chamber was continuously evacuated. The supporting frame could accommodate a Weissenberg film cassette, a spectrometer table for mounting a counter detector, or a Debye–Scherrer film cassette.

Renton and Baun (1963; Baun and Renton, 1964) describe a cryostat mounted on a modified goniometer head. A miniaturized dewar vessel (Figure 4–8a) is mounted directly on a goniometer head base with translational motions only. It consists of a glass-walled vacuum-insulated dewar, with a cold finger on which the specimen is mounted. The top is covered with a beryllium dome (0.25 mm thick and 80 mm in diameter) forming an enclosure which is evacuated, filled with dry nitrogen, and sealed off.

Another miniature cryostat, which can also be mounted on a goniometer head, has been designed by Sharon (1975). This device consists of a vacuum-insulated tube with a porous plug in the middle of the tube (Figure 4–8b). Liquid nitrogen enters through one end of the tube and cools the plug; gaseous nitrogen exits from the other end. A mounting pin, attached to the plug and lying perpendicular to the tube, cools the crystal by conduction. The crystal is covered with a thin plastic dome which can be evacuated. This device has been used on a precession camera.

A simple low-temperature cryostat can be made by circulating a cooled liquid through coils attached to a cylindrical cassette or located in a chamber that fits over the cassette. The lowest temperature is limited to the temperature of the liquid. At least two such cameras suitable for oscil-

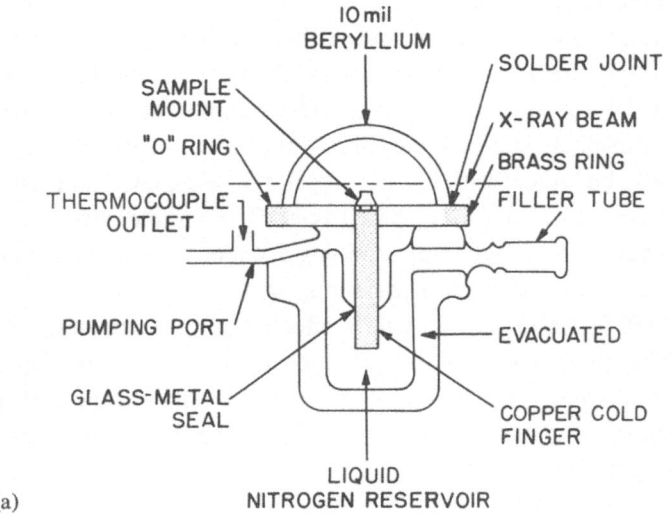

(a)

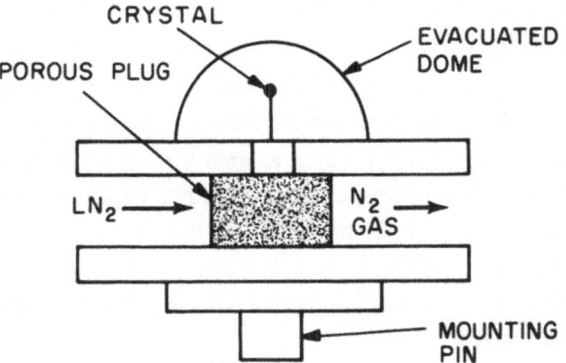

Fig. 4–8. Cross sections of low-temperature adapters for single-crystal goniometer head. (a) Renton and Baun (1963); (b) Sharon (1975).

lation–rotation photographs have been reported (Barnes, 1929; Zubenko and Umanskii, 1957). A gas-stream adapter for Zubenko and Umanskii's camera, capable of reaching 90 K, has been described (Boiko and Ovchinnikov, 1968).

Hovi et al. (1966) enclosed a Laue camera inside an evacuated glass container. The collimator, covered with a beryllium window, was inserted through the wall of the container.

A device that permits low-temperature and/or high-pressure studies of compressed solid gases was described by Bronsveld et al. (1973). An interesting feature of this device is that the sample is cooled by conduction paths originating at both the top and bottom of the sample chamber. The tube connecting the two liquid-nitrogen chambers restricts the viewing angle to 300° in the horizontal plane. The sample chamber is machined out of 1.5-mm-thick sintered beryllium; the space between the sample chamber and the outer wall of the cryostat is evacuated and the outer window is covered with 0.1-mm-thick Mylar. The sample chamber is pressurized by connection to a compressed-gas cylinder containing the material being studied. It can withstand pressures up to 2700 bars. Temperature stability is 0.03° and crystal growth is effectuated by controlling the temperature. The presence of a crystal and its alignment are determined entirely by film techniques, with the single-crystal rotation method used to measure lattice constants as a function of pressure and temperature with an estimated precision of 0.01%.

4.4.3. Powder-Sample Cryostats for Use with Liquid Helium or Liquid Nitrogen

Cryostats used for powder diffraction require fewer moving parts than those constructed for single-crystal intensity-data collection because the powder sample does not require as many degrees of freedom. However, a number of applications, such as thermal expansion studies, phase transition investigations, and precision lattice constant measurements, require exceptional mechanical and thermal stability. Several cryostats capable of being used to collect zero-level single-crystal data are also described in this section.

A list of cryostats constructed for low-temperature powder investigations is found in Table 4–2 along with short comments describing the most interesting features of each device.

4.4.3.1. Camera Apparatus

One of the earliest methods of studying frozen gases by Debye–Scherrer photography was to condense the material on a cold substrate in a previously evacuated camera. This method has been used with several variations, but the basic approach was to cool a narrow-diameter tube or wire by filling it with a liquefied gas, by allowing the liquefied gas to flow through it, or by maintaining it in contact with a cold bath.

Table 4–2.   Conduction-Cooling Powder Cryostats

---

a. Diffractometer, below 78 K

Barrett (1956)	Provision for cold-working of metals
Barrett and Meyer (1964)	Modification of Barrett (1956) for use with solidified gases
Barret et al. (1967)	Modification of Barrett (1956) for use with reactive systems (Teflon sample holder)
King and Preece (1967)	For precision lattice parameter measurement; discusses calibration of temperature at liquid-helium temperatures
Abell and King (1970)	Modified version of King and Preece (1967) to allow cold-working of specimen
King et al. (1969)	For powder diffractometry in high magnetic fields; specimen can be removed without disturbing the vacuum insulation
Linkoaho (1968)	Describes alignment procedure for MRC (Materials Research Corporation) cryostat for horizontal powder diffractometer
Smith and Leider (1968)	Describe technique for aligning cryostat mounted on horizontal diffractometer
Black et al. (1958)	For use on the G.E. horizontal diffractometer
Mauer and Bolz (1961)	Improved version of Black et al. (1958)
Bolz and Mauer (1963)	Further details on Mauer and Bolz (1961)
Stochl and Ullman (1963)	For Norelco vertical diffractometer, supports weight of dewar with U-shaped bracket
Lytle (1964)	For precision lattice constant measurements; range in 2θ of 220°
Gruber et al. (1969)	For precision lattice parameter measurements
Himmler et al. (1969)	For precision lattice parameter measurements
Peterson and Simmons (1965)	For back-reflection studies up to pressure of 20 atm
Losee and Simmons (1968)	Modified version of Peterson and Simmons (1965)
Balzer and Simmons (1974)	Samples in cylindrical Lucite pressure cell to several hundred bars
Zakharov (1969)	Temperature control to 0.02°; beryllium window
Sears and Klug (1962)	Sample deposited from vapor
Sampson (1970)	Can reach 1.3 K
Boiko et al. (1972a)	As cold helium gas leaves the cryostat, it passes over the sample
Flinn et al. (1961)	For studying large crystals
Sears (1974)	Gas deposited on cold substrate in evacuated chamber

Table 4–2.   Conduction-Cooling Powder Cryostats (*cont.*)

Penfold (1971) (Oxford Instrument Co.)	Continuous-flow cryostat
Air Products and Chemicals, Inc.	Continuous-flow cryostat
Calvarin and Berar (1975)	Sample holder contains heat-transfer gas; temperature stability to 0.03°
Etourneau et al. (1975)	Special sample holder with heat transfer gas
Sayetat (1975b)	Dewar weight is balanced with counterweight

b. Camera, below 78 K

de Smedt et al. (1930)	Gas condensed on cold copper rod
Ruhemann (1932)	Gas condensed on cold tube through which cold helium is circulated
Keesom and Kohler (1934)	Gas condensed on cold rod
Keesom and Taconis (1938)	Pressurized for study of solid helium; sample in 0.04 mm thick aluminum tube
Gunther et al. (1939)	Gas condensed on cold finger
Cheesman and Soane (1957)	Gas condensed onto wire held between two dewars
Kogan and Bulatova (1969)	Gas condensed on substrate
Kogan et al. (1960)	For radioactive tritium; condenses gas on outside of copper rod
Kogan et al. (1964)	Condenses sample inside Be tube; compares results to Kogan et al. (1960)
Atoji et al. (1959)	Flat-plate cassette; dewar supported in front of vertical X-ray tube
Bogoyavlenskii and Bereznyak (1967)	Pressurized to 150 atm for study of helium; flat-plate cassette placed inside evacuated specimen chamber
Figgins et al. (1956)	Forms liquid hydrogen with miniature liquefier
Gränicher et al. (1959)	Vertical sample axis
Rühl (1954)	For study of thin films to 7 K
Queisser (1958)	Similar to Rühl (1954)
Schuch (1958)	Can reach pressure of 175 atm; sample contained in 0.3-mm-thick Be cell
Schuch and Mills (1962)	Improved version of Schuch (1958)
Mills and Schuch (1974)	Used Al high-pressure cell to 4500 atm
Reeber and Powell (1967)	Back-reflection camera
Eeles (1960)	Precision camera
Girt et al. (1974)	Back-reflection camera; can take film at reference temperature and at low temperature
Girt et al. (1975)	Modification of Girt et al. (1974) to allow three temperatures to be measured

Table 4–2.   Conduction-Cooling Powder Cryostats (*cont.*)

c. Diffractometer, 78–200 K

Claus (1931)	Used with Bragg ionization spectrometer; crystal in cavity of cold copper block; crystal is cooled by cold air and is not in contact with the copper
Harvey (1933)	Used with Bragg ionization spectrometer to measure diffuse scattering from large crystals
Obsieger (1963)	For Debye–Waller temperature factor measurements on large platelike single crystals
Kagan and Umanskii (1960)	Cools by conduction from back of flat specimen
Goldak (1964)	Automatic powder diffractometer and cryostat that operates *inside* a vacuum chamber; details of operating detector and X-ray tube inside a vacuum chamber are given
Calhoun and Abrahams (1953)	Specimen-holder adapter through which liquid flows; can be used on Norelco diffractometer without modification of existing instrument
Abrahams and Kalnajs (1954)	Improved version of Calhoun and Abrahams (1953)
Butters and Meyers (1955)	Circulate cold gas through specimen plate on Norelco vertical diffractometer; rigid double-walled transfer line replaces shaft of specimen holder
Smith and Heady (1955)	Sample holder for Norelco vertical diffractometer through which cold gas circulates; has plastic film canopy
Jetter et al. (1957)	Cryostat replaces shaft on Norelco vertical diffractometer; can cold-work the specimen by removing cover while blowing cold gas over specimen
Hutchinson and Miller (1958)	Modification of Jetter et al. (1957)
Gould and Gerold (1965)	Sample holder for Norelco vertical diffractometer
Smith (1961)	Circulates cold gas through sample holder for Norelco vertical diffractometer
Dowell and Rinfret (1960)	Replaced sample holder and shaft on Norelco vertical diffractometer with vacuum-chamber attachment through which liquid nitrogen or cold nitrogen gas is circulated; a 0.009-mm-thick Ni foil doubles as a heat shield and beta-radiation filter
Baun (1959, 1961)	Teflon sample holder for Norelco vertical diffractometer; cold gas or liquid is circulated through sample holder and then sprayed over sample surface

Table 4–2.   Conduction-Cooling Powder Cryostats (*cont.*)

Honeywell et al. (1964)	Cryostat for Norelco vertical diffractometer; can be cooled to 80 K at pressure of 100 atm, liquid sample held in Be sample cell
Ghislain et al. (1965)	Sample holder for Norelco vertical diffractometer to which a small cryostat is attached; can be used with no modification of diffractometer
Weltman (1962)	Sample holder for Norelco vertical diffractometer; cold gas circulates through coils in sample holder; covered with radiation shield of polyurethane foam with Mylar-covered window
Stammler et al. (1963)	Weltman (1962) device adapted to G.E. horizontal diffractometer
Masson (1960)	Sample holder for G.E. diffractometer; copper rod extends into LN_2 bath; X-rays pass through windows in styrofoam chamber
Roessler and Bolling (1964)	Simple specimen holder for horizontal diffractometer; sample *immersed* in cold bath and held in place inside polystyrene foam container adjacent to a thin-walled section
Brady and Reuth (1966)	Circulated liquid nitrogen through copper coil in intimate contact with the sample holder; for G.E. horizontal diffractometer
Baun and Renton (1963) Renton and Baun (1963) Baun and Renton (1964)	Cryostat and accessories for horizontal diffractometer
Trut et al. (1973)	Similar to Baun and Renton (1963), with provision for pumping liquid nitrogen to reach 65 K
Intrater and Appel (1961)	Gas, cooled by standard heat exchange methods, flows through hollowed-out copper specimen block; designed to fit into MRC high-temperature attachment
Davis and Eby (1975)	Modification of Norelco-MRC high-temperature, high-vacuum attachment for low-temperature (to 100 K) use
Peisl and Waidelich (1959)	Forms vacuum chamber by spreading Mylar film over aluminum screening (with 15-mm-diameter holes)
Shimura (1960)	Diffractometer can scan 2θ at fixed temperature or vary temperature over short 2θ scan range (for thermal expansion measurements)
Potapov (1962)	Uses helium transfer gas inside sample chamber
Bochkarev and Egorov (1971)	For investigating thin films

Table 4–2.   Conduction-Cooling Powder Cryostats (*cont.*)

d. Camera, 78 -200 K

Dennison (1921)	Sample in special dewar flask that is rotated continuously during exposure
de Smedt and Keesom (1924)	Gas condenses on cold finger
Simon and Simson (1924a,b)	Cold gas blown through tube in vacuum chamber; gaseous sample condenses on outside of tube
McLennan and Wilhelm (1925)	Gas condenses on cold finger
Terrey and Wright (1928)	Small capillary extending from dewar into camera; sample condenses on outside of capillary
Vegard (1931)	Improvement of Terrey and Wright (1928)
Ruhemann and Simon (1931)	Vertical dewar with thin tube at lower end inside camera; thin tube is inside glass sphere; gaseous sample introduced into evacuated sphere condenses on cold thin tube
Natta (1933)	Gas condenses on cold finger
Burton and Oliver (1936)	Sample condenses in cold tube; heaters at either end of sample adjust temperature of tube to ensure uniform sample temperature
Pohland (1934)	Sample axis vertical, dewar placed on top of evacuated camera chamber
McFarlan (1936)	For study of metastable phases; sample prepared and mounted at low temperatures
Helmholz (1935)	Sample support attached to inside of dewar; sample covered with thin-walled glass tube which is evacuated
Frazer and Pepinsky (1950)	Helmholz device extended by incorporating adjustable arcs and translations
Shevelev and Balakina (1960)	Sample suspended from cryostat
Kogan and Omarov (1965)	Holds two specimens; for easily oxidized materials
Kan and Lazarev (1951)	Sample in capillary tube attached to copper block extending into LN_2
Gervais et al. (1966)	Sample in contact with copper rod and also is bathed in escaping cold gas to prevent ice formation
Golob and Horn (1973)	Utilizes a movable inner dewar to quench samples which are then warmed to desired temperature; incorporates a movable film cassette to obtain photographs at different temperatures
Ray (1964)	Liquid oxygen circulates through hollowed-out copper plate to which sample is attached

Table 4–2.   Conduction-Cooling Powder Cryostats (*cont.*)

Kruner (1926)	Sample condensed on flat surface with cold copper block
Simon and Vohsen (1928)	Cools back of flat specimen
Clifton (1950)	Basic cryostat can be used with diffractometer or for front-reflection or back-reflection photograph; level of liquid nitrogen in contact with cold copper specimen holder is controlled by styrofoam-covered copper piston
Viswamitra and Ramaseshan (1963)	Flat-cassette back-reflection modified Unicam rotation camera; liquid air drips through hypodermic needle into brass cup, to which crystal is attached, inside thin polystyrene chamber
Bronsveld et al. (1973)	LN_2 in contact with both top and bottom of high-pressure sample cell
Dumbleton and Bowles (1966)	Liquid nitrogen flows through holes drilled in Al block bolted to collimator stand; sample covered with Mylar; warm gas blows across face of sample; for Polaroid photographs of polymer samples
Weigle and Saini (1936)	Vertically placed Seeman–Bohlin back-reflection camera; X-rays enter from bottom; dewar placed on top cools sample by conduction
Mascarenhas and Mascarenhas (1967)	Adapter for standard parafocusing symmetrical Seeman–Bohlin back-reflection camera
Malyushitskaya et al. (1975)	High-pressure (to 130,000 atm), low-temperature (to 100 K) camera

e. Cryostats for above 200 K

Cox (1932)	Cold alcohol circulated between the double walls of a cylindrical camera; dry air blown over camera on humid days; also used for single-crystal studies
Barnes and Hampton (1935)	Copper block is cooled by cold acetone flowing through a closed circulation system
Hanson and Halverson (1948)	Cold acetone circulating through copper sample holder; flat-film cassette
Guengant (1958)	High pressure applied to sample in tube
Kamb and Davis (1964)	Regulated stream of LN_2 passing through cylindrical jacket covering a high-pressure cell
Gerard and Pernolet (1973)	Low-temperature ($-60°C$), high-pressure (30 atm) cell for θ/θ diffractometer; cooled by circulating cold alcohol

Keesom and de Smedt (1922; de Smedt and Keesom, 1924) used the cold-finger method while Simon and Simson (1924a,b) used a narrow tube through which cold gas flowed. The thin tube used by Ruhemann and Simon (1931) was the end of a vertically placed dewar, while Cheesman and Soane (1957) employed a wire cooled from both ends by suitably placed dewars. In another variation, liquid air circulates through a thin copper tube which is kept at constant temperature over its entire length by adjusting two heaters (one above and one below the sample) until two thermocouples (one on either side of the heater) are at the same temperature (Burton and Oliver, 1936). In the case of a compound with a high vapor pressure, the substrate was enclosed in a glass bulb so that the sample chamber was separate from the evacuated film chamber (Ruhemann and Simon, 1931).

The use of this cylindrical sample can lead to some errors in the position of the observed powder lines. This factor has been discussed by Kogan et al. (1960). Cheesman and Soane used the copper-wire substrate to calibrate the pattern. Their attempts to use a platinum wire as a sample support, calibration standard, *and* resistance thermometer failed when the resistance of the necessarily thin wire was not suitable for temperature measurements.

Kogan et al. (1964) condensed hydrogen on the inside of a beryllium tube and found that the texture of the powder was different from that observed when the sample was condensed on the outside of the cold finger (Kogan et al., 1960).

A standard commercial parafocusing symmetrical Seeman–Bohlin back-reflection powder camera was adapted for low-temperature use by constructing a copper sample support attached to a liquid-nitrogen reservoir (Mascarenhas and Mascarenhas, 1967). This sample support and attached reservoir are part of a large metal cover that fits over the standard camera. The cover can be evacuated to prevent ice formation on the sample.

4.4.3.2. Liquid-Helium Cryostats for Powder Diffractometers

A continuous-flow liquid-helium cryostat was first built in 1971 (Penfold) and later modified for commercial production (King, 1975, personal communication; Oxford Instrument Co.). The sample is suspended in an evacuated enclosure with aluminized Mylar windows and is cooled by a continuous flow of liquid helium from a reservoir attached to the cryostat by a flexible metal transfer line (Figure 4–3). The rate of liquid-helium transfer and the spent helium gas removal are controlled by a pump at the exhaust end of the cryostat. The temperature can be maintained continuously at any value from 4.2 to 300 K, and temperatures below 4.2 K can be achieved by stopping the flow and pumping on the liquid helium collected in a cavity above the sample. A temperature of 3.5 K can be held for about twenty minutes. The cryostat can be used on either horizontal or vertical powder diffractometers, using various adapter plates. An alignment slit is built into the specimen block, which can be moved into position without disturbing the operation of the cryostat. A straining device [similar to one reported by Abell and King (1970)] is under development.

A similar device is also manufactured by Air Products and Chemicals, Inc. In one version, the shaft of a vertical diffractometer is replaced by a new shaft assembly to which the cryostat and sample holder can be attached. Liquid helium is transported to the cooling block through a flexible metal transfer line and the helium gas is exhausted through a tube extending away from the cryostat. The cryostat, sample chamber, and transfer line are supported on roller bearings so that over a range of 60° in θ (or 120° in 2θ) they are rotated no more than ±30° about their common axis. This system has been used by Roberge (1975) in a study of the lattice parameter of niobium between 4.2 and 300 K.

Peterson and Simmons (1965) describe a rigid-tail helium cryostat for X-ray back-reflection diffraction studies requiring temperature control within a few hundredths of a degree for extended time intervals in the range above 2.3 K. The specimen chamber permits the growth, visual study, thermal etching, and annealing of crystallized gases. It may also be pressurized to 20 atm for X-ray studies of compressibility.

The specimen chamber is fitted with a single 13-mm-diameter thin beryllium window. Only one window is required for the back-reflection studies at $\theta > 65°$. A glass window for visual observation of the sample is constructed of Pyrex and sealed with epoxy resin to a molybdenum insert so as to minimize cracking due to differential thermal expansion.

Crystals are grown by a temperature gradient that can be controlled to 0.1 K during the several days needed for crystal growth. The consumption rate for liquid helium is approximately 0.13 liter per hour. This was later modified by Losee and Simmons (1968).

A diffractometer-mounted cryostat capable of measuring lattice changes of 1 part in 6500 using a silver-target X-ray tube was described by Lytle (1964). An aluminized Mylar vacuum-tight window encircles the cryostat for 220°. The liquid-helium consumption rate is 0.2 to 0.4 liter per hour.

Metallurgical specimens have been studied with several specially designed cryostats that incorporate provisions for cold-working, strain-testing, or applying a magnetic field.

Barrett (1956) designed a liquid-helium cryostat with a rod aimed at the specimen and attached to a sylphon bellows and ball-and-socket joint. The sample is cooled to 5 K and cold-worked with the rod by hammering and rubbing the surface in order to investigate phase changes that are initiated by cold-working. The basic instrument was later adapted by Barrett and Meyer (1964) for work with solidifed gases and by Barrett et al. (1967) for use with reactive systems.

The liquid-helium cryostat (Figure 4–9) described by King and Preece (1967) was later modified for the study of strain deformations at cryogenic temperatures (Abell and King, 1970). The cryostat is designed for use with a horizontal diffractometer, and specimens can be scanned over a range of 2θ from $-8°$ to $+165°$. The specimens can be aligned by translation and rotation at all operating temperatures and can be removed without disturbing the alignment. The modified apparatus contains a tension straining jig with provision for hammering (cold-working) the specimen surface while at cryogenic temperatures. Samples can only be mounted by breaking the vacuum seal.

A liquid-helium cryostat with provision for changing the specimen probe *without* disturbing the vacuum insulation of the cryostat was reported by King et al. (1969). They suggest that with this method, X-ray diffractometer cryostats can be kept cold for indefinite periods by using closed-cycle liquid-helium refrigerators (see Section 4.5).

Boiko et al. (1972b) describe a cryostat which can be used for diffractometer studies in the 5–300 K range. This cryostat directs the cold helium gas over the sample as it leaves the dewar. The sample is thus cooled by conduction and convection. The consumption rate is 0.12 liter per hour.

The majority of liquid-helium cryostats are mounted on horizontal diffractometers. However, Stochl and Ullman (1963) designed a cryostat for use on the Norelco vertical diffractometer. The weight of the dewar assembly, which would otherwise misalign the goniometer, is supported by

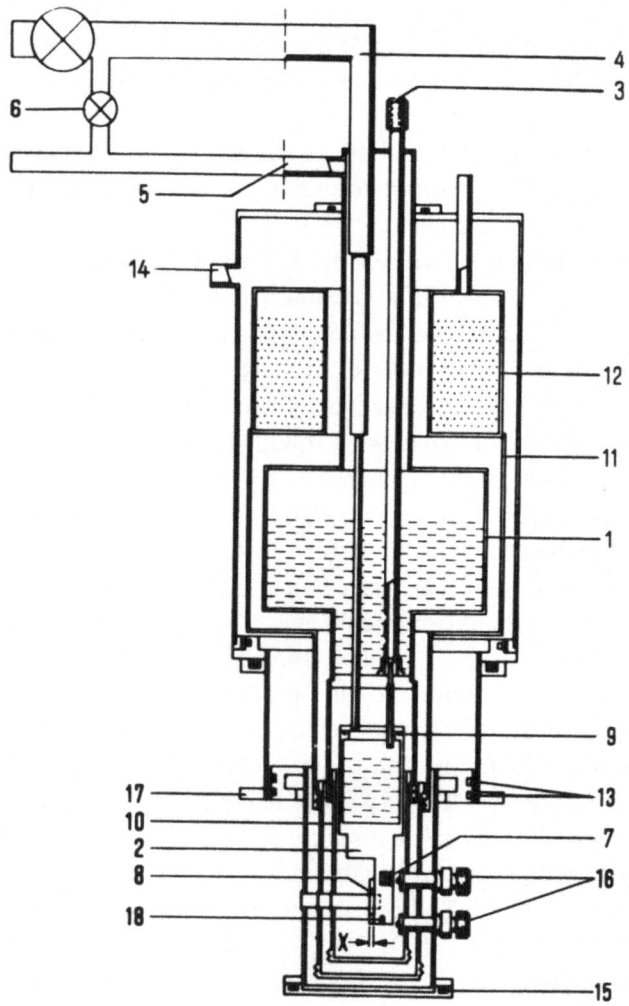

Fig. 4–9. Schematic diagram of liquid-helium powder cryostat (King and Preece, 1968). (1) Liquid-helium reservoir;
(2) hollow copper block; (3) needle valve; (4) pumping line
from 2; (5) helium boil-off line; (6) valve for isolating 4
from 5; (7) resistance heater; (8) powder specimen in mount;
(9) indium O-ring seal; (10) liquid-helium-cooled radiation
shield; (11) liquid-nitrogen reservoir; (12) double O-ring sliding
seal; (13) pumping line for evacuating cryostat; (14) end-flange
O-ring seal; (15) plungers for rotating multiple specimen
mounts; (16) flange to connect cryostat to support frame;
(17) thermometer or thermocouple; (18) liquid-nitrogen-cooled
radiation shield.

an inverted U-shaped bracket that passes over the top of the diffractometer and is attached to the dewar assembly on one side and the goniometer shaft assembly on the other side. Any misalignment caused by contraction of the sample holder can be compensated for by using a built-in slide adjustment. The authors recommend the use of 0.05-mm-thick Mylar windows.

4.4.3.3. Liquid-Nitrogen Cryostats for Powder Diffractometers

Conduction cooling of powder samples at liquid-nitrogen temperatures is fairly routine and can be accomplished without using an evacuated chamber. Suitable protection against ice formation can be provided by a Mylar cover and a stream of warm, dry gas (Calhoun and Abrahams, 1953). However, the use of an evacuated specimen chamber is quite common.

Baun and Renton (1963, 1964) have described a cryostat for a horizontal diffractometer; a number of accessories are described in a separate publication (Renton and Baun, 1963). Samples fixed to one of several mounting plates are placed in the chamber, which is then evacuated, and a bath at any temperature between that of liquid nitrogen and $+100°C$ is placed in the dewar. A 0.25-mm beryllium window is used to transmit the diffracted beam.

A similar device, with provision for pumping on the liquid and thereby lowering the minimum temperature to 65 K, has been described by Trut et al. (1973).

4.4.3.4. Diffractometer Adapters

Diffractometer adapters are particularly useful because they either replace or cover the sample holder and can be used without having to modify the instrument in any way.

A specially designed sample holder for the Norelco vertical diffractometer is cooled from below by a liquid-nitrogen reservoir surrounded by styrofoam insulation (Figure 4–10). The goniometer shaft is kept warm by mica strips inserted between the sample holder and flat of the shaft and by a resistance heater that is wound around the shaft. A Mylar cover and flowing warm dry nitrogen gas prevent ice formation on the sample (Calhoun and Abrahams, 1953; Abrahams and Kalnajs, 1954).

Weltman (1962) described a similar sample holder that is cooled by

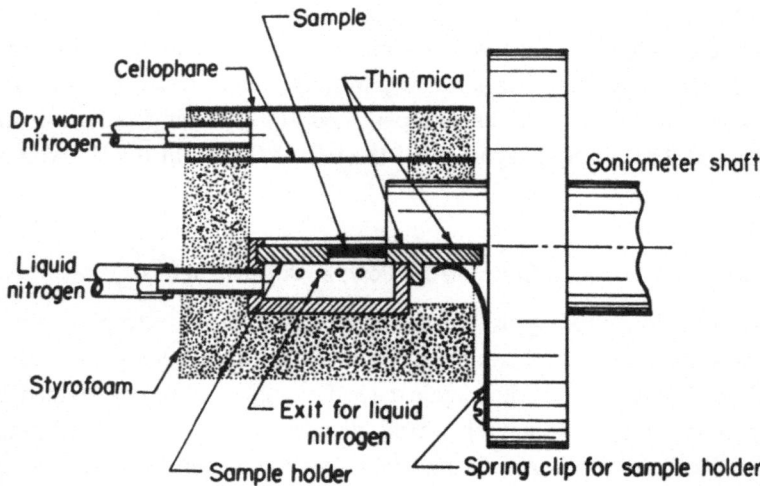

Fig. 4–10. Low-temperature adapter for vertical powder diffractometer (Calhoun and Abrahams, 1953). (Reproduced with the permission of the American Institute of Physics.)

passing cold gas through coils in the sample holder. Smith (1961) and Gould and Gerold (1965) also described similar devices.

An extremely useful, self-contained low-temperature sample holder for the Norelco vertical diffractometer was reported by Ghislain et al. (1965). The device was designed to fulfill the following requirements:

a. To allow the specimen (prepared under liquid air) to be put on the specimen holder without intermediate reheating; this necessitates the specimen holder's being cooled in advance and kept cool when the specimen is mounted on it and when the attachment is mounted on the diffractometer.

b. To maintain the specimen at a temperature very near that of liquid air.

c. To allow diffraction peaks to be recorded in the whole range of angles permitted by the diffractometer.

d. To make the device extremely easy to construct, with a minimum of joints, and able to be fastened simply but rigidly on the diffractometer.

Figure 4–11 shows the attachment. The main piece is a plastic tube, 25 cm long, 50 mm o.d. and 45 mm i.d. After being softened in an oven,

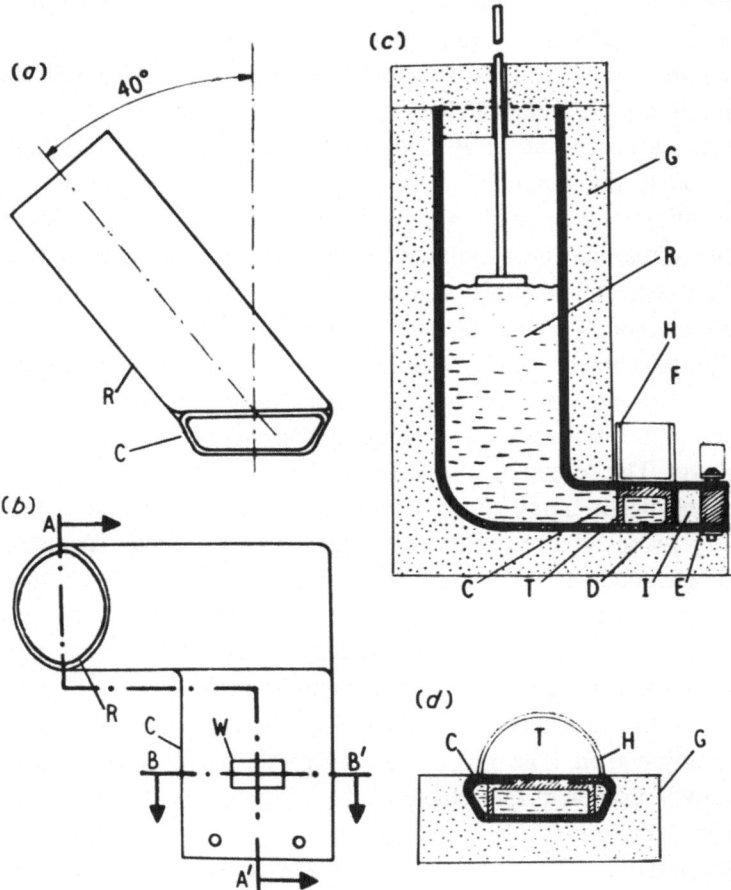

Fig. 4–11. Low-temperature sample holder for vertical powder diffractometer. (a) Elevation of the shaped tube; (b) plan of the shaped
tube; (c) section AA′ through the attachment; (d) section BB′ through
the attachment. Details are given in the text (Ghislain et al., 1965).
(Reproduced with the permission of the Institute of Physics.)

the tube was bent and shaped in two parts: (1) a cooling chamber C with
a nearly trapezoidal cross section and a flat upper surface, to be coplanar
with the flat on the goniometer shaft; and (2) a roughly cylindrical reservoir R whose axis lies in a plane perpendicular to the axis of the cooling
chamber and makes an angle of 40° on each side of the vertical direction
when 2θ increases from 0 to 160°.

A rectangular window W (2 cm × 1 cm) is cut out of the top of the chamber C and a four-legged aluminum table T is placed inside the chamber. In the middle of the table top is a rectangular elevation which fits into the window W and supports the specimen. The chamber is closed at one end by a plastic plate D backed by an insulating layer of plastic foam I and an aluminum block E. This block allows a rigid fastening of the attachment to the goniometer; clamp F fits on the semicylindrical surface of the diffractometer shaft and is fastened by two screws through the chamber walls and the aluminum block. The joints around the table top and the plastic plate D are sealed with silicone grease. The chamber C and the reservoir R are insulated by plastic foam G.

The attachment is filled with liquid nitrogen and allowed to cool. The reservoir is then covered with a plastic foam cover. A wooden rod attached to a float protrudes through a hole in the cover and marks the liquid level. The specimen is mounted and a hood, H, consisting of a semicylindrical plastic frame covered with Mylar foil, is attached to the chamber with silicone grease to prevent icing on the specimen. The attachment is then fastened to the diffractometer, and a gentle flow of dry nitrogen gas is directed on the Mylar film to prevent ice formation on it. The reservoir and chamber hold about 250 ml of liquid nitrogen, which lasts about half an hour. Replenishing can be easily accomplished.

The diffractometer shaft is well insulated from the liquid and is not appreciably cooled. The rigidity of the specimen chamber is sufficient to avoid any noticeable displacement of the diffraction peaks when the degree of filling of the reservoir is varied.

Stammler et al. (1963) constructed a device similar to Weltman's (1962) but for the G.E. horizontal diffractometer.

Masson (1960) built an adapter for a horizontal diffractometer out of polystyrene foam. A copper rod in contact with LN_2 supports the sample, and X-rays pass through the walls of the styrofoam chamber. The use of polystyrene foam is recommended since it is a thermal insulator, relatively transparent to X-rays, and is easily machined and joined to other parts. It can also be immersed in liquid nitrogen if the sample is to be loaded in that manner. Similar adapters employing immersion cooling are described in Section 5.2. This method of constructing a chamber with low-density walls through which the X-rays pass was initially used by Rinne (1917), who placed the sample in a cork chamber.

Many X-ray equipment manufacturers produce cryostats designed to fit their instruments. For example, Rigaku-Denki sells a dewar that fits over their horizontal powder diffractometer and allows up to 50 hours of operation at 85 K with one filling. Enraf-Nonius offers an adapter which modifies their Universal Low-Temperature Device (see Section 3.4.2.2) for use with their Guinier–Lenné camera so that photographs of d-spacing vs. temperature can be obtained at low temperatures.

4.5. Conduction Cooling via Mechanical Refrigeration

Mechanical or closed-cycle refrigerators can be adapted for use in conduction cooling in two ways. The simplest approach is to use a standard cryostat similar to those already described, but to replace the liquefied gas with a liquid bath that is cooled by a mechanically refrigerated probe (see Appendix 3). However, the lowest temperature attainable by this method is, currently, approximately 135 K.

The second method is to mount a mechanical refrigerator directly on the X-ray instrument and to use it to cool the sample by conduction. A vacuum-jacketed, multiple-exposure back-reflection camera utilizing a closed-circuit pressurized helium cryostat has been described by Woodard and Straumanis (1971). It can operate at any temperature between 25 K and 180 K without the need for any liquefied gases. No sample motion of any kind is permitted by the original design of the instrument, so that powder specimens with particles smaller than 20 μm are used to obtain uniform diffraction lines. Details of construction and the use of this apparatus to obtain lattice parameters correct to within ± 0.00003 Å at temperatures as low as 4 K are given in the original article.

This camera was redesigned and improved by Johnson et al. (1974) to permit the use of both Laue and front-reflection, as well as back-reflection, techniques. Temperatures between 1.4 K and 30 K can be maintained; the sample can be rotated during exposure; and the film is outside the evacuated chamber.

Recently, Simmons (1976) has described a cryostat which incorporates a liquid-helium dilution refrigerator. This closed-cycle refrigerator employs the mixing of ^{3}He and ^{4}He in a chamber attached to the sample to cool the crystal by conduction. The cryostat is constructed with five heat shields and contains a plastic sample chamber that can be pressurized to 400 atm. It is designed for the study of single crystals of helium. A minimum temperature of 30 mK can be attained.

This complex cryostat is rigidly fixed in place and is supported so that the sample is in the center of a θ/θ diffractometer. The diffractometer is mounted on two other circles (bearings) resulting in a three-circle single-crystal θ/θ diffractometer in which the X-ray tube and counter both move around the fixed helium single crystal.

4.6. Conduction Cooling via Thermoelectric Cooling

The use of thermoelectric cooling (see Appendix 5) has certain advantages: the sole source of "cooling power" is electricity; there are no expenses beyond the initial cost and that of the electricity consumed; and no effort need be expended in moving and changing compressed-gas cylinders or dewars of liquefied gases. However, there are also serious restrictions to the use of thermoelectric cooling devices. The cooling capacity (an important consideration if crystals must be grown *in situ*) is limited. In addition, the lowest temperature that can be attained is a function of the number of thermoelectric stages used in the device. Thus the size of the apparatus becomes a controlling factor, especially if it is designed for use with a single-crystal specimen. Finally, both electric and cooling-water connections must be made to the device, thereby introducing some restrictions on the rotational freedom allowed the apparatus. These factors, together with the need for a covered sample chamber, suggest that thermoelectric cooling be restricted to powder samples and single crystals that are solid at room temperature and that do not have to be cooled greatly. With currently available units, a practical lower temperature limit is $-45°C$, although one device for neutron diffraction, which permits larger apparatus and, therefore, more cooling stages, could be cooled to $-80°C$ (Khan and Erickson, 1970). The use of miniaturized multistage modules coupled with specially designed goniometer heads would permit the extension of these devices for use to $-100°C$ and lower. In this case it might be necessary to consider placing the thermoelectric cooler over the crystal, so that the crystal is between the cooler and the goniometer head base.

In general, a thermoelectric cooling device requires an evacuated chamber for the sample and good thermal conductivity between sample and cooler. Only a few operating devices have been reported in the literature.

Horne et al. (1959) described a single-stage thermoelectric cooler for the Norelco vertical powder diffractometer capable of reaching a minimum temperature of only 0°C, which was still lower than the temperature reached by Petz (1963).

Zickert and Guttzeit (1966) reported fabrication of three sample holders for use with Debye–Scherrer, rotating-crystal, and small-angle scattering techniques, but no details were given.

Hartoulari and Dufour (1970) describe a sample holder for use on a powder diffractometer. A special feature of this cooling device is that an ammonia atmosphere can be maintained inside the sample chamber. A similar device in which the humidity over the sample could be controlled was capable of reaching a minimum temperature of −22°C (Johnson, 1975).

Agron, Levy, and Bogardus (1972) designed cooling devices for use with neutron and X-ray single-crystal samples. The X-ray unit is surrounded by a Mylar window and is used on a precession camera. Cohen et al. (1971) cooled a protein crystal in a capillary tube to 3°C.

4.7. Conduction Cooling via Joule–Thomson Expansion

Joule–Thomson refrigerators achieve cooling through the rapid expansion of a gas at an orifice (see Appendix 4). These devices have been extensively used in spectroscopy, where size and accessibility requirements are less stringent than in LTXRD. A successful cooling device for neutron diffraction was first described by Coppens, Godel, and Sabine (1967). The gas expands at the orifice, which is adjacent to the crystal mounting pin, cools the crystal, and then flows back past the heat exchanger to precool the incoming high-pressure nitrogen gas (Figure 4–12). A modified version incorporating a self-regulating device was reported by Ebdon and Wheeler (1971).

Air Products and Chemicals (1973) has designed a refrigerating unit called the Cryo-Tip, which is available in several models and is capable of operation at several temperatures. Their unit for single-crystal investigations at liquid-nitrogen temperature has been used in several crystal-structure analyses (Iwasaki and Iwasaki, 1972; Frenz and Ibers, 1972; Einstein et al., 1972; Drew and Einstein, 1973; Guttormson and Robertson, 1973; Ito and Sakurai, 1973; Wheeler and Colson, 1975).

This device consists of a subminiature, single-fluid, open-cycle Joule–Thomson cooler mounted on a 64-mm-high eucentric goniometer head (Figure 4–13) and is covered with a beryllium shroud. The cooler is

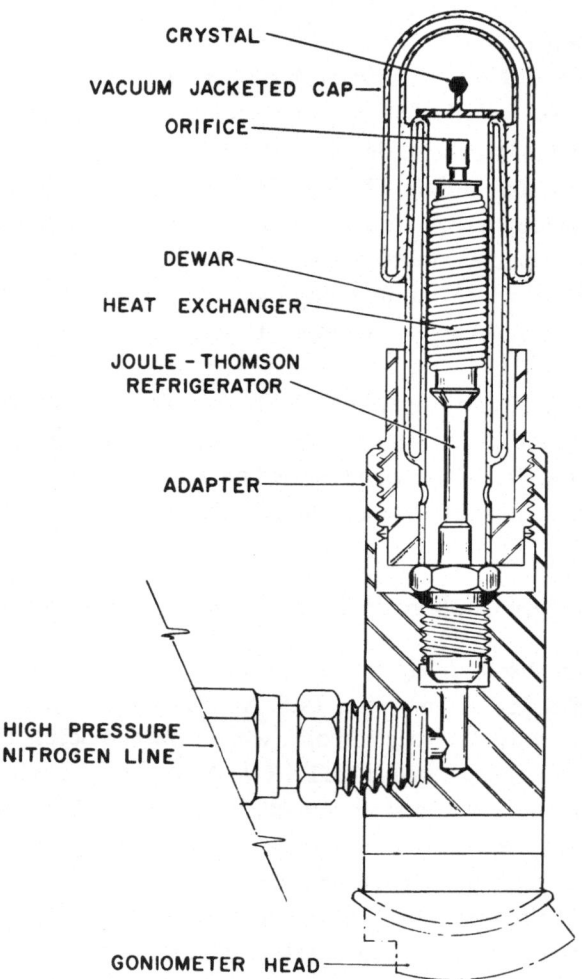

Fig. 4–12. Joule–Thomson refrigerator for use with neutron diffraction (Coppens et al., 1967). (Reproduced with the permission of the American Institute of Physics.)

mounted in a heat exchanger so that the cooled, expanded gas cools the incoming gas. This allows the use of minimum refrigerator size and warms the exiting gas to nearly room temperature. The temperature is varied by adjusting the flow rate of the nitrogen gas. A large cylinder of compressed gas lasts approximately ten hours while operating at maximum flow rate (i.e., at minimum temperature).

The beryllium shroud is transparent to X-rays and forms a vacuum-

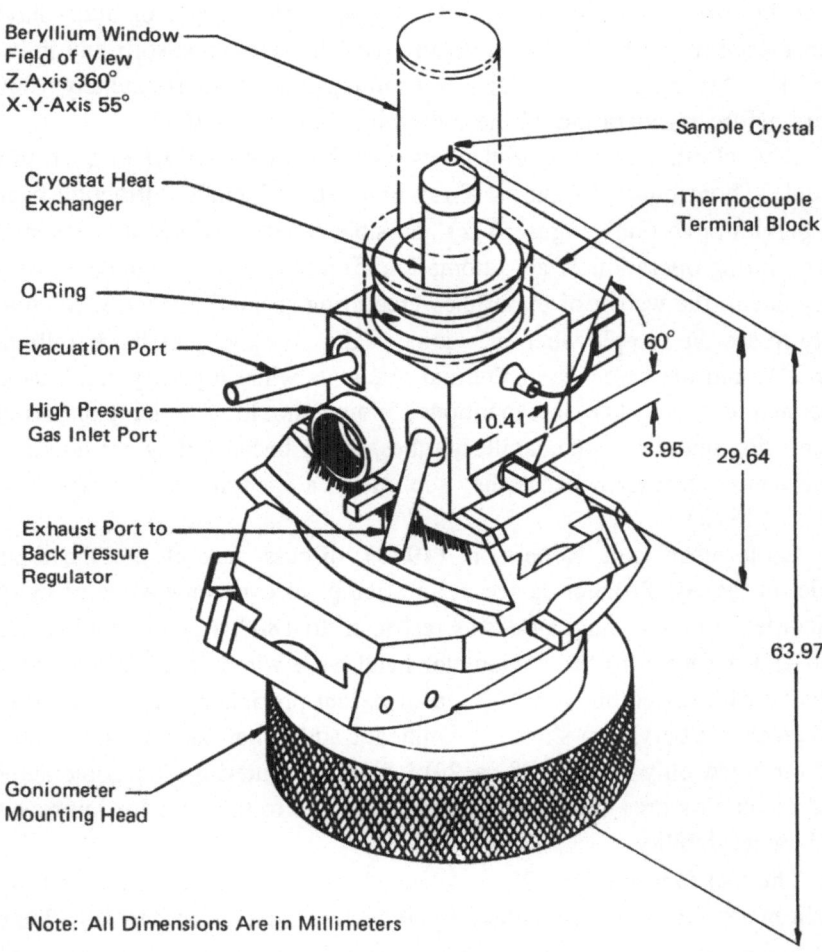

Fig. 4–13. Schematic diagram of Joule–Thomson refrigerator for low-temperature X-ray diffraction. (Courtesy of Air Products and Chemicals, Inc.)

tight enclosure about the crystal so that it can be evacuated. A built-in copper–constantan thermocouple indicates the temperature of the copper tip; any temperature differential between this tip and the crystal must be determined. For example, Ito and Sakurai (1973) calibrated the temperature and found that the temperature of the sample was 13° higher than the thermocouple reading at 240 K and 33° higher at 110 K.

A basic problem with the use of Joule–Thomson expansion coolers is that any moisture in the supply tank or gas lines may condense at the expansion orifice and block the flow of gas. Various types of filters have been placed in the lines to remove any particles and/or adsorb any moisture. The Air Product unit claims a running time of approximately 1000 hours before regeneration of the molecular sieve is required.

Several other potential difficulties must be considered when operating a Joule–Thomson device on an X-ray unit. These include diffraction from the beryllium shroud (Figure 4–2), possible tangling of gas and vacuum lines (during operation of an automatic diffractometer), mechanical instability due to the weight of these lines (requiring special brackets), the inability to see the sample when it is cooled (due to opacity of the beryllium shroud), and the presence of thermal gradients when a poorly conducting specimen is used. The last-mentioned point should also be considered when using samples inside capillary tubes, either because they are liquid at room temperature or because they sublime in a vacuum (see Iwasaki and Iwasaki, 1972).

Guttormson and Robertson (1973) discuss several modifications made on the Air Product device in an attempt to overcome some of these difficulties: (1) the range of ϕ was restricted to 180°; (2) the goniometer head and the leads to the goniometer head were wired so as to prevent a collision with the collimator; (3) a transparent plexiglass shroud was used to replace the beryllium shroud (significant additional background scatter was observed only below 10° in 2θ); (4) after centering, the goniometer head (including arcs and translations) was glued to the base for improved mechanical stability.

The fact that the Cryo-Tip is attached to a flexible high-pressure hose has been used to good advantage by Wheeler and Colson (1975). They mounted a crystal on the Cryo-Tip inside a large freezer, then activated the Cryo-Tip and transferred the cooled crystal to the diffractometer.

The crystal structures determined with this apparatus were studied at temperatures ranging between $-65°C$ and $-100°C$. It is evident that this device is useful for many routine crystal-structure determinations that can be carried out in this temperature range. However, special requirements, such as accurate temperature control and measurement or growing of single crystals *in situ,* are not easily met when using this device. Its main advantage is that it can be used with little or no changes in the X-ray instrument, aside from some modifications to the software if an automatic diffractometer is used.

A final word of precaution is that liquid nitrogen will not form if the pressure drops below a certain minimum [e.g., Ebdon and Wheeler (1971) report 120 kg/cm²; Coppens et al. (1967) report 105 kg/cm²]. Thus, provision should be made for disposing of approximately half a cylinder of compressed nitrogen gas every 10 hours. Coppens et al. suggested using an auxiliary compressor to fully utilize the nitrogen gas from the cylinder. This, of course, will raise the price of a system considerably and will become economically sound only if large quantities of gas are used and no other research teams in the laboratory can use the partly full tanks. The temperature stability of the system seems to be about ±3°.

Bonilla et al. (1970) mention using the Cryo-Tip to attain temperatures as low as 20 K, but do not give any further details.

CHAPTER 5

Immersion-Cooling Apparatus

5.1. General Principles

The temperature of a sample which is cooled by immersing it in a cold medium is restricted to the temperature of that medium. This cold medium is usually a liquid or a gas. The most commonly used coolant is LN_2, but other materials, such as cold alcohol and sand, have also been used. One of the advantages of using LN_2 is that it evaporates without interacting with the crystal or camera. If alcohol is used, the possible interaction of the coolant with the sample must be considered, and provision must be made for removing the liquid from the vicinity of the sample and diffraction instrument.

Cold-gas cooling is accomplished by conducting the investigation in a cold room, by placing the X-ray apparatus in a cold chamber, or by using a stream of cold gas.

Some of the earliest low-temperature studies were conducted in a cold environment by opening windows during the winter or conducting the experiments outdoors. More recently, refrigerated cold rooms have been used. If the sample is enveloped by a *stream* of cold gas, we have the special case of immersion that was discussed in Chapter 3.

5.2. Immersion of Sample in Cold Liquid

Three methods have been described in the literature for cooling a sample with a cold liquid. They are spraying, dripping, and dipping.

5.2.1. Spraying

Santos and West (1933) described a Debye–Scherrer camera in which the specimen axis was coaxial with a copper tube immersed in liquid air. The dewar containing liquid air was below the specimen and the pressure differential between the atmosphere and the boiling liquid air inside the copper tube caused the liquid air to rise and spray the sample which was suspended above it. The sample could be rotated about its axis *and* moved up and down vertically to insure random orientation of the crystallites in the beam.

In another variation, a dewar of liquid air was pressurized, causing the liquid air to spray over a sample placed near the nozzle (Goetz and Hergenrother, 1932).

5.2.2. Dripping

Dripping is the most common of the sample-immersion methods. A stream of cold liquid, usually a liquefied gas, drips onto the specimen from above (Figure 5–1). To prevent ice formation, Hengstenberg and Mark (1928) passed a stream of dry air through the camera. Other investigators claim that the continuously moving liquid stream is sufficient to keep the sample clear of ice. However, the presence of the liquid stream in the X-ray beam produces diffuse rings on the X-ray photograph (Lonsdale and Smith, 1941; Francombe, 1957; Biswas, 1958). Liebling and Marsh (1965) also cooled their sample by dripping LN_2 over it, but arranged the flow rate so that the liquid vaporized before reaching the crystal. Since the crystal was bathed only in cold gas, no halo due to liquid nitrogen was observed.

If the cooling liquid is a liquefied gas, there will be no problem with disposing of the liquid after it cools the sample. However, if a material that is still liquid at room temperatures is used, provision must be made for disposing of it before it collects in the camera. Tombs (1952) described a powder camera in which the sample was studied at $-70°C$ and $-10°C$ by allowing cold alcohol to drip over the sample. The alcohol was removed through a small hole in the bottom of the camera. The use of such viscous liquids requires a wider than usual outlet from the dewar tube (a diameter of 1 mm is satisfactory). Anhydrous coolant should be used to prevent formation of ice crystals which can block the outlet.

Fig. 5–1. Schematic diagram of device for dripping liquid coolant over sample. The orifice size can be changed by replacing tube B in plug A. (Steward, 1960). (Reproduced with the permission of the Institute of Physics.)

The practical aspects of this method were reviewed by Lonsdale and Smith (1941) and by Clifton (1950). A double-walled dewar flask reservoir is provided with an opening at the bottom of approximately 3 mm diameter into which a piece of metal tubing of about 0.3-mm bore is fitted. Tape can be wrapped around the metal tubing to effect a good fit between it and the glass outlet. Various jets and outlets can be used with the different cameras and diffractometers.

Taylor (1931) used a hypodermic syringe and showed that if a very small hole is used, gas forms and stops the flow. Kuznetsov (1956) described a nozzle consisting of just such a needle. However, a small hole in the tube just above the needle is used to vent any excess gas that might otherwise interfere with the steady flow of liquid. Clifton (1950) described a nozzle in which the size of the opening is adjusted by means of a lever-controlled insert.

The reservoir can be automatically filled since it is open to the atmosphere and any changes in pressure that occur during refill periods have a negligible effect on the sample temperature. This method is very good for cooling samples to the temperature of the cold bath. It is not recommended for higher temperatures or for growing crystals *in situ*.

Other cameras utilizing this method have also been described. Biswas (1958) described a Debye–Scherrer camera in which liquid air drips on the sample. The camera is continuously evacuated to remove the cold gas and prevent ice condensation.

Hirshfeld and Schmidt (1956) cooled a single-crystal specimen by using a dewar (similar to the gas generators described in Chapter 3) in which the delivery tube was placed in the liquid nitrogen rather than in the cold nitrogen gas which forms above the liquid level. A heater boils the LN_2 and maintains a pressure which delivers a steady stream of LN_2 to the crystal. A similar device for a powder diffractometer was described by Zevin and Kheiker (1958).

Several investigators have designed instruments which can be used to deliver a stream of liquid nitrogen *or* cold nitrogen gas. These include Owen and Williams (1954), Pearson (1954), Kuznetsov (1956), and Viswamitra (1962). Kogan and Bulatova (1969) cooled the sample by conduction and immersion.

A beryllium cell that can be used either on a precession camera or a single-crystal diffractometer is cooled by blowing cold gas or dripping LN_2 over the cell inside a plastic bag (Figure 2–7). This cell, which transmits 70% of Mo K_α radiation or 17% of Cu K_α radiation, can also be run at high pressures (Morosin and Schirber, 1974).

5.2.3. Dipping

Several devices have been described in which the X-ray beam passes through a sample which is completely immersed in liquid nitrogen or helium. A 1-cm path length through liquid helium was used by McDonald et al. (1966) in a study of metastable phases. The sample was prepared in a press and the pressure was released while the press was immersed in liquid helium. Copper powder was used as an internal standard.

Grushko et al. (1970) used tungsten radiation to penetrate a 2-mm thickness of LN_2 (in addition to the glass windows of the container). Interestingly, a *conical* film cassette was used to obtain the Laue diagram. Rather recently, Segmüller et al. (1974) designed a dewar with Mylar windows 2 mils in thickness for use with liquid helium.

An interesting adapter for a horizontal powder diffractometer was constructed by Roessler and Bolling (1964). The flat sample is immersed in a container made of high-density foamed polystyrene. A spring pushes the sample up against one wall of the container. This wall is thick near the edges in order to support the pressure of the spring-loaded sample but is very thin in the middle (1.5 mm). The thin portion of the wall absorbs less than 15% of the X-ray beam and allows nearly normal diffractometer patterns to be obtained. Above 14° in 2θ, the background scattering from the foam is no problem. This is a simple device to fabricate and use, but some distortion due to contraction of the sample holder at low temperatures does occur (Figure 5–2).

The special characteristics of several devices are summarized in Table 5–1.

5.3. Immersion of Camera in Cold Liquid

The technique of immersing the camera in a cold liquid is not commonly used for several obvious reasons. Cooling a camera can interfere with its alignment and/or with the operation of moving parts due to thermal contraction. The X-ray path must be considered and absorption must be minimized. In addition, sample changing is complicated by the need for removing the camera from the bath. However, this method is a logical extension of the cryostat conduction-cooling method and excellent stability and temperature control of the specimen (by the use of a heater) can be accomplished.

Pearson (1955) described a powder camera that is placed in a liquid-helium bath. The X-ray tube is placed horizontally and a long brass collimator extends down into the dewar. In this manner, there is no absorption of the X-ray beam by the liquid helium.

Beck et al. (1973) described a single-crystal camera utilizing modified Weissenberg geometry in which the translational movement of the film is replaced by rotation of the cylindrical cassette about its axis. This compact camera can be placed in a dewar 115 mm in diameter and can be immersed in liquid helium.

A powder camera which was cooled by surrounding it with one of several different cold baths, including sand cooled by LN_2, was designed by Hawes (1959).

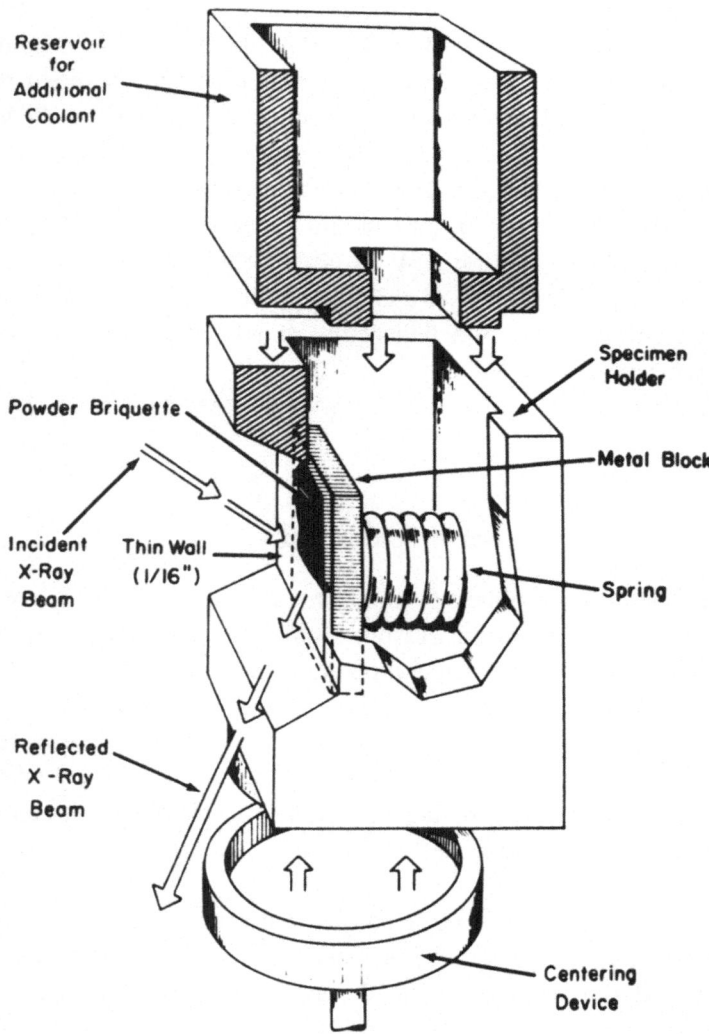

Fig. 5–2. Low-temperature sample holder for horizontal diffractometer (Roessler and Bolling, 1964). (Reproduced with the permission
of the American Institute of Physics.)

5.4. Use of Cold Room

Some of the earliest low-temperature investigations were conducted in
laboratories kept cold simply by opening the windows. St. John (1918)

Table 5–1.   Apparatus for Cooling by Dipping Sample in Cold Bath

Aronova et al. (1959)	Sample at the bottom of a liquid-nitrogen dewar is continuously washed by liquid nitrogen
Donde et al. (1967)	Metallic dewar filled with liquid helium; outfitted with beryllium windows; specimen rotated with magnetic actuator
Grushko et al. (1970)	Device to grow large argon crystals with provision for obtaining transmission Laue patterns while crystal is immersed
Golik (1960)	Metallic sample is partly immersed in liquid nitrogen; also, mechanically rotating scoops lift liquid nitrogen and pour it over the sample
Kiseleva and Mikhalenko (1962)	Lower portion of sample dips into liquid nitrogen; by adjusting height of sample can control its temperature at the point where the X-ray beam strikes it
Ling and Wagenfeld (1965)	Crystal suspended in liquid nitrogen inside dewar equipped with X-ray transparent windows; absorption due to the liquid nitrogen is measured and accounted for
McDonald et al (1966)	Sample, in metastable phase, released from high-pressure press in liquid-helium bath
Roessler and Bolling (1964)	See text
Sandler and Akhmechet (1958)	Flat sample on end of rotating rod is dipped into liquid nitrogen, but it is photographed above the liquid level
Segmüller et al. (1974)	Dewar with Mylar windows is used

studied the crystal structure of ice in a cold laboratory; Broomé (1923) studied benzene at $-2°C$ and Owston (1949) investigated the diffuse scattering from ice in a similar manner. In all three cases the results were not particularly satisfying, and other methods of cooling were used subsequently.

More recently, specially designed cold rooms or converted large, walk-in freezers have been used. Wheatley (1960) studied pyrimidine at $-2°C$, while Gieren et al. (1973) reported a crystal-structure analysis at $-50°C$.

A variation of the cold room is the cold chamber in which only the X-ray camera is placed. This method was used by St. John (1918), who put cans of ice and salt inside a chamber containing the camera, and by Bacon et al. (1964), who used a foamed polystyrene box with dry ice inside it. The X-rays passed through the walls of this box. More recently, a chamber has been cooled by connecting it to a freezer (Tachez and Theobald, 1975).

Currently, it is possible to purchase a mechanically refrigerated cold chamber which contains the evaporator, air circulator, and defrost heaters carefully shielded to prevent turbulence over the equipment placed in the chamber. The X-ray tube is mounted in a cutout in the rear wall of the insulated, stainless steel chamber. The hinged front door provides access to the X-ray equipment via several ports in a protective plexiglass plate. This refrigerated chamber can be cooled to −50°C (Cincinnati Sub-Zero Products).

In a few instances, the sample has been cooled by placing dry ice (Oda et al., 1943) or liquid air (Rinne, 1917) into the chamber which contained the sample.

Bertolucci and Marsh (1974) describe the use of a cold room maintained at −17°C. They discuss techniques for preparing samples and growing crystals in the cold room (see Sections 6.2.1 and 7.2).

PART III

Low-Temperature X-Ray Diffraction Techniques

CHAPTER 6

Sample Preparation

6.1. Sample Preparation: Samples Solid at Room Temperature

Samples that are solid at room temperature can be divided into two groups: normal, well-behaved solids and solids which are volatile, reactive, or otherwise unstable unless cooled. The latter group includes materials with high vapor pressure, those that decompose at room temperature, and those that decompose rapidly when exposed to X-radiation (at room temperature).

6.1.1. Well-Behaved Solids

When dealing with normal solids, very few *special* procedures are necessary.

Polycrystalline samples are prepared in the usual way. A conduction-cooled powder-diffractometer sample must be in good thermal contact with the sample holder and cooling block. This can be accomplished by using a thermally conducting grease, which can be prepared from vacuum grease and fine copper powder or can be purchased (see Appendix 11). Care should be taken to keep the copper–grease mixture out of the path of the X-ray beam. (See Section 4.1.3 for limitations of this method.)

Several people have suggested using collodion as a means of securing the sample to the sample holder. Penfold (1971) prepared a collodion film of the sample which was attached to the sample holder by vacuum grease.

With single-crystal specimens, the primary consideration is the proper choice of mounting adhesive. Some adhesives shrink or become brittle during cooling and may affect the crystal alignment. Ideally, the adhesive should have the same thermal characteristics as the fiber and crystal. Several good adhesives are available, including low-temperature epoxies and cyanoacrylates (e.g., Eastman 910). Viswamitra (1962) recommends silicophosphate dental cement.

Good results have also been obtained with vaseline (Vonnegut and Warren, 1936; Milledge, 1969), which becomes rigid and holds the crystal in place when cooled. One advantage of using vaseline is that when the specimen is rewarmed the crystal can be adjusted or removed.

Some crystals have been mounted by freezing a liquid in which they are placed. For example, water has been used as a mounting medium for ice (Owston, 1949) and mercury for crystals of tin (Bilderback and Colella, 1975).

The selection of the crystal does not entail any special considerations, and it should be treated as if the sample were to be studied at room temperature. However, if the crystal is the least bit soluble in water, it should be protected by mounting it in a capillary tube or by covering it with a coating of shellac. Otherwise, any ice forming on the surface of the crystal may dissolve it (Dickens, 1966). A detailed discussion of cleaning, handling, and sealing thin-walled capillary tubes is found in Section 6.2.3.1 (see also Sections 6.1.2 and 6.4).

6.1.2. Unstable Samples

Several methods have been described for mounting a single crystal inside a capillary tube. One of the easiest and most efficient of these is to mount a crystal in the normal manner at the end of a thin glass fiber and then to cover both crystal and fiber with a capillary tube 0.7 to 1.0 mm in diameter. The capillary tube can be sealed to the top of the goniometer head sample holder (nib) with epoxy or other adhesive (see also Section 6.4).

Dickens (1966) suggested another method. The crystal is removed from the microscope stage with a 0.2-mm-diameter capillary tube that has been electrostatically charged (by rubbing it on a cloth) and inserted into the capillary tube in which it is mounted. The crystal is then wedged firmly into this capillary tube with a fine glass tube or straight wire. The capillary tube is sealed 1 or 2 mm away from the crystal with a very small torch, using a heat sink to protect the crystal. (Using the torch directly on the open end of the capillary tube introduces appreciable amounts of water vapor, which can dissolve the crystal in the tube or form ice when cooled.)

If the capillary tube is over 0.5 mm in diameter, the expanding gas may blow a hole in it while it is being sealed. This can be prevented by keeping one end of the tube in liquid nitrogen while the other end is sealed with the torch. If the crystal-mounting process takes place inside a dry box, the open end may be sealed with a small heater or plugged with epoxy or grease and then resealed with a torch when removed from the dry box. Finally, the capillary tube containing the crystal is mounted with cement or epoxy to a short length of fine wire which is affixed to a 3-mm-diameter rod set in the goniometer head nib. The wire should be bent so that the arc settings are 10° or less.

In order to minimize the portion of the crystal holder which is in the beam, Verschoor and Keulen (1971) mounted a crystal in a short glass capillary tube fused onto a thin glass fiber.

If absorption is not a serious problem, it is possible to mount crystals in a short length of tube by removing the tip of the capillary tube and placing a small amount of vaseline–beeswax mixture on the open end. A heating element can be used to warm the wax so that it is drawn into the tube, which is then sealed with the same heating element (see Section 6.2.3.1). The crystal is inserted in the large end of the capillary tube with a fine glass rod or a wire (ca. 0.1 mm diameter) until it is embedded in the soft wax. The capillary is cut, dressed with a metal probe, and sealed. It can then be cemented onto a metal wire or glass fiber.

The use of vaseline–beeswax may not be practical for all crystals at low temperatures since the rates of contraction of sample and wax may be quite different. However, a little experience will determine what mixture works best. Milledge (1969) states that vaseline is a satisfactory mount down to liquid-helium temperatures.

Others secure the crystals by wedging them into place inside the capillary tube or by holding them in position with glass wool (e.g., Guttormson and Robertson, 1973).

The use of standard, cylindrical capillary tubes, even though they are thin-walled, introduces definite errors into the measurement of relative intensities due to absorption of the X-ray beam by the cylindrical walls of the capillary tube. Several methods have been developed to correct for these errors (see Section 8.4). The major advantages of using these tubes are that (1) the capillary tubes are readily available, and (2) the sample is in direct or very close contact with the walls of the tube so that it is cooled evenly.

However, Nelmes (1970) describes thin-walled transparent bulbs manufactured from pure, fused silica which have relatively low X-ray absorption and quite high physical strength. These bulbs are spherical, about 15 mm in diameter, with walls less than 0.05 mm thick. They have a neck tube of internal diameter 3–4 mm. For a specimen located at the center of a bulb, all rays pass normally through the same thickness of silica and so suffer the same small absorption (excluding small areas close to the neck and apex of the bulb). If a specimen is mounted on a fiber positioned at the center of the bulb and the fiber is sealed in place with wax, the problem of calculating absorption corrections can be eliminated. Prior to sealing, the bulb can be flushed out with dry nitrogen or helium gas through hypodermic needles inserted through the wax. The use of helium gas, which has a greater thermal conductivity than nitrogen gas, will improve the cooling efficiency at the crystal. A thermocouple could also be sealed inside the bulb just below the crystal. Although these bulbs were developed for room-temperature studies, their application to the field of LTXRD appears worth investigating. (See Appendix 12 for the manufacturer of these bulbs.)

6.1.3. Protein Samples

The study of proteins and other large biologically interesting molecules presents special problems. Crystals of these molecules contain large quantities of water, and in order to maintain the crystals, they must be in equilibrium with the mother liquor. The use of normal low-temperature techniques results in the destruction of the crystals, presumably due to the expansion of water on freezing. However, because such crystals are highly susceptible to radiation damage when studied at or near room temperature and are still partially affected at $0°C$, it is desirable to study them at as low a temperature as possible.

Recently, several successful approaches to this problem have been reported. Haas (1968) succeeded in cross-linking the surface of lysozyme crystals so that they were insoluble in water–glycerol mixtures. The cross-linked crystals were then maintained in a water–glycerol mixture which formed an amorphous glass, rather than a crystalline solid, on cooling. Photographs were obtained at $-50°C$.

However, the cross-linking procedure did not work for lactate dehydrogenase (LDH) (Haas and Rossmann, 1970). Instead, these crystals were soaked in a concentrated acrose solution, which also forms a glass when cooled. LDH crystals soaked in sucrose solution were cooled to $-75°C$ and showed a rate of radiation damage that was one-tenth that of crystals at room temperature. These crystals have to be frozen quickly and isotropically. The technique found to be most efficient consisted of placing a single LDH crystal on a strip of filter paper, waiting until the liquid had been drawn up by the paper, scooping the crystal up on the end of a 0.25-mm glass fiber, and immediately plunging it into, and retaining it in, liquid nitrogen. The crystal was now frozen to the fiber, which was itself mounted on a goniometer head. The mounted crystal was then quickly transferred to a diffractometer, where it was placed in a stream of cold nitrogen gas. The gas stream was maintained at a temperature below $-60°C$ in order to keep the sucrose solution in a glassy state.

One serious drawback to the use of this technique is that the unit-cell dimensions change with the sucrose concentration. This indicates that the low-temperature crystals are not necessarily isostructural with the room-temperature crystals.

Another approach to the problem was reported by Thomanek et al. (1973). The difficulty in freezing crystals, as has already been mentioned, is that water expands as it is cooled below $4°C$. However, this is not true for the high-pressure modification known as ice III. The transition to this state (at 2100–3500 atm) is accomplished by a volume contraction. This modification remains in a metastable state at atmospheric pressure at low temperatures. In this method, crystals of myoglobin were subjected to hydrostatic pressure using isopentane as the liquid, cooled to 77 K, and depressurized to atmospheric pressure at 77 K. The isopentane was easily removed and the crystals were transferred to a precession camera while constantly at liquid-nitrogen temperature. Photographs taken of the crystals under a steady flow of dripping liquid nitrogen showed no evidence of damage due to the high-pressure, the low temperature, or the immersion in isopentane.

Petsko (1975) has developed a method which appears to be the most general method yet described for maintaining the integrity of protein crystals at low temperatures. The technique is to replace the normal crystal mother liquor with a salt-free aqueous–organic solvent mixture of low freezing point. Crystals treated in this manner have withstood several temperature cycles between room temperature and $-120°C$. 2-Methyl-2,4-pentanediol is the organic additive that has been found most suitable for many proteins. However, if a flow cell is to be used and viscosity is a problem, ethylene glycol or isopropanol can also be successfully employed in some cases.

It is necessary to determine the proper percentage of alcohol to use for each crystal. If few crystals are available, some preliminary studies can be made in solutions of the protein. In practice, the protein crystal is placed on filter paper to remove the mother liquor, and the crystal is then plunged directly into the alcohol–water mixture maintained at a temperature close to the freezing point of the mother liquor.

If the crystal shatters or cracks it is an indication that the alcohol concentration is too high, while softening or dissolving of the crystal indicates too low an alcohol concentration.

Advantages of this method include ease of sample preparation, wide range of crystals which can be cooled to low temperatures, the fact that the solvent mixture remains fluid even at low temperatures, and the absence of significant unit-cell dimension changes.

Disadvantages include the necessity of finding the exact working conditions for each protein individually and the high viscosity that is observed for some solvent mixtures at low temperatures. The exact extent of any changes in protein conformation is not yet known and may possibly turn out to be significant.

6.1.4. Phase Transitions

When solid–solid phase transformations occur, the high-temperature phase is often formed without any difficulty. However, there are a number of well-known cases in which the crystal shatters as it passes through the transition point. In this case it is necessary to try some of the suggestions found in Section 7.3.

Difficulties in initiating a phase transition at low temperatures are quite commonly encountered in studies of metals. Barrett (1956) designed a liquid-helium cryostat for a powder diffractometer with a metal rod aimed at the specimen and attached to sylphon bellows with a ball-and-socket joint. The sample could be cold-worked by hammering and rubbing its surface with the rod while being maintained at 5 K. In this manner, phase changes initiated by cold-working could be studied.

6.2. Sample Preparation: Samples Liquid at Room Temperature

A material which is liquid at room temperature can be crystallized by cooling the sample prior to, simultaneously with, or after mounting in the sample holder. All three methods, for both single-crystal and powder specimens, have been described in the literature and will be discussed here. (Single-crystal growth and mounting are discussed in Section 7.2.1, while the preparation of a randomly aligned polycrystalline sample is described in Section 6.5.)

6.2.1. Crystallization Prior to Mounting

Single crystals of several materials have been grown from solution at low temperatures. The use of this method is usually restricted to cases where (a) single crystals cannot be grown from the liquid enclosed in a capillary tube, (b) it is desired to grow crystals in a specific phase (at a specific temperature), and (c) it is desired to use a small well-formed crystal in place of the cylindrical crystal of uncontrolled size that would occur if the crystal were grown from a liquid enclosed in a capillary tube. Extreme care must be exercised in handling single crystals grown in this manner as they must be collected, selected, mounted, and transferred to the X-ray unit while being maintained below their melting or transition points. The experimental difficulties encountered in utilizing this method, as well as the tendency of many molecules crystallizing at low temperatures to have high vapor pressures, are the basic reasons for the popularity of enclosing samples in capillary tubes.

An interesting, early use of this method was reported by Vonnegut and Warren (1936). Bromine crystals (which melt at $-7.3°C$) were deposited from the vapor onto the surface of a round-bottomed flask filled with dry ice and placed over the mouth of a beaker containing liquid bromine. The crystals were mounted by being placed upon a smooth surface of dry ice and cemented to a fine glass rod with soft vaseline, which froze upon cooling. The crystal was maintained at about $-150°C$ during the investigation. (The vapor pressure of the bromine crystal was sufficiently reduced by cooling, so that a single crystal would last through a twelve-hour oscillation photograph.)

A number of investigators have grown crystals in cold baths or refrigerators and then collected and mounted them at low temperatures. Gopalakrishna and Cartz (1972) mounted the crystals in precooled glass capillary tubes which were then cemented shut and mounted on the goniometer head. All work was done in a refrigerated area. Wheeler and Colson (1975) mounted crystals on a Cryo-Tip goniometer head (see Section 4.7) inside the freezer where they were grown. The Cryo-Tip was activated and cooled down prior to removing it from the freezer.

Crystals of hexafluorobenzene were grown in a cold room at $-17°C$ by sublimation of a polycrystalline mass. The thermal gradient was supplied by a 40-watt bulb. A glass capillary tube containing a small amount of the liquid sample in its tip was allowed to freeze, a single crystal was then wedged into place, and another, larger crystal was wedged into the large portion of the capillary tube. Thus the sample crystal was surrounded by material which maintained vapor equilibrium such that the sample crystal did not change its size during the period of data collection. The capillary tubes were sealed by fitting a brass pin into the funnel-shaped end of the tube and cementing it in place with lacquer. Other crystals were grown by sublimation from material sealed in a capillary tube (Bertolucci and Marsh, 1974).

McFarlan (1936) was faced with the problem of preparing randomly aligned polycrystalline samples of various high-pressure modifications of ice which were stable at atmospheric pressure only at low temperatures. He successfully ground these samples to a fine powder while they were under liquid air. This technique has also been used to grind samples that are glassy or tacky at room temperature.

6.2.2. Crystallization Simultaneously with Mounting

This approach is limited to the preparation of polycrystalline samples, although in at least one case, a single crystal was later grown from the powder.

In powder diffractometry, a glass plate, mounted on the diffractometer, is cooled by any convenient method. The liquid sample is sprayed from an atomizer or similar device and condenses onto the cold surface (Miksic et al., 1959; Weltman, 1962). If a metal plate is used, the thickness of the sample should be sufficient to mask any diffraction pattern from the metal. If this is not possible, the lines should be accounted for by running a blank at the temperature of the investigation.

Another often used method is to condense vapors onto a conduction-cooled wire substrate so that Debye–Scherrer photographs can be obtained.

Since this method is more commonly used with gaseous samples, it will be discussed in Section 6.3.

6.2.3. Crystallization after Mounting

The popular method of crystallization after mounting has many advantages (the primary one being ease of handling and mounting) and a few disadvantages (primarily absorption of the X-ray beam).

Special sample holders are used for powder diffractometry. For a vertical diffractometer, the θ angle is set so that the sample holder is horizontal. This holder consists of a depression in the surface of a flat aluminum plate. The liquid is placed in the depression in the horizontal holder, frozen, and then studied in the usual manner.

Alternatively, a standard sample holder is adapted by sealing the upper surface with a liquid-tight cover of thin Mylar film or similar material. The liquid is placed in the sample holder and held in place by covering the back. This type of sample holder can be used for either horizontal or vertical powder diffractometers.

6.2.3.1. Filling and Sealing of Capillary Tubes

The most common method used to prepare liquids and gases for Debye-Scherrer and single-crystal low-temperature studies is to enclose the samples in thin-walled glass capillary tubes. Solids which are deliquescent, reactive, or volatile at room temperature can also be enclosed in sealed tubes. The techniques used to grow single crystals from materials contained in capillary tubes which are mounted on the X-ray instrument are discussed in Section 7.2.2, while correction factors for absorption of the X-ray beam by the tube walls are described in Section 8.4.1.

A sample contained within a glass capillary tube is cooled most efficiently by the gas-stream method. If conduction cooling is employed, the sample should be mounted on a good conducting material, such as a metal wire.

Source of Capillary Tubes. It is most convenient to use the commercially available tubes which are usually used to hold Debye–Scherrer samples. These tubes are manufactured from soft glass and from quartz in a variety of diameters (see Appendix 12).

Several methods for preparing capillary tubes have been described in the literature, and if the commercial tubes are not available, the article by

Tanaka and Amma (1964) can be consulted. Capillary tubes have also been made from nonreactive plastic for use with highly reactive materials (Atoji and Lipscomb, 1954).

Shaping. The commercially available thin-walled capillary tubes are available in a variety of sizes. If the sample is to be obtained from a vacuum line, the sample tube must be attached directly to the vacuum line. Several investigators reported difficulties in forming a successful glass-to-glass seal between the soft-glass tubes and the Pyrex glass in the vacuum line. For this reason, thin-walled tubes (0.01–0.02 mm thick) have been drawn from standard Pyrex tubing (Abrahams et al., 1950; Tanaka and Amma, 1964). These tubes were drawn using a glass lathe and can be attached directly to the vacuum line. An all-glass system is often necessary for reactive samples.

Alternatively, an adapter is made in which one end of a metal tube is furnished with an O-ring seal which fits the large end of the capillary tube. The other end of the adapter can be tapered to fit a ground glass taper in the vacuum line or it can also have an O-ring seal large enough to fit onto a glass nipple on the vacuum line (after Mueller et al., 1958; Figure 6–1).

Wax seals have also been used to attach the capillary tubes to a vacuum line, but these are not always satisfactory.

As will be described in Chapter 7, crystals grown in capillary tubes often grow in a preferred orientation. This direction of growth may not be that preferred by the investigator. In order to overcome this problem, specially shaped capillary tubes may be used.

Reed and Lipscomb (1953) used a straight capillary tube with a relatively large bulb at one end. The seed crystals were grown in the bulb with the hope that their orientations would vary, thus allowing a range of crystal orientations in the narrow portion of the tube that is fixed in the X-ray beam.

Shallcross and Carpenter (1958) used a variation of this approach when they employed an L-shaped capillary tube. The seed was formed in the short end and proceeded to form a crystal with an orientation of 90° to that which was normally obtained.

Cleaning. The commercially available soft-glass capillary tubes are made from "alkali glass." They should be soaked in acid if there is any danger of contamination due to residual alkali (King, 1954). After soaking, the tubes should be rinsed thoroughly and dried in an oven. In order to ensure that the insides of the tubes are properly washed, the rinsing solution can

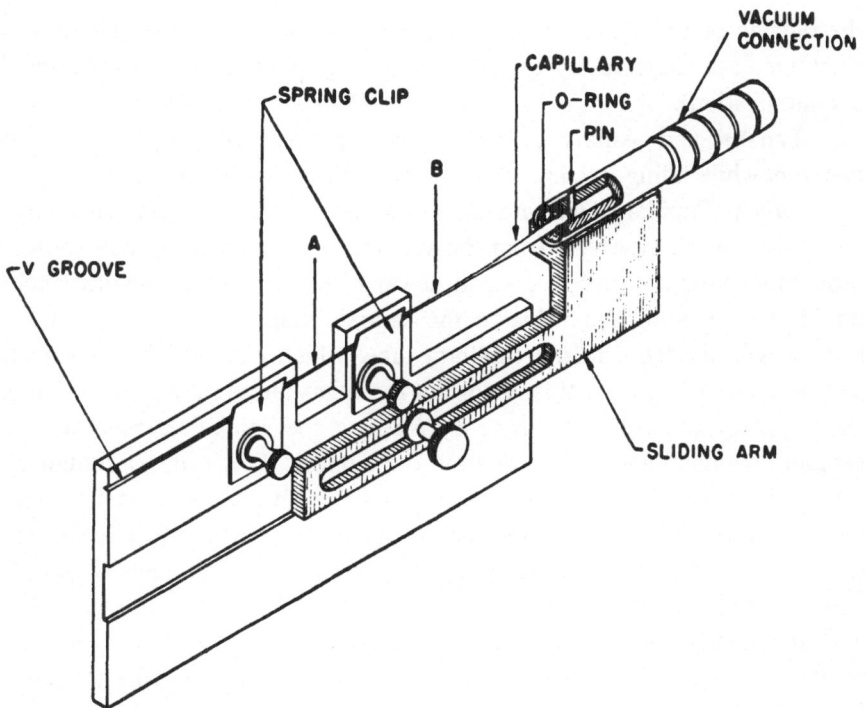

Fig. 6–1. Capillary-tube sealing jig with vacuum attachment. The capillary tube may be sealed at A or B depending on the length of the sample tube that is desired (Mueller et al., 1958). (Reproduced with the permission of the American Institute of Physics.)

be introduced via a microsyringe or by cutting off the tips of the tubes and resealing them after cleaning. A final rinse with the compound to be studied will ensure a clean capillary tube.

Handling. King (1954) suggested cementing a short length of metal rod to the end of the sample tube. This handle facilitates the mounting of these fragile tubes in cameras and goniometer heads. Mueller et al. (1958). describe a "capillary sealing jig"—a metal supporting frame into which the capillary tube can be fixed. The open end of the tube fits into an O-ring adapter through which the tube can be attached to a vacuum system. The fragile end of the tube is supported in a V-groove in which provision has been made for access at several points for sealing with a small flame (Figure 6–1). With this device it is not necessary to seal the capillary tube directly to the vacuum line, although in certain cases an all-glass system is preferred. Another suggestion is to place rings of sealing wax around the outside of the capillary tube on either side of the sample. These wax rings increase the stability of the tube during handling (Petsko, personal communication).

Tanaka and Amma (1964) have suggested using sponge-padded tweezers while filling, sealing, and mounting these fragile tubes.

Filling. Capillary tubes attached to a vacuum line can be filled by isolating the sample reservoir from the vacuum line, evacuating the capillary tube, immersing the tube in a small dewar of liquid nitrogen or other cold liquid, and opening the valve to the sample chamber. Vapors from the sample will condense in the capillary tube, which can then be sealed off with a torch or hot wire. It is important to outgas the sample prior to filling the capillary tube. This is done by freezing the sample, evacuating the sample chamber, and then remelting the sample. This process should be repeated several times until the sample does not bubble as it melts under vacuum. The cold bath over the sample should be removed just before the sample is allowed to distill over into the capillary tube. The melting sample will then distill slowly into the sample tube.

It is also possible to fill a capillary tube without using a vacuum line. In most cases, the small amount of gas absorbed in the sample will not interfere with the crystalline modification that is formed when the liquid is frozen. The capillary tubes normally used are of small diameter and special techniques are used to fill them. Several methods have been described.

1. If a large quantity of sample is available, a longer and very fine capillary tube is inserted into the sample tube (which should be about 1–2 cm long, with one end sealed off), and this assembly is immersed in the liquid sample. The sample tube will readily fill, after which the open end can be sealed (Abrahams et al., 1950).

2. The sealed end of a commercial capillary tube is snipped off and dipped into the sample. Liquid will be drawn up the tube by capillary action, after which the tube is sealed on both sides of the sample. This method has two advantages. The sample tube can be rinsed with the sample several times by adroitly shaking the liquid out of the tube. Secondly, the amount of liquid in the tube can be easily controlled.

3. A method that is quite useful if a small amount of liquid is available is to insert the sample directly into the capillary tube by using a microsyringe.

4. The final suggestion is to wrap the large open end of a standard commercial capillary tube with a small piece of cotton and fit it firmly into

a test tube so that the fragile end of the capillary tube is protected by the walls of the test tube. A drop of sample is placed into the opening of the capillary tube, and the test tube is fitted into a centrifuge. A few minutes of gentle centrifuging will force the liquid down the capillary tube, which can now be removed and sealed just above the sample.

Sealing. A properly sealed capillary tube can hold a sample indefinitely, even if the sample has a significant vapor pressure. The key to properly sealing the glass capillary tube is to gently draw the glass apart as it is melted by a small flame. The glass is thus drawn to a fine point which can be melted into a secure seal.

An attempt should be made to seal the capillary as close as possible to the liquid, i.e., to leave as little empty space as possible inside the capillary tube. If there is too much empty space, the sample will tend to spread itself over the length of the tube (e.g., by sublimation), making it difficult to obtain a good single crystal or powder. However, if a torch is brought too close to the liquid, it will usually boil the liquid, causing (1) a loss of sample and/or (2) a poor seal as the boiling liquid will blow pinholes in the soft, molten glass. For this reason, a heat sink should be used while the capillary tube is being sealed. If the tube is attached to a vacuum line, it can be immersed in liquid nitrogen so as to freeze the liquid. Resting the capillary tube on a smooth block of dry ice has also proved helpful. In other cases, covering the tube with a piece of wet filter paper has been suggested (King, 1954).

Mueller et al. (1958) suggested inserting a thin glass rod inside the capillary tube to fill the empty space that would otherwise form when the tube is sealed a short distance away from the sample. The combination of this technique with methods 1, 3, and 4 of filling should result in a small sample held securely in place near the professionally sealed end of a commercial capillary tube. The glass rod also serves to reinforce the sample tube and acts as a mounting rod.

A very small torch should be used to seal the tube. There are several commercial torches available which can be adjusted to an extremely small flame. A very small flame can also be obtained by drawing a piece of Pyrex tubing to a fine point and attaching it to the gas source via a piece of rubber tubing. The size of the flame is adjusted by a screw clamp attached to the tubing. This adjustment is rather sensitive, and it is helpful to have available a lit standard microburner for relighting the fine flame.

If the open end of the capillary tube is sealed with the direct flame, water vapor may enter the tube and contaminate the sample. To prevent this, the capillary tube should be sealed by heating from the side of the tube.

The capillary tube can also be sealed by heating it with a hot wire. Meyer (1973) suggests using a loop of nichrome wire (15–20 gauge), 1 cm in length, silver-soldered to a short length of common electrical cord. The temperature at the tip of the heating element is controlled by a rheostat (Variac), with fine control obtained by using a stepdown transformer or two rheostats in series. With this heating loop, the soft-glass capillary tube can be successfully sealed by heating and drawing the glass, even in a non-oxygen environment (dry box).

The sample tube can also be sealed with wax and epoxy, but these seals are subject to leaks due to the variation in rates of contraction between sealant and glass as the sample is cooled. Low-temperature epoxy should be used and the epoxied end should be fitted into the nib so that no unwanted diffraction or absorption occurs (see also Section 6.1.2).

The overall length of the sealed sample tube depends on the type of X-ray instrument used and the type of investigation being conducted.

In most cases, a sample 1–2 cm in length is sufficient. However, if an automatic diffractometer with a very limited height adjustment (z-translation) is being used, a shorter sample should be prepared. Otherwise, the region of interest in the capillary tube may not fall within the range of the z-translation. If a precession camera or Weissenberg goniometer with a translation of several centimeters is used, the sample could be longer. However, if the sample tube extends too far into the gas stream, the orientation of the sample may be restricted, since the sample tube may come in contact with the nozzle at large arc corrections (Figure 3–10).

It should be noted that many cooling devices utilizing conduction cooling cannot be used to grow single crystals from liquids, or, if they are used for powder samples, they have special sample holders. The few exceptions to this rule have very specific procedures for sample preparation (e.g., Smith, 1966).

Mounting. The metal pin cemented to the capillary tube can be used to mount it in the nib. Alternatively, the glass capillary tube itself can be fitted into a standard sample support on the camera or goniometer head.

The use of a specimen enclosed in a glass capillary tube is a standard technique for Debye–Scherrer photographs. The only special attention required when low-temperature techniques are used is to ensure a random distribution of crystallites (see Section 6.5).

However, when accurate single-crystal intensity data are collected, the data must be corrected for absorption by the glass capillary tube (see Section 8.4). In addition, the orientation and size of the crystal (grown from a liquid) are often not under the control of the investigator. In particular, with some systems in which single crystals are particularly hard to grow, more than one crystal may be in the beam or the crystal may be larger than the beam, resulting in different volumes of the crystal diffracting at different orientations. Special correction factors have been derived for this latter case and are described in Section 8.4.4.

6.3. Sample Preparation: Samples Gaseous at Room Temperature

The problems and techniques involved in preparing samples from materials that are gaseous at room temperature are very similar to those described for liquids. Usually capillary tubes are filled from vacuum line assemblies (e.g., Collin and Lipscomb, 1951; Post et al., 1952), as described in Section 6.2.3. In the case of easily condensed materials, the sample may be cooled to the liquid state and treated as a standard liquid sample (Collin, 1952).

In the case of gaseous samples, it has been found advantageous to use a capillary tube with a relatively large bulb blown at one end. This allows more material to be contained in the sample tube (Reed and Lipscomb, 1953; Siemons and Templeton, 1954; Huffman, 1974).

Single crystals can be grown from the condensed gases using the techniques described in Section 7.2.2.

The most common method for preparing polycrystalline samples is to condense the gas directly onto a cold surface in the camera or cryostat. This method has also been used to prepare a polycrystalline sample from which single crystals were grown (Keesom and Taconis, 1935). The instruments in which this approach is used include both Debye–Scherrer cameras (Simon and Simson, 1924a,b; de Smedt et al., 1930; Ruhemann and Simon, 1931; Taylor, 1931; Vegard, 1931; Ruhemann, 1932; Gunther et al., 1939; Cheesman and Soane, 1957; Kogan et al., 1960; Kogan and Bulatova, 1969) and diffractometer-mounted cryostats (Mauer and Bolz, 1961; Barret and Meyer, 1964; Bol'shutkin et al., 1970).

One apparatus requiring special mention is that described by Smith (1966), which can be used for a three-dimensional structure analysis of single crystals grown from condensed gases at liquid-helium temperatures.

It has been successfully used for the investigation of γ-oxygen, CH₄, β-borane, and other materials. The gas is condensed inside a capillary tube about 1 mm in diameter which is attached to a copper block in contact with liquid helium. A single crystal is grown by careful control of a heater attached to the copper block. Reflections from the crystal, which grows in an arbitrary orientation, are detected by film or a scintillation counter. The unit cell is calculated from the counter setting angles and is refined as data collection proceeds.

6.4. Preparation of Reactive or Radioactive Samples

Many materials studied at low temperatures are unstable or corrosive and require special care. Some sources of difficulty include sensitivity to air (easily oxidizable compounds), sensitivity to moisure (easily hydrolyzed compounds), and sensitivity to temperature (compounds that are thermodynamically unstable at room and higher temperatures). Such substances should be enclosed in a container for ease of handling and/or temperature control. The techniques are essentially the same as those described above and depend on whether the material is solid, liquid, or gas at room temperature.

Often the sample must be loaded into a capillary tube inside a dry box. Johnson and Anderson (1966) and Lange and Haendler (1972) have described techniques and apparatus for use in filling, sealing, and protecting thin-walled glass capillary tubes inside a dry box. The former suggest fixing the capillary tube inside the stem of a funnel with soft-melting wax, while the latter describes a clear plastic support for holding the capillary tube inside a piece of rigid, clear plastic tubing. Once the sample is in the capillary tube, the top of the tube can be sealed with wax. The sample is vibrated to the bottom of the tube and/or permanently sealed outside the dry box.

Single crystals can be mounted on a fiber and sealed within the capillary tube as described previously (see Section 6.1.2) or they can be wedged into place inside the tube (Ford and Powell, 1954). Nelmes (1970) describes the use of spherical, rather than cylindrical, sample tubes (see Section 6.1.2). He mounted highly deliquescent crystals that showed no measurable deterioration eight months after mounting.

Meyer's method (1973) of using a nichrome wire to seal the capillary tube can also be used inside a dry box. If the sample decomposes as the tube is sealed, the seal may have to be made some distance from the sample. The use of a thin glass rod inserted inside the capillary tube has already been described as one method of filling the otherwise empty volume and of improving the structural stability of the fragile capillary tube (Mueller et al., 1958).

If the material is particularly corrosive, then fluorethene or Teflon capillary tubes (Atoji and Lipscomb, 1954) and powder diffractometer sample holders (Barrett et al., 1967) can be used (Burbank, 1973). The fluorethene tubes can be sealed with heated pliers while still on the vacuum line. Some waxes, glues, and epoxies may also be attacked by various samples and should be used with care.

In many cases, the use of an evacuated cryostat is the only protection needed for easily oxidized samples (Kogan and Omarov, 1965). Mauer and Bolz (1961) describe a vacuum cryostat in which the sample chamber is separated from the evacuated chamber by a Mylar window so that highly volatile substances can be studied.

Another class of special samples arises when quantitative mixtures of gases must be obtained inside the sample tube. An example is the study of various hydrates of ammonia (Siemons and Templeton, 1954; Olovsson and Templeton, 1959; Olovsson, 1960). Quantitative mixtures were prepared using a gas burette attached to a capillary tube or by condensing the gases directly into the capillary tube on a vacuum line, and then determining the composition by weighing the filled capillary tube.

The investigation of radioactive samples at low temperatures requires additional precautions to filter out radiation due to the sample, which would otherwise interfere with the detection of the diffracted X-ray beam. A powder camera (Kogan et al., 1960) and a powder diffractometer (Skaba and Krivy, 1970) have been described.

The use of a modified Guinier camera sample holder to support a series of sealed glass capillary tubes (Veith, 1975) has been described in Section 3.5.4.

6.5. Techniques for Preparing Randomly Aligned Powder Samples

A difficulty commonly encountered during the study of polycrystalline samples at low temperatures is the tendency of many substances which are liquids or gases at room temperature to crystallize with preferred orientations. In many cases, and particularly with powder diffractometer samples, this tendency is enhanced by the shape of the sample holder. In this latter case, too, it is more difficult to recognize the presence of preferred orientation. On the other hand, in some situations (e.g., the deposition of powder sample from vapor onto a cold wire or plate) random distribution of crystallites is easily accomplished. If preferred orientation is suspected the sample may be annealed (Sears and Klug, 1962).

However, Bol'shutkin et al. (1972) reported that best results were obtained by condensing *small* portions of CF_4 onto the substrate at 8 K and annealing at 45 K. They observed that a rapid, single-stage condensation of the gaseous sample results in the formation of a texture in the sample that is not removed by annealing. On the other hand, preparation of the sample by condensing several layers on the sample holder results in a finely dispersed, randomly oriented polycrystalline sample.

McFarlan (1936) prepared randomly oriented powder samples by grinding the sample under liquid air before placing it in the sample holder (see Section 6.2.1), as did Bertie et al. (1963) and Wheeler (1968).

In another variation of low-temperature grinding, Tompa (1968) mounted a vibrating grinder (manufactured by Crescent Dental Manufacturing Co.) in an inverted position. The stainless steel vial (containing the sample and steel grinding ball) and the mounting clip were suspended in a styrofoam chamber. Cold nitrogen gas was used to cool the vial to any desired temperature down to 100 K. The grinder's ball bearings were warmed by a stream of warm air.

When the sample is already contained in a capillary tube, two other methods can be used. Dennison (1921) plunged the room-temperature sample into liquid air. The rapidly freezing material formed only minute crystals. Eastman (1924) melted and refroze the sample several times during the course of the exposure. Over a period of several hours, a normal powder diagram will be obtained. This latter method, of course, is restricted to the use of a gas-stream cooling process, where the sample can be melted and refrozen very quickly. In order to maintain a uniform temperature during the exposure, the most efficient method of melting the sample is to squeeze the rubber or soft plastic tubing carrying the gas to the heat exchanger immersed in the cold bath. If the gas flow is halted by shutting a valve, it is not always possible to readjust the valve opening to the same position; if the complete air supply is shut off, even momentarily, ice will form on the cold surface of the sample.

Unfortunately, some substances tend to supercool and will not refreeze at the temperature at which the photograph is being taken. In this case, it may be necessary to touch the capillary tube with a swab of cotton or glass wool dipped in liquid nitrogen or dry-ice-cooled acetone in order to initiate freezing of the sample. Care should be taken in this case since it is possible to bypass thermodynamically unstable, but easily formed, crystalline phases (Carpenter and Richards, 1962; Lundgren, 1970; Rudman, 1970b).

If the material is melted for only a few moments, the X-ray beam may be left on. Otherwise, the shutter should be closed to prevent the broad bands characteristic of the liquid from being superimposed on the powder pattern.

Powder samples that are mounted using a binder are subject to strain effects at low temperatures as a result of the different thermal expansion coefficients of the sample and binder. This effect was noted by Cocks et al. (1966) and was systematically investigated by Etourneau et al. (1975), as described in Section 4.1.3. Cogan and Cocks (1975) suggest mounting a fine powder specimen (300–400 mesh, i.e., a particle size of approximately 40 μm) with a lightweight machine oil on a glass tray. The glass slide is prepared by using epoxy to cement sections from a microscope slide cover onto another slide cover to form a well 0.13 mm deep by approximately 1 cm^2 in area. The powder is sprinkled into the well and covered with an extremely thin layer of oil by smearing an oil-covered slide over the powder. Several layers are deposited until a satisfactory sample is obtained. Cogan and Cocks did not observe any strain effects for tin and zinc down to 100 K, but did not test this method at lower temperatures.

CHAPTER 7

Crystal-Growing Techniques

7.1. Samples Solid at Room Temperature

No Phase Transition. Crystal growth is accomplished prior to mounting the sample using standard techniques.

Below Phase Transition. If the material undergoes a phase transition as it is cooled, it is not unusual for the crystal to shatter at the transition point. Various methods of crystal growth by slow cooling, by annealing, or by growing the crystal from a solution maintained below the transition temperature are described in Section 7.3.

7.2. Samples Liquid or Gaseous at Room Temperature

Single crystal growth by vapor deposition has not been used too often, although Burbank (1973) states that it is easier to obtain single crystals with this method than by growing them from a liquid. He found that crystals grown from the vapor (in capillary tubes) are often well-formed and separated from their neighbors, are firmly attached to the capillary wall, and can be readily isolated in the X-ray beam. In liquid-to-solid crystal growth from samples contained in capillary tubes, the crystals do not have a well-defined shape or size and often have inconvenient crystallographic orientations. Special capillary tubes to control the crystal orientation have been designed (see Section 6.2.3.1); methods for calculating corrections for absorption by the sample container or for crystals larger than the beam have been reported in the literature (see Section 8.4).

7.2.1. Crystal Growth at Low Temperatures Prior to Mounting the Crystal

As was described previously (see Section 6.2.1), Vonnegut and Warren (1936) grew and mounted single crystals of bromine which were grown by deposition from the vapor onto the surface of a flask filled with dry ice and placed over the mouth of a beaker. Since this material melts at $-7.3°C$, the method was successful.

Several other crystals, crystallizing at considerably lower temperatures, have also been prepared. Nordman and Reimann (1959) grew single crystals of ammonia-triborane by the evaporation of a diethyl ether solution of the compound in a vial immersed in a chlorobenzene slush bath at $-45°C$. A slow stream of dry nitrogen prevented ice formation in the vial. The crystals were mounted and sealed in thin-walled glass capillary tubes while on a microscope cold stage.

Since crystals of hydronium perchlorate shatter when cooled below the transition point, single crystals were grown directly from solution below the transition temperature (Nordman, 1962). Crystals were grown at dry-ice temperature from a water–anhydrous perchloric acid slush. After being kept at dry-ice temperature for a week or more, small portions of coarsely crystalline slush were transferred to a microscope cold stage cooled from below and blanketed with cold, dry nitrogen gas. A crystal was selected and inserted in a glass capillary tube held in a hole in a small block of dry ice. The tube was then sealed off and mounted on a precession camera by transferring it from the dry-ice block directly into the stream of cold nitrogen gas used for cooling the X-ray unit. This method required much patience and many attempts before it was successful. The present availability of flexible metal transfer lines may be helpful in this case. The cold stream could be directed over the mounted crystal and moved with the crystal as it is placed on the X-ray unit, just as Burbank (1973) has done while moving a sample from one unit to another (Figure 7–1). One of the major advantages of this method is that the size, shape, and eventual orientation of the crystal are under the control of the investigator.

A similar approach was used by Scheuerman and Sass (1962). Crystals of valeric acid were grown from several milliliters of the acid placed in a 30-ml pear-shaped flask placed in a beaker filled with glass beads. The beaker was set in a dewar flask filled with a dry-ice slurry. When the acid had frozen into crystals, the crystals were broken free from the flask and shaken onto a metal plate cooled by a block of dry ice. While still on the cold plate, a crystal was inserted into a capillary tube, glued on a brass

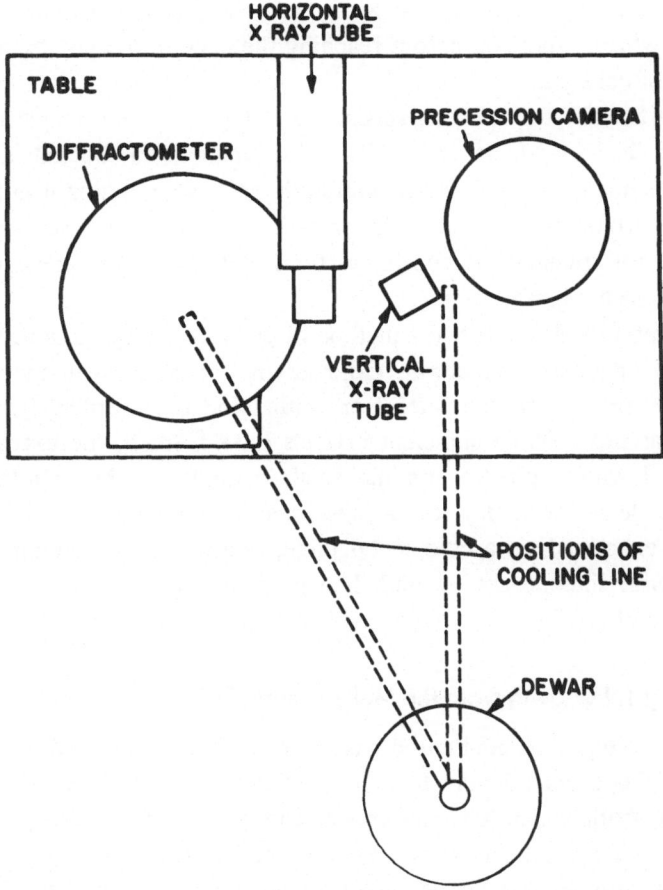

Fig. 7–1. Schematic diagram of Burbank (1973) apparatus. (Reproduced with the permission of the International Union of Crystallography.)

pin, placed on a piece of dry ice, and transferred to a goniometer head in a stream of cold gas.

Several examples of growing and mounting crystals in a cold room or cold chamber have been discussed in Section 6.2.1.

Liebling and Marsh (1965) slowly lowered a sealed tube of cyclopentadiene from a dry-ice acetone bath into liquid nitrogen. The tube was broken open under liquid nitrogen and clear crystals were extracted from the frozen mass. All further operations, including mounting and orientation of the crystals, were carried out *under* liquid nitrogen. The capillary tube was then mounted on the X-ray unit in a steady stream of liquid nitrogen, which vaporized before reaching the crystal, thereby bathing it in a stream of cold gas.

A system for transporting specimens at liquid-helium temperature has been described by Maeta et al. (1975). A small dewar bucket is used to lift the specimen from a liquid-helium bath while maintaining it in a liquid-helium environment. Several variations of the device are described; it has been used for transporting specimens for X-ray diffraction measurements, but no details are given.

Luzzati (1953) described a method of growing crystals in a beaker suspended in an unsilvered dewar. Cold gas is blown between the walls of the beaker and dewar; the rate of crystallization is controlled by the rate of cold gas flow. When sufficient crystals have formed, the excess liquid is removed with a pipette and the single crystals can be maintained by filling the dewar with dry ice. A specimen is removed with refrigerated tweezers, examined on a cold metal plate, attached to an aluminum wire filament with silicone grease, and then mounted on the X-ray instrument in a stream of cold gas.

7.2.2. Crystal Growth from the Melt; Sample in Capillary Tube

When a liquid or condensed gas enclosed in a sealed capillary tube is frozen, a single crystal may form immediately or a polycrystalline mass may form, from which a single crystal can be grown. Growing a crystal inside a capillary tube is most easily accomplished with the gas-stream method of cooling, but, occasionally, it has been used successfully when the sample is cooled in a cryostat by conduction.

Several critical factors must be considered when growing a single crystal inside a capillary tube. While the sample should be pure, it has been found that a high-purity compound can be more difficult to work with because of the sharp melting point that is encountered. Often a slightly impure sample, melting over a range of several degrees, will yield better crystals than the pure compound, which immediately forms a polycrystalline mass as soon as one seed crystal forms.

Crystals can be formed in a capillary tube that is completely filled with the sample. However, when the length of the capillary tube is greater than the diameter of the X-ray beam, and if the single crystal fills the capillary tube, the volume of material diffracting X-rays will vary with the angle between the X-ray beam and the capillary tube. In such a case absorption corrections are required (see Section 8.4). On the other hand, if several small single crystals are formed, it may be difficult to sort out the reflections due to one of these crystals. Huffman (1974) suggests using small droplets of sample in the capillary tube, with a distance at least 1½ times the diameter of the capillary tube separating adjacent droplets. The maximum sample length should be no more than about twice the diameter of the capillary tube.

When gas-stream cooling is employed, the normal cold-stream cross section of 1 to 2 cm means that within a distance of 0.5 to 1 cm the temperature varies from a low of approximately $-160°C$ to room temperature (Figure 3–15). This very sharp thermal gradient may result in gross local variations in temperature and consequent difficulties in growing the crystal. It is best to maintain the temperature of the cold stream only 5–15° below the melting point of the sample while growing the crystal. Burbank (1973) has pointed out that, in those circumstances where it is desirable, the thermal gradient across the sample tube can be increased by placing the tube at a 60° angle to the cold stream.

A serious, commonly encountered problem is that many samples tend to supercool and may remain in the liquid state to 30–50° below the freezing point. This difficulty can be overcome by touching the capillary tube with a piece of dry ice or a cotton swab dipped in liquid nitrogen or dry-ice-cooled acetone (Abrahams et al., 1950; Post et al., 1951). However, if phase transitions are involved, this procedure might lead to certain difficulties (see Section 7.2.2.4).

7.2.2.1. Growth Not on the X-Ray Instrument

The most convenient method is to grow the crystal *in situ*, that is, while it is mounted on the X-ray instrument. However, in those cases in which the instrument cannot be tied up during the crystal growth process (e.g., on an automatic diffractometer) or in which the crystal is to be studied on more than one instrument, it is possible to grow the crystal in one location and then transfer it to the X-ray instrument. A simple means of accomplishing this is to place the goniometer head (with the crystal) inside a dewar (containing liquid nitrogen) during the transfer process. This method has been utilized successfully for transferring a crystal from a precession camera in one building to a neutron diffractometer in another building (S. LaPlaca, personal communication).

However, if the crystal undergoes a phase transition when cooled to

liquid-nitrogen temperatures, more accurate temperature control is necessary during the transfer process. Bouttier (1949) described a device for
growing a crystal from a liquid enclosed in a capillary tube inside a
temperature-controlled cryostat. After the crystal was grown, the capillary
tube (attached to a long-handled rod) was retracted into a *cavity* in an
aluminum block, through which the rod passed. This block was cold and
kept the crystal from melting while it was brought over to the X-ray unit.
This method was used by Luzzati in several investigations (1951, 1953).

A slightly different approach has been described by Burbank (1973).
Two X-ray instruments, a precession camera and a single-crystal diffractometer, are located on the same generator adjacent to each other. The
crystal is grown and examined on the precession camera. A special clamp
then locks the goniometer head to the nozzlelike end of the flexible metal
transfer line, and the crystal is moved over to the diffractometer, fixed in
place, unclamped, and studied. In this manner, the temperature of the
sample remains constant during the transfer (Figure 7–1).

7.2.2.2. Growth by Controlling the Temperature of a Gas Stream

(a) Controlled Rate of Cold Gas Flow. Two conflicting factors are encountered when the gas flow rate is reduced: the cooling gas remains in
the heat exchanger for a relatively long period of time and is cooled to
nearly the temperature of the bath; and the heat leaks normally encountered in the transfer line tend to warm the gas (unless a vacuum-insulated
line is used). Furthermore, if the flow rate is very slow, the cold gas
stream may not bathe the crystal uniformly. For a reasonably efficient
copper-coil heat exchanger, it is found that the temperature at the outlet
of the nozzle, and, therefore, the temperature at the crystal, will vary inversely with the flow rate up to an optimum flow rate. Greater-than-
optimum flow rates will not further cool the crystal, and an excessively
high flow rate may even cause the crystal to vibrate in the beam.

For the optimum control of the growth process, the direction of the gas
stream should be parallel to the axis of the capillary tube. By carefully
adjusting the flow rate, a temperature gradient can be established along
the length of the capillary tube so that part of the crystal is melted and
part is frozen. As the rate is increased, crystallization proceeds until the
entire sample is a single crystal. This method has been used successfully
by many investigators (Kaufman and Fankuchen, 1949; Collin and Lips-

comb, 1951; Post et al., 1951). Post et al. used the small nichrome heater incorporated at the base of their specimen holder (Figure 3–8) to establish a temperature countergradient along the length of the capillary tube.

Olovsson (1960) used a slight variation of this method. The cooling temperature of the gas stream is first adjusted to a point just below the melting point. (If the sample supercools, crystallization is initiated, as described above, by using a cold metal wire.) The polycrystalline mass obtained in this way is melted (by touching with some warm object) except for a tiny piece at the end of the capillary. From this seed the crystal is allowed to grow, and, if the result is not a single crystal, most of the mass is remelted and the crystal is allowed to grow again. If the sample consists of two components (e.g., a mixture of gases) care should be taken that the sample is homogeneous before it is frozen. This can be done by freezing the liquid quickly.

(b) Controlled Temperature of Gas Stream. The temperature gradient required for successful crystal growth can also be obtained by adjusting the temperature of the gas stream. The temperature can usually be adjusted by mixing warm gas with the cold gas or by adjusting a heater in the gas stream (see Section 3.1.3). The former method is somewhat slower since it takes time for the temperature along the length of the nozzle to equilibrate. The latter approach attains equilibrium more quickly (when the heater is near the end of the nozzle) and has been used by many investigators. For example, Henshaw (1957) used two heaters, one for "rough-warming" and the other for fine control of the gas-stream temperature. Single crystals were grown by changing the resistance in one branch of an ac bridge in which the fine-control heater was another branch.

(c) Moving Crystal in Gas Stream. If the gas-stream conditions are held constant, single crystals can also be grown by moving the crystal slowly along the temperature gradient in the gas stream. Parkes and Hughes (1963a) coupled a clock motor to the translation mechanism of a precession camera. As the capillary tube is driven into the jet of cold nitrogen gas, a single crystal is grown. A similar device, which can also be used to convert a precession camera into a flat-plate oscillation–rotation camera, has been constructed by Lippman and Rudman (1976; see Section 3.5.2).

Burbank (1973) grew single crystals on a precession camera while the camera was in continuous motion at $\bar{\mu} = 10°$ in order to even out thermal variations.

7.2.2.3. External Heating

A single crystal can be grown by cooling under fixed conditions while heat is supplied from an external source.

(a) Cooling Methods Not Using a Gas Stream. Temperature gradients in the vicinity of the sample can be established through the use of a heater built into a cryostat. Smith (1966) grew single crystals near liquid-helium temperatures using this method (see Section 6.3). Optically transparent heat-absorbing windows should be included in the cryostat for visual observation of the crystal during growth and subsequent centering.

Viswamitra and Ramaseshan (1960) describe a chamber which is attached to a conventional goniometer head. A thin capillary tube filled with the sample is mounted at the center of a brass pin which is surrounded by a small electric heater. The whole arrangement can be cooled to liquid-air temperature by conduction, by allowing liquid air to drip continuously onto a brass receptacle which has two concentric grooves. A second heater, supported by two legs 1 cm above the lower heater, has a central hole through which a movable copper rod (in good thermal contact with the heater) passes. The copper rod has a hole at the bottom so that, in its lowest position, it completely encloses the capillary tube. With the copper rod at its maximum height, the heaters are adjusted so that the lower end of the capillary is 5° below and the upper end is 10° above the freezing point of the liquid. After the liquid has frozen, the copper rod is lowered so as to melt the sample. As the rod is slowly raised, a single crystal starts to grow. The authors explain that since the isothermal planes in the temperature gradient are not only strictly normal to the direction of growth, but also are maintained perfectly parallel during crystallization, the growing of single crystals becomes an easy and controlled process. Crystals 1 mm in diameter and 6 mm long have been grown in 2 to 3 minutes. A similar device has been used by Deiseroth et al. (1975).

(b) Cooling Methods Using a Gas Stream. An external source of heat is used to melt a small portion of the polycrystalline material in the capillary tube. As the molten zone passes through the sample, a single crystal can be grown on the low-temperature side of the zone. This method embodies the concept of zone refining, and several passes of the molten zone through the sample can be used to purify the sample while growing a single crystal. The heater can remain stationary while the sample is moved or vice versa.

Wire Loop Heater: Singh and Ramaseshan (1964) attached a small

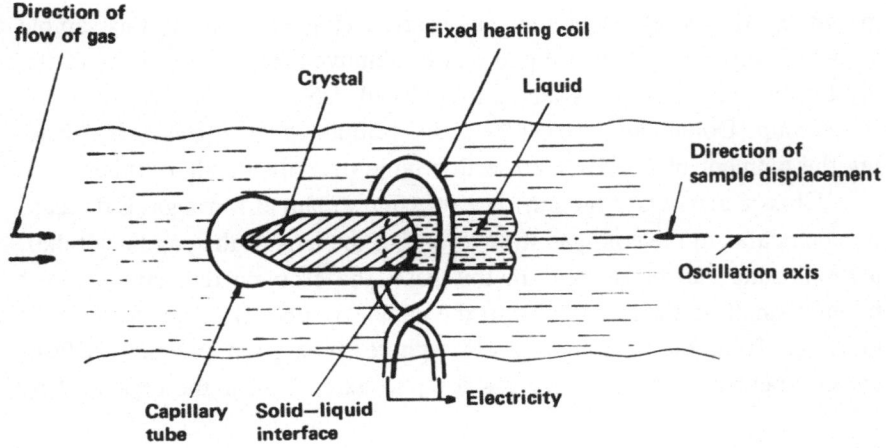

Fig. 7–2. Schematic diagram of method used to grow crystal from liquid contained in a thin-walled capillary tube. The sample is moved slowly through the heated wire loop maintained in a cold gas stream (Renaud and Fourme, 1966).

heater (consisting of a loop of nichrome wire) to the carriage of a Weissenberg goniometer and moved it slowly past the capillary tube containing the frozen material. Renaud and Fourme (1966) fixed the loop over the sample, coupled a motor to the translation drive of the Weissenberg goniometer, and moved the sample through the heating loop (Figure 7–2).

A specially shaped nichrome wire loop (Figure 7–3) that warms both ends of a sample has been described by Huffman (1974). The sample is initially cooled to form a polycrystalline mass and is then warmed slowly until only one seed crystal remains in the liquid. The single crystal is

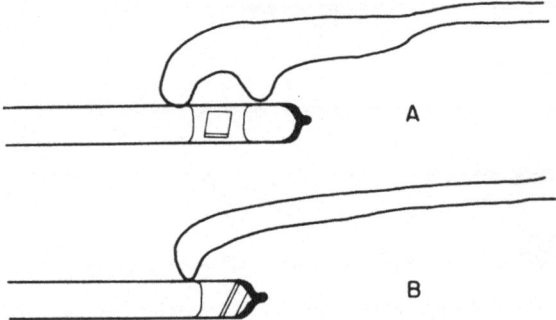

Fig. 7–3. Use of nichrome heaters to grow crystals in capillary tube (Huffmann, 1974). (a) Double-loop heater to warm both ends of a sample; (b) single-loop heater to warm one end of a sample.

grown by slowly lowering the temperature (Figure 7–4). If the sample volume is too large, excess liquid can be removed from the vicinity of the crystal by using a cotton swab saturated with LN_2.

Lamp: Dohlen and Carpenter (1955) obtained local melting by focusing the filament of a 100-W projection lamp onto the capillary tube.

Cross-Current Gas Stream: If a very fine stream of warm gas is directed perpendicular to the cold gas stream (and to the axis of the capillary tube), a small zone can be melted. In this case, the cross-current stream must be very small and located close to the capillary tube so as not to draw in moist air from the surrounding atmosphere. It is good practice to flush the cross-current line with dry gas prior to fixing it near the crystal. This

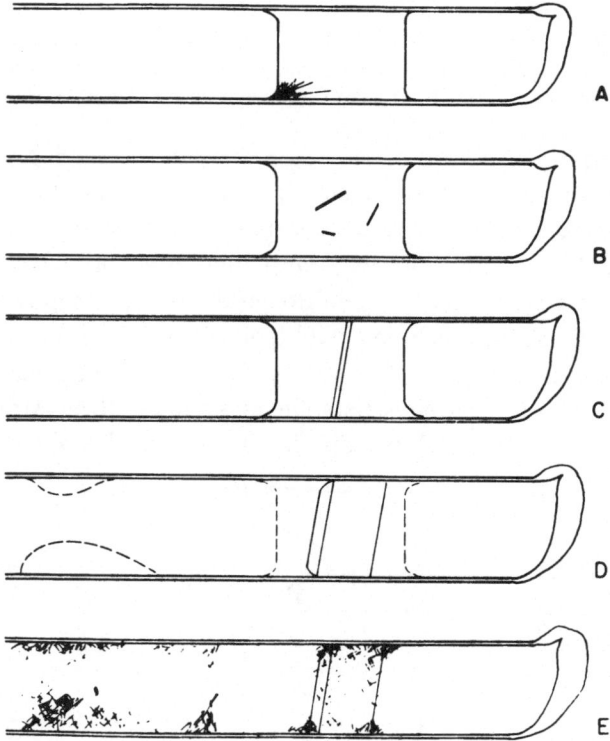

Fig. 7–4. Stages in growing single crystal from sample in capillary tube (Huffman, 1974). Polycrystalline material (A) is melted until a single seed crystal (B) is left. This is allowed to grow to fill the diameter of the tube (C). It continues to grow (D) until desired size is attained, at which time excess liquid is removed (E).

method has been described by Abrahams et al. (1950), Kreuger (1955), and Rudman (1966a). Rudman used a small, hollow needle that was squared and flattened at the end and attached to a length of rubber tubing and a valve. The crystal was then translated past the warm air stream emanating from the tip of the needle.

7.2.2.4. Special Problems

Glassy State. If the sample forms a glass when it is cooled, single crystals cannot be obtained. André et al. (1972) found that diethyl ether forms a glassy state whether it is cooled slowly or rapidly. A crystalline powder was induced through thermal shocks administered to the glassy state using a small, electrically heated wire loop. Single crystals were then grown from the powder by the method of Renaud and Fourme (1966), as described in Section 7.2.2.3.

Viswamitra and Ramaseshan (1960) found that their method of crystal growing (see Section 7.2.2.3) proved useful for growing crystals from liquids which are viscous or which have a tendency to form glasses.

Metastable Phases. In many materials, more than one crystalline modification can be obtained from solutions maintained at a given temperature, with the controlling factor being the solvent, the rate of crystallization, or the concentration of the solution.

Similar effects have been noted for the growth of crystals at low temperatures from the pure liquid. Some very subtle changes in the technique of crystal growth can result in different crystalline modifications being formed.

Iodine monochloride can be grown in either of two forms (Carpenter and Richards, 1962). In both cases, the material is contained in a glass capillary tube and crystallization is initiated by touching the tube lightly with a piece of dry ice. However, if contact is made while the cooling gas is flowing, one form is obtained; if the cooling gas is turned on immediately after crystallization begins, then the other form is obtained. The former is the more stable of the two.

During a study of polymorphism in the methylchloromethane compounds, Rudman observed the appearance of metastable phases in several of these compounds (Rudman and Post, 1968; Rudman, 1970b). In the case of carbon tetrachloride, crystals of Phase Ia are formed by carefully freezing the liquid. As the temperature is lowered, they transform to a different modification, Phase Ib. Upon further cooling, a major transition occurs and the crystal forms Phase II. Warming the Phase II crystal above the transition temperature induces a transformation back to Phase Ib, which persists to the crystal's melting point (which is a few degrees higher than that of Phase Ia).

The initial low-temperature X-ray diffraction investigation of the methylchloromethane compounds did not indicate similar behavior for the closely related methylchloroform and 2,2-dichloropropane. However, a differential scanning calorimetry (DSC) study indicated an effect closely related to that observed for carbon tetrachloride. A reinvestigation of these compounds using X-ray diffraction revealed identical behavior for all three compounds. Of interest is the reason why Phase Ia was not observed during the initial study of methylchloroform and 2,2-dichloropropane. Both these compounds have a tendency to supercool to a large extent. In the initial investigation, crystallization was initiated by touching the capillary tube with a cotton swab dipped in a dry-ice–acetone mixture. The thermal shock induced the formation of Phase II, which immediately warmed to form Phase Ib; Phase Ia was never observed by this method of crystallization. During the reinvestigation, the temperature was lowered until crystallization occurred, without any external interference or assistance. Phase Ia crystals grew from the supercooled liquids.

It is good practice to investigate any samples showing signs of phase transitions with low-temperature DSC, to learn how many unique crystallographic phases to expect. Although it is often easier to initiate crystallization by sudden cooling of the capillary tube, it is also important to be sure that no phases are being by-passed when this method is used. This is especially important during examinations of simple molecules in which polymorphism is a common occurrence.

7.2.3. Crystal Growth from the Melt; Sample Not in Capillary Tube

Single crystals of argon have been grown in a beryllium cell with an inner diameter of 1 mm and an outer diameter of 2.5 mm. Strain-free crystals are grown from prepurified argon in the following manner. The cold cell, under constant pressure, is filled with argon and an oscillation photograph is taken. Since there is no visual access to the cell, the photograph is used to reveal the presence of crystalline material. If the developed film indicates that a solid is present, the temperature is increased a few tenths of a degree and another photograph is taken. This process is repeated until the diffraction spots disappear. According to the construction of the cryostat, this means that the solid–liquid interface is then located just below the equatorial plane in which the X-ray beam lies. By slowly lowering the temperature, a single crystal is grown in the path of the X-ray beam (Bronsveld et al., 1973).

A cryostat with a heat-absorbing window for observation of the sample during crystal growth and alignment was described by Losee and Simmons (1968). The sample of krypton was contained in a 6.4-mm-diameter cylindrical tube of 0.025-mm-thick Mylar sheet sealed at top and bottom with copper plugs.

With the specimen chamber slightly above the triple-point temperature, krypton gas was admitted to the specimen tube through a capillary tube and liquefied. The sample chamber was cooled a few tenths of a degree below the triple point and a single crystal was grown from the resulting solid by moving an electrically heated coil over the cylindrical specimen tube at a rate of about 0.4 mm/h.

By carefully controlling the vapor pressure over the solid and the temperature, it was possible to completely separate the single crystal from the Mylar tube. Several small electromagnets were used to hold and then to release spherical steel markers (50 μm diameter) which were embedded at several points in the crystal during the crystallization process. These markers were used as reference points for length measurements and to detect any rotation or translation of the specimen.

A technique requiring a great deal of patience and skill has been described by H. Hope (1975, personal communication). A small drop of liquid is placed on the tip of a fine fiber held in a stream of cold gas. The drop, which is initially held in place by surface tension, is frozen by the cold gas stream. A single crystal can be grown from this drop by careful manipulation of the temperature gradient. In this manner, many of the disadvantages associated with a sample held in a capillary tube in a cold gas stream (e.g., formation of crystals larger than the beam, absorption by the capillary tube) are absent.

7.3. Crystal Growth below the Temperature of Phase Transition

Many compounds undergo crystallographic phase transitions as the temperature is lowered. A comparison of the crystal structures of these phases often suggests the most probable mechanism for the phase transformation. While it is normally a routine matter to cool a single crystal through the phase transition, it is not unusual for a single crystal of the high-temperature modification to shatter immediately upon transformation to the low-temperature modification.

Alternatively, the phase transition may display kinetic, as well as thermodynamic, effects. For example, Lundgren (1970) found that the low-temperature phase of $HBr \cdot 3H_2O$ (which occurs at approximately 90 K) did not return to the high-temperature phase even when held at 170 K for several hours.

The possible occurrence of crystallographic phase transitions should be investigated whenever a crystal is cooled. A survey using low-temperature DSC is recommended. However, the possible occurrence of metastable phases, kinetically slow transitions, irreversible phase transitions, and other similar effects should be considered if the data indicate any unusual occurrences. In many cases, twinning (reversible or irreversible) as a function of temperature may also be present. Whenever possible, a low-temperature survey using film techniques should precede data collection on a diffractometer. It is much easier to identify the above-mentioned effects on a photograph than from counter data.

Several techniques have been developed to obtain single crystals of the low-temperature forms.

7.3.1. Slow Cooling

Shattering of the crystal may be caused by sudden changes in the temperature of the crystal. Very slow cooling of the crystal, particularly in the vicinity of the transition temperature, may alleviate any thermally induced strains which would otherwise cause the crystal to shatter.

7.3.2. Annealing

Pseudospherical molecules, e.g., methylchloroform, form high-temperature crystalline phases, known as plastic crystals, which are characterized by a large degree of dynamic molecular reorientation (see Section 2.7.1). Attempts to prepare crystals of the low-temperature phases by slow cooling of the plastic crystals through the transition are nearly always unsuccessful. The single crystals of the plastic phases shatter as they are cooled, even when the cooling rate is one or two degrees per day.

After many attempts to find a simple, reliable technique for preparing single crystals of the low-temperature phases suitable for routine X-ray diffraction studies, the following method was developed (Rudman, 1970a).

A single crystal of the high-temperature phase, grown from the melt, is cooled until it shatters to form an opaque powder of the low-temperature phase. (In cases where there is substantial supercooling, the temperature is lowered to just below the transition point and the sample tube is touched with a cotton swab dipped in a dry-ice–acetone mixture.) The sample is maintained at a temperature of 2–4° below the temperature of transition. After a few hours, observation through crossed polaroids and X-ray photographs of the sample will indicate that the specimen has some preferred orientation. Finally, after a period ranging from eight hours to five days, sections in the capillary tube, ranging in length from 0.5 to 2 mm, anneal to form suitable single-crystal specimens.

Burbank (1973) suggests annealing the transformed microcrystals into a single crystal by letting the capillary tube oscillate through the precession motion, thus varying the thermal gradients across the sample tube.

In a variation of this approach, Fourme (1972) grew single crystals of the low-temperature form of furan by carefully cycling the sample through the transition temperature several times.

7.3.3. Growth from Solution

Single crystals of the low-temperature form can also be obtained by growing crystals from a solution maintained at a temperature below that of the phase transition, using techniques decribed in Section 7.2.1.

7.4. Test for Centrosymmetric Crystal

During the course of a study of polymorphism of hydrogen cyanide, Dulmage and Lipscomb (1951) developed a pyroelectric test to determine whether or not the crystallographic phase under investigation is centrosymmetric.

A capillary tube containing the sample was suspended by a thread approximately 5 mm from the tip of a thermocouple. A dewar of liquid nitrogen was slowly raised, cooling the capillary tube as the nitrogen surface approached it. At a temperature of approximately 168 K (as measured by the thermocouple), the capillary tube suddenly jumped to the tip of the thermocouple. The dewar was lowered and when the sample temperature reached approximately 174 K, the capillary tube fell away from the tip. This phenomenon was repeated several times and indicated strong pyroelectricity in the low-temperature form. This procedure was also used with a stream of cold nitrogen gas (Dulmage and Lipscomb, 1952).

CHAPTER 8

Data Collection and Reduction

8.1. Temperature of Data Collection

8.1.1. Choice of Temperature

The fundamental reason for using LTXRD techniques is that the thermal motions of the atoms in the sample are reduced as the temperature is lowered. As a result, gases and liquids crystallize, phase transitions occur, diffraction intensities are increased quantitatively and improved qualitatively, and thermal contraction is observed. However, the "proper operating temperature" is one which is low enough for the desired effect to be observed but not so low as to cause unnecessary experimental and economic difficulties. The complexity of the apparatus, the experimental difficulty of collecting data, and the costs involved all rise in inverse proportion to the lowering of the temperature. Therefore, the temperature at which data are collected should be as high as possible without adversely affecting the fundamental research objectives.

The benefits of collecting data at low temperatures for crystal-structure analysis have been reported by several authors (see Section 2.2).

Cruickshank (1956) described the variation of vibration amplitudes with temperature in some molecular crystals. He concluded that a minimum worthwhile temperature for collecting single-crystal data would be about 25 K. At 90 K, the thermal motion is about one-third that at room temperature, while at 25 K it is reduced to one-third that at 90 K. There is little to be gained by using temperatures below 25 K.

In a similar analysis, Coppens (1972) concluded that for charge-density analysis, temperatures of or below 30 K are strongly suggested. However, for the determination of accurate atomic coordinates, liquid-nitrogen studies will often be adequate, since a further increase in precision of coordinates may be meaningless unless effects of charge asymmetry on the least-squares refinement are properly accounted for.

Thermal diffuse scattering (TDS) effects may also be large at room temperature and are generally not corrected for. Cooling to LN_2 temperatures greatly reduces the ratio of TDS to Bragg scattering by a factor of 5 with a similar increase in the intensity of the higher-order reflections. Further cooling to liquid-helium temperatures results in another reduction by at least a factor of 5 (Figure 2–4) in the TDS-to-Bragg scattering ratio, with little gain in Bragg intensity (Reid, 1973).

Hanson (1965a), in a comparison of room-temperature counter data and low-temperature (140 K) film data from the same compound, found that the standard deviation was about three times larger for low-temperature film data than for the room-temperature counter data, but that the magnitude of anisotropic thermal motion is smaller at low temperatures. In addition, the bond lengths were somewhat larger for the low-temperature data, indicating that less of a correction is needed for the thermal motions of the atoms.

On the other hand, a recent comparison of three sets of counter data obtained at 298, 219, and 95 K, showed similar standard deviations for all three data sets, but the thermal parameters at 95 K were about 30% of those at 298 K (Thomas et al., 1974). This is in good agreement with Cruickshank's (1956) previously mentioned calculations.

These conclusions are also in general agreement with Huber-Buser (1971), who analyzed the results of a number of crystal structures in which light atoms were refined in the presence of heavier atoms. The analysis of these data indicates that if hydrogen and nonhydrogen atoms are to be distinguished, a minimum signal-to-noise ratio is required. Thus, if one is to improve structure determinations by using low-temperature techniques, one must also improve the quality of observations, so that the signal-to-noise ratio is raised. Otherwise, the increased number of high-angle reflections will not be very useful in locating hydrogen atoms.

A final interesting point is that the characteristics of the sample itself may sometimes restrict the temperature of data collection. For example, Hanson (1965b), during a study of the azulene–trinitrobenzene complex, found that *some* reflections are extremely sensitive to slight temperature variations at $-140°C$, but not at $-95°C$ (Figure 8–1). For this reason,

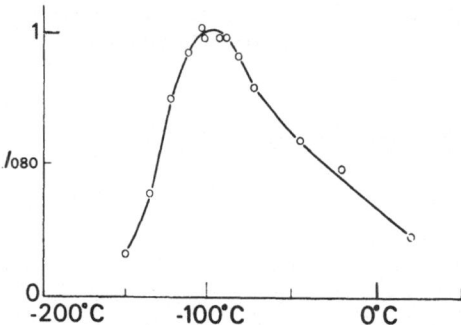

Fig. 8–1. Temperature dependence of I_{080} of the azulene–trinitrobenzene complex (Hanson, 1965b). (Reproduced with the permission of the International Union of Crystallography.)

all the data were collected at $-95°C$. This very interesting observation deserves further investigation.

8.1.2. Temperature Calibration

The principles and use of temperature-measuring instrumentation are discussed in Appendix 1. Under most experimental conditions, the thermometer or thermocouple is not in direct contact with the sample. A temperature calibration curve (thermocouple reading vs. actual sample temperature) must be obtained if the temperature is to be determined accurately. For many studies, e.g., routine crystal-structure analysis, the temperature need be known only to within 5°. However, if X-ray and neutron data are to be compared, if thermal expansion data are to be obtained, or if temperatures of transition are being studied, the temperature may have to be known to within 0.1–0.03°. [For the latter case, Axon et al. (1953) discuss several possible sources of error for powder-camera apparatus.]

One method of temperature calibration in common use is to prepare a calibration curve of "crystal temperature" vs. "thermocouple temperature." If the crystal can be seen (e.g., in a gas stream or a cryostat with a window), the following method can be used (Olovsson, 1960; Renaud and Fourme, 1967; Abowitz and Ladell, 1968).

An organic liquid is sealed in a glass capillary tube and the temperature is adjusted until liquid–solid equilibrium is established, as verified by visual

Table 8–1.   Melting Points and Transition Temperatures Useful for
Temperature Calibration

Compound	Melting point (K)[a]
p-Dioxane	284.8
Benzene	278.6
Water	273.1
t-Butyl chloride	248.2
Pyridine	231.1
Trichlorethylene	200.1
Toluene	178.1
Acetone	177.7
Ethanol	155.8
Methyl cyclopentane	130.7
cis-2-Pentene[b]	93.1
	Transition temperature (K)
TbVO$_4$[c]	33
TmVO$_4$[d]	2.15

[a]After Abowitz and Ladell (1968).
[b]Coppens (1972).
[c]Coppens et al. (1974).
[d]Segmüller et al. (1974).

observation. This is done for several materials, and a calibration curve of real vs. observed temperature is then drawn. Abowitz and Ladell suggest using the materials listed in Table 8–1. They state that in all cases it was first necessary to supercool the liquids to achieve crystallization, and that reliable temperature calibration could not be carried out below 120 K because they did not find any liquids which could be crystallized even by quenching the capillary tube with liquid nitrogen. Coppens (1972) suggested using cis-2-pentene, which has a melting point of 93.1 K.

It is also possible to use solid–solid phase transitions between optically birefringent and isotropic phases. The sample is placed between crossed polaroids (see, e.g., Lippman and Rudman, 1976), and the sharp transition observed on *warming* through the phase transition is noted. The temperature of transition observed on cooling should not be used since these materials have a tendency to supercool. Suitable materials include carbon tetrachloride (225.5 K), t-butyl chloride (219.2 K), and neopentane (140.0 K).

In a variation of this method, Burbank (1973) made careful temperature measurements for all conceivable operating conditions, i.e., for a wide variety of power settings on the heating elements in the gas generator, using a standard thermocouple configuration placed in the crystal position, prior to doing any crystallographic work. He then assumed a one-to-one correspondence between calibrated Variac settings and desired temperatures and did not employ a thermocouple during the actual low-temperature data collection.

One could also calibrate a working thermocouple vs. a thermocouple placed in the sample position and use the resulting calibration curve to obtain the real temperature during the data collection. One must ensure even cooling of the thermocouple leads so as to minimize errors in temperature measurement. Robertson (1965) placed a very small thermocouple junction, comparable in size to the crystal, inside a thin-walled capillary tube and mounted it in exactly the same position as the crystal specimen. The leads of the thermocouple were coiled both within the capillary tube and outside it around its base to minimize errors caused by heat conduction in the wires.

The presence of a window in a cryostat can lead to poor temperature control due to absorption of radiation by the sample. Small (Okazaki and Kawaminami, 1973) or heat-absorbing (Losee and Simmons, 1968) windows should be used.

Inasmuch as few cryostats include optically clear windows for visual observation of the sample, it is often impossible to calibrate the instrument by observing phase transitions. In this case, it is necessary to calibrate the thermocouple by making use of materials whose lattice constants are known at low temperatures. Pollock (1955) suggested the use of pure lead since it is a material that has a large coefficient of thermal expansion and a fairly intense reflection at a high Bragg angle. He presented data down to 90 K, which would allow one to estimate temperature to approximately $0.5°$.

King and Preece (1967) described a double-scanning method which combined techniques for diffractometer alignment and temperature measurement down to 18 K, using tungsten and silicon as standards.

Woodard and Straumanis (1971) developed a method for the precision determination of lattice parameters at temperatures to 40 K. Thermal expansion data for tungsten (Shah and Straumanis, 1971), silicon (Shah and Straumanis, 1972), and also aluminum, silver, and molybdenum (Straumanis and Woodard, 1971) have been determined and could be used as standards for checking temperature calibration and instrument alignment.

Any material used for temperature calibration should have a reasonably large thermal expansion coefficient in the temperature region of interest. For temperatures below 40 K, europium oxide (EuO) appears to be a satisfactory standard which exhibits a 0.1% contraction between 100 K and 4.2 K (Lévy, 1969; Etourneau et al., 1975).

Eubig and Tomizuka (1972), who were interested in measuring lattice parameters to 1 ppm, determined that a temperature variation of 0.1° would result in lattice parameter changes greater than 1 ppm. Since their cryostat could not maintain the temperature to within 0.1°, they monitored the temperature constantly and applied corrections to the data.

Girt et al. (1974) described a back-reflection camera in which the film can be placed in two positions so that photographs can be obtained at a reference temperature and at the working temperature. The reference temperature data are used to check the calibration of the temperature and of the material. This is important in studies of metastable states obtained by quenching, since one must be sure that the desired metastable state is being cooled. They later modified the camera to take exposures at three temperatures (Girt et al., 1975).

Finally, there is the question of the temperature of samples irradiated by X-rays. At very low temperatures, where the heat capacity of the crystalline sample becomes very small, the heat produced by X-ray absorption may increase the temperature of the substance's outer layers considerably. Straumanis and Woodard (1971) examined this effect for silver powder and, although the results were not conclusive, suggest that at 40 K the heating effect may result in a difference of about 0.0004 Å in the observed lattice parameter. Calculations based on their data for the thermal expansion coefficient of silver at 40 K show that this corresponds to a $\triangle T$ of 16°! This phenomenon requires further investigation.

8.2. Apparatus to Be Used for Data Collection

The type of *X-ray* apparatus that should be used for data collection, which depends on the type of investigation being contemplated, is generally independent of the low-temperature system and will not be discussed here. The type of *cooling* apparatus to use depends on the desired temperature, as described above. The relative merits of the various cooling systems have

Table 8–2. Comparison of Gas-Stream, Cryostat, and Immersion Methods

	Method		
	Gas stream	Cryostat	Immersion
Consumption of cryogen	0.5–8 liters/h	0.1–0.5 liters/h	0.2–0.5 liters/h
Absorption of radiation	None	Present	Present
Temperature stability	0.5°–5°	0.1°–0.02°	0.1°
Special problems	Frost condensation	Mechanical stability	Absorption of radiation
Relative cost	Low to moderate	Moderate to high	Low to moderate

been described in Part II, and are compared in Table 8–2 (adapted from Coppens, 1972).

At ultra-low temperatures, the low rate of cryogen consumption and the insulation of the sample from external heat sources result in the cryostat being the preferred instrument. At intermediate temperatures, the choice is less obvious. For single-crystal structure analyses, the conduction method is recommended only for use at temperatures below those attainable with liquid nitrogen. Above 77 K, the cost of the instrument and the potential absorption errors (caused by the windows) act to favor gas-stream techniques. Also, if ready accessibility of the sample is a must (e.g., to grow single crystals from a liquid), the gas-stream method is highly recommended.

For powder diffractometer studies the conduction method is generally preferred at all temperatures, with the complexity of the cryostat design being a function of the temperature required, the frequency of use, and the accuracy of data needed. However, the gas-stream method is quite satisfactory for many purposes, particularly for preliminary studies.

8.2.1. Protection of the X-Ray Instrument

The primary function of low-temperature apparatus is to cool the sample and not the X-ray diffraction instrument. Usually the sample volume that must be cooled is very small and cooling can be accomplished easily. However, when the gas-stream method is utilized, special precautions should be taken to prevent the cold gas stream from cooling the camera or diffractometer. In order to prevent this from happening, heat shields which deflect the cold gas stream away from the parts that could be adversely affected by cooling have been employed by several investigators. These areas include the sample rotation axis of a Debye–Scherrer powder camera, the arcs and translations of a goniometer head, the rotation axes of Weissenberg and precession instruments, and the ϕ and χ circles of diffractometers; in other words, all movable parts of the instrument.

The most obvious adverse affect is the formation of ice on a cold surface, making it impossible, for example, to adjust the arcs on a goniometer head. A second problem is the possible misalignment of the instrument. Such difficulties are unavoidably encountered with many instruments, and methods have been developed to realign instruments at low temperatures when this problem cannot be surmounted in any other manner (see Section 8.3.3.). However, for most cooling applications it is possible to shield critical parts of the X-ray apparatus and so bypass any alignment difficulties. A third problem, which is more subtle and takes longer to recognize, is the condensation of water vapor, rather than ice, on the X-ray apparatus. In this case, portions of the apparatus at some distance from the sample may become cool enough to condense water. This does not interfere with the immediate operation of the instrument, but over a long period of time, the surface or internal parts may begin to corrode, eventually ruining the instrument.

It is a useful precaution to coat any surfaces which are likely to become cool with water-resistant spray or grease (e.g., silicone spray or vacuum grease). The use of a plastic tent over the apparatus also aids in preventing the condensation of moisture on the instrument as well as on the sample itself.

A very common method that is utilized in several of the instruments described in Part II is to blow a stream of warm air over parts of the X-ray instrument that would otherwise be cooled by the low-temperature system (see Sections 3.1.8 and 4.2).

When a goniometer head is used, two factors must be considered: the prevention of ice formation on the goniometer head itself and the minimizing of any cooling of the rest of the apparatus by conduction through the goniometer head.

Examples of heat shields for goniometer heads range from the easily made devices described by Post et al. (1951) and Abrahams et al. (1950) to the more elaborate redesigning of the goniometer head described by Renaud and Fourme (1967; Fourme, 1975).

Bolhuis (1971) suggests using a copper rod to connect the glass rod on which the crystal is mounted to the goniometer head. Thermal conduction maintains the copper rod above 0°C and prevents ice from forming on the support rod or goniometer head. However, this technique will work only for a very finely controlled cold stream such as the one he describes. Another approach is the use of a plastic base for the goniometer head, which reduces any cooling-by-conduction of the remainder of the X-ray apparatus. Such a goniometer head has been incorporated by Stoe in their low-temperature apparatus.

These preventive measures are necessary only when the gas stream is directed at the goniometer head. When it is dircted perpendicular to the sample axis (so that the goniometer head itself is not cooled), special modifications of the goniometer head are not needed.

The above comments are, in general, limited to cooling with the gas-stream method. If other methods of cooling are used, specially designed sample holders and cameras are an integral part of the low-temperature apparatus design and/or all cold parts are contained within an evacuated chamber so that icing problems are minimized or absent. For powder diffractometry, the use of the gas-stream method and certain simple conduction-cooling adapters require careful shielding of the goniometer shaft in order to prevent serious misalignment problems. The shielding consists of electric resistance heaters or strategically placed insulation.

8.2.2. Film at Low Temperature

Jauncey and Richardson (1934) and Reekie (1939) investigated the sensitivity of photographic film to X-radiation at low temperatures. Their results, which were limited to the films tested, showed that the sensitivity decreases to a minimum and at that point the films are more sensitive to X-rays than to light. The techniques used in carrying out these tests are described in the original articles.

Lonsdale and Grenville-Wells (1956) also tested some X-ray films and found that Ilford Industrial G film shows a large increase of light sensitivity at low temperatures while Industrial A does not show a similar increase, and that the G film is equally sensitive to X-rays at 20° and −95°C. All these investigators stress the importance of *uniform* cooling of the X-ray film so as to ensure uniform intensity distribution when collecting data at low temperatures.

Another requirement is that the pressure on the film should be uniform without any stresses in the film, since the sensitivity of the film at points of stress and pressure is not the same as that at other points on the film. Thus, nonuniform cooling and mounting of the film can lead to errors in relative intensities caused by nonuniform recording of diffraction intensities across the surface of the film.

In an extension of these data to lower temperatures, Pearson (cited by Cruickshank et al., 1956) found that the sensitivity of X-ray film exposed at 4 K was not greatly different from that at room temperatures.

In some of the cameras discussed in Part II, the designers were careful to keep the film at room temperature, even though the sample was cold. One way in which this was done was to circulate warm water through coils attached to the outside of the film cassette. A second method is to place the film cassette outside the cooling chamber and have the X-rays pass through a window.

8.2.3. Alignment and Calibration of Apparatus

Several techniques have been developed for aligning both single-crystal and powder-diffractometer cryostats. Inasmuch as the details of alignment are specific to the instruments used, they will not be discussed here.

At LN_2 temperature or higher, gross misalignment is not expected. However, it has been shown that sample displacements of the order of 1.5 mm can occur on cooling the cryostat from room temperature to 18 K (King and Preece, 1967). This is the result of cryostat design in which the helium reservoir and the specimen are suspended within the cryostat to reduce heat leaks. Uneven contraction of the supports results in sample displacement and may be inherent in cryostats of similar design. The surface specimen has to be realigned at low temperatures; for accurate work it should be realigned after every 10° change in temperature.

Procedures for aligning cryostats mounted on horizontal powder diffractometers have been described by King and Preece (1967, 1968) and Smith and Leider (1968). Smith (1966), Woodley et al. (1971), Coppens (1972), and Rees and Coppens (1973) have describe procedures for aligning single-crystal diffractometer cryostats.

The approach of Rees and Coppens (1973) is typical. They measured twelve strong reflections (of high order so that Mo $K\alpha_1$ and Mo $K\alpha_2$ peaks are well separated) in different regions of reciprocal space. Each reflection was manually centered in the counter by approaching the reflection from both positive and negative 2θ, and the value considered for each setting angle was the average between the two corresponding measurements. A least-squares refinement of all four setting angles was finally performed.

It is suggested that restrictions on the setting angles be made part of the angle setting program when an automatic diffractometer is used, so that transfer lines and electrical and gas connections to the low-temperature device are not fouled up. Such restrictions have been incorporated in the Syntex P2$_1$ diffractometer programs.

Once the apparatus is aligned mechanically, it should be calibrated. Weltman (1962) advised mixing aluminum powder into the powder sample to act as an internal standard. If the sample can be removed without affecting the alignment, pure materials such as those mentioned in Section 8.1.2 can be used.

Low-temperature cell parameters were obtained from a Weissenberg camera by superimposing room- and low-temperature exposures on the same film and using the room-temperature cell parameters as a calibration standard (Thomas et al., 1974).

8.3. Data Collection

8.3.1. Strategy of Data Collection

A major source of concern during a LTXRD investigation is the possible failure of the low-temperature system. If the sample is not affected by such a failure, data collection may be continued from the point at which it was stopped and will proceed in a normal manner. However, if sample preparation is difficult and dependent on a continuous source of cooling (e.g., when a crystal is grown from a liquid), a minimum set of low-temperature data should be collected as quickly as possible, so that if the measurements are interrupted, the crystal structure can still be solved. This approach is discussed by Smith (1966), who suggests measuring the most intense inner sphere of reflections first, and increasing the maximum value of d^* in steps of 0.1 or 0.2 until all available data are collected.

Furthermore, each reflection should eventually be measured two or three times to provide a check for possible measuring or recording errors. Further checking by measuring symmetry-related equivalent reflections is also recommended. These checks are particularly important when counter measurements are made of crystals grown from liquids or gases, since regrowing crystals to recheck data during the course of the subsequent structure refinement can only be done with extreme difficulty.

Data collection for a single-crystal structure analysis can last from several days to a few weeks. If the data are collected on film, minor variations in temperature are usually averaged out over the course of a single exposure and any long-range temperature variations can be noticed easily enough from the thermocouple readings. However, if data are collected on a diffractometer, accuracy demands that all data be collected at the same temperature or that suitable corrections be applied. Constant monitoring of the temperature can be done electronically or by using the crystal itself.

Hanson (1965b) plotted the b dimension of his crystal as a function of temperature. The value of b (derived from an 080 reflection measured at approximately 135° 2θ) was easily measured whenever necessary, giving a sensitive indication of the sample temperature.

Verschoor and Keulen (1971) investigated the change of intensity of the 800 reflection of cyanuric acid as a function of temperature and found it to be 1.2% per degree Celsius. During the course of the data collection, they checked the stability of the crystal and X-ray system by measuring the usual standard reflections every hour, and they also measured the temperature stability of the low-temperature apparatus by measuring the 800 reflection once every 6 hours. From the 19 measurements obtained during the course of the data collection, it was concluded that the standard deviation of the temperature was 0.5°.

Strouse (1975) used the intensity of a diffracted beam at the inflection point on one *side* of the maximum intensity as a sensitive indication of the temperature stability.

Ross and Williams (1974) printed the temperature at which each reflection was measured using a physical-interrupt meter tied to a proportional temperature controller.

Long-range fluctuations can be measured by using a recording potentiometer connected to the thermocouple. Since the output of a copper–constantan thermocouple is approximately -5 mV at LN_2 temperatures, increased sensitivity to any temperature fluctuations can be obtained by introducing a bucking voltage of approximately $+4.5$ mV on the recorder and running it on a 1-mV full-scale range, with the zero-adjust offset as much as necessary.

Hanson (1965a) was faced with the choice of collecting room-temperature counter data or low-temperature film data. He collected two sets of data, a set of counter data at room temperature and a visually estimated film set at 140 K. In order to correct for any systematic errors in visually estimating the film data, one zone of film data was collected at room temperature and compared with the same zone collected using the scintillation counter. The resulting error vs. Bragg angle curve was used to correct the low-temperature data.

In a low-temperature precession camera study of pentaborane, Dulmage and Lipscomb (1952) found it more convenient to collect all the data on zero-level photographs.

Woodley et al. (1971) describe a three-circle diffractometer for use at liquid-helium temperatures (see Section 4.4.1), which they explain is easier to design than a four-circle instrument. However, one cannot identify multiple diffraction errors on a standard three-circle diffractometer, and so they suggest mounting the detector on an arc that will allow a departure from equatorial geometry (Hine et al., 1975).

8.3.2. Check for Single Crystal

When single crystals are cooled, twinning may occur or the crystal may undergo a phase transition. It is always advisable to test a crystal prior to data collection by low-temperature optical microscopy or differential scanning calorimetry or by examining it on a low-temperature camera.

When data are collected on a counter diffractometer, much unnecessary work can often be avoided by making certain that the crystal is in fact a single crystal. This check is recommended whether the sample material is a solid or a liquid at room temperature. Although the presence of twinning is often indicated by the appearance of a split peak profile, in many cases the twinning is more subtle and not easily recognized.

A simple adapter to hold a set of crossed polaroids on a camera or diffractometer can be constructed (Post et al., 1951; Lippman and Rudman, 1976). This device can be used for verifying the presence of a single crystal at low temperatures and for rough alignment of this crystal. It is especially helpful when the crystal has been grown from a liquid *in situ*.

The use of film as a means of checking the crystal quality is also recommended. A film cassette can be adapted for use on a diffractometer (Smith, 1966; Syntex, 1974); Polaroid film is especially recommended because of the very short exposure times required.

The Polaroid film cassette offers a particular speed advantage in a frequently used method for assessing the single-crystal character of a specimen. The method applies to crystals of known cell dimensions and symmetry and makes use of the fact that all reflections occurring with the same Bragg angle will fall on a circle on a symmetrically placed flat-plate film cassette. No matter what the orientation of the specimen, if it is a single crystal, the multiplicity of spots occurring on a given circle cannot exceed a given maximum. By counting the number of spots on a circle and comparing it with the expected multiplicity for that set of reflections, the presence of a single crystal can be verified quickly. Since each reflection will, in general, appear twice (once as it enters the sphere of reflection and once as it leaves), the number of spots on each circle will actually be twice the theoretical multiplicity (Rudman, 1968).

A precession camera can be converted to a flat-cassette rotation camera by attaching a slow motor (1 rpm) to the spindle axis by a pulley or simple gear arrangement, with the precession angle set at 0° (Lippman and Rudman, 1976). The use of the Polaroid cassette is recommended (see Section 3.5.2).

A flat-plate adapter can also be constructed for the Weissenberg goniometer. This was first mentioned by Collin and Lipscomb (1951). Later modifications included an adapter to take oscillation–rotation photographs using a Polaroid cassette (Rudman, 1968) and an adapter to take flat-plate, Polaroid Weissenberg photographs (Hope, 1969). The former is useful for aligning a crystal, as well as for determining its quality, while the latter can be used to conduct a rapid survey of the lattice constants of a crystal (see Section 3.5.1).

Huffman (1974) has presented explicit instructions for analyzing twinned or multicrystalline samples on a diffractometer. In many cases it was found that as crystal growth nears completion in a capillary tube, small fragments will occur which can be significant scatterers. While the presence of such satellite crystals is easily recognized and corrected for using film techniques, it is more difficult to identify from diffractometer data. Once it has been found impossible to index the lattice properly or to find a consistent set of symmetry elements, it can be assumed that one or more of the diffraction maxima located belong to a second crystal, or that a twin has occurred.

Relative intensities for a given lattice row will be identical to symmetry-related rows and from crystal to crystal. Thus if two pairs of symmetry-related reflections occur at the same Bragg angle (within experimental error), it can be shown that, if they are due to *one* crystal, they must be one of the following: (1) two general reflections in a monoclinic system, (2) two zonal reflections of an orthorhombic system, or (3) two zonal or axial reflections of a tetragonal cell. In each of the three cases the symmetry element relating the two reflections of one pair must be either perpendicular to or coincident with the symmetry element relating the other pair of reflections. If the experimental results do not comply with this condition, the axial row for each reflection should be checked, and the relative intensity of each higher-order peak recorded. If the relative intensities are identical, this indicates that more than one crystal is present or that a twin has occurred.

At this point, it is necessary to determine which reflections are due to which crystal. This can be accomplished by inspecting the pairs of reflections with identical 2θ values and assuming that the approximate *ratio* for relative intensities will remain constant.

Should this method fail, it is usually possible to determine from which crystal in the capillary a particular reflection has originated, assuming the centers of the various crystals are separated by several tenths of a millimeter. In order to accomplish this, the instrument must be well aligned and the reflections must be carefully centered in both the positive and negative regions of 2θ. Huffman has shown that crystal centering errors can be expressed as

$$t_a = \sin \Delta 2\theta \cdot R/2 \cos \theta$$

$$t_b = R \cdot \tan \Delta\omega/\sin \theta$$

$$t_c = R \cdot \tan \Delta\chi \sin \theta \cos \omega$$

where t_a, t_b, and t_c are crystal translational errors in the direction of the diffraction vector, χ axis, and instrument axis respectively; $\Delta 2\theta$, $\Delta\omega$, and $\Delta\chi$ are the discrepancies between the respective angles when measured in the positive and negative 2θ region, and R is an instrument constant. Transforming these equations to a coordinate system fixed with respect to the goniometer head with t_1 along with the goniometer head axis and t_2 and t_3 perpendicular to this axis and 90° apart yields

$$t_1 = (t_a^2 + t_c^2)^{1/2} \cdot \sin\left(\arctan \frac{t_a}{t_x} + \chi - 90\right)$$

$$t_2 = (t_b^2 + t_x^2)^{1/2} \cdot \sin\left(\arctan \frac{t_b}{t_x} - \phi\right)$$

$$t_3 = (t_b{}^2 + t_x{}^2)^{1/2} \cdot \cos\left(\arctan\frac{t_b}{t_x} - \phi\right)$$

where

$$t_x = (t_a{}^2 + t_c{}^2)^{1/2} \cdot \cos\left(\arctan\frac{t_a}{t_c} + \chi - 90\right)$$

Thus by carefully centering each reflection in the positive and negative region of 2θ, one can determine the relative location of the crystallite from which the diffraction occurred.

By retaining only the largest set of reflections which appear to be definitely from one crystal, it is possible to assign indices and characterize the crystal. The orientation matrix thus obtained can be used to determine which reflections located in the hemisphere belong to that particular crystal, and they can be eliminated from the list. The remaining reflections can then be treated. If in fact a twin is present, one or more reciprocal lattice vectors will coincide, and if no overlap is present it can be assumed that separate crystallites are present.

Although data collection is not complicated by more than one crystal being present, in the data reduction it is necessary to determine which reflections are likely to overlap from one crystal to the next, and either determine a correction factor or simply eliminate these reflections. The presence of twinning must also be taken into account when applying absorption corrections.

8.3.3. Alignment of Crystal

When film methods are used, any of the standard methods of aligning crystals can be used, but if data are being collected on a Weissenberg goniometer, the use of a Polaroid adapter can greatly reduce the time necessary to align the crystal. When studying crystals that have been grown at low temperatures, time is of the essence.

The method recommended for aligning crystals using the Polaroid adapter on the Weissenberg goniometer is as follows (Rudman, 1968): Point R (Figure 8–2), which is a reciprocal-lattice point located on the surface of the sphere of reflection, can be brought into the equatorial plane $(MOPQ)$ by an adjustment through angle α or angle β. The three points used to define the plane which is to be brought into coincidence with the equatorial plane are R, O (the origin of the reciprocal lattice), and any

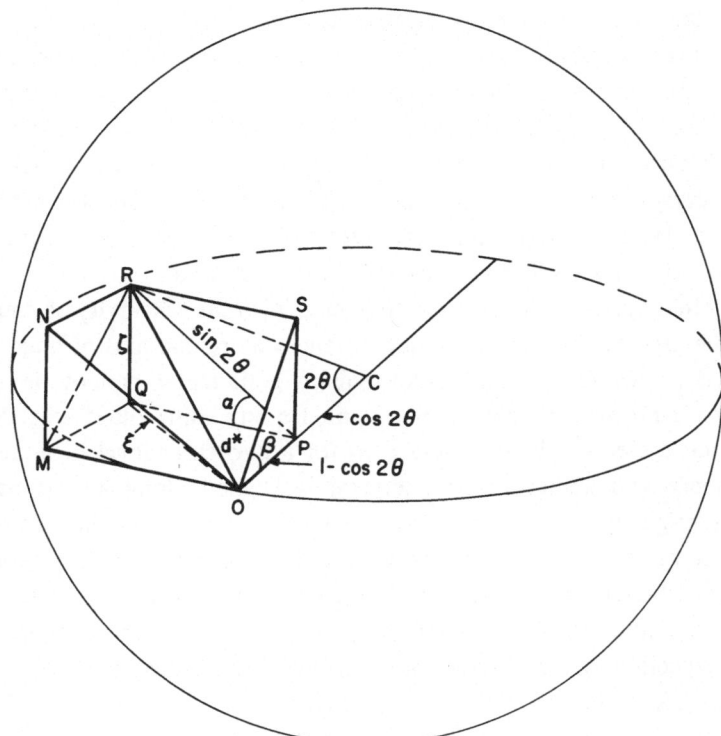

Fig. 8–2. A Weissenberg goniometer Polaroid adapter, which can be used for aligning a single crystal as described in the text. Reciprocal-lattice (R.L.) point R located on the surface of the sphere of reflection can be brought into the equatorial level (plane $MOPQ$) by an adjustment through $\alpha°$ or $\beta°$. In practice, three R.L. points are used to define the plane which is to be brought into coincidence with the zero layer. They are R, O (the origin of the reciprocal lattice), and any other R.L. point. It is this third point which determines whether the required correction is an α or β correction or a combination of the two. The origin of the sphere of reflection of radius 1 is located at the crystal, C (Rudman, 1968). (Reproduced with the permission of the International Union of Crystallography.)

other reciprocal-lattice point. The required arc adjustments are determined by the choice of this third point. From the figure it is clear that $\sin \alpha = \zeta/\sin 2\theta$ and $\tan \beta = \zeta/(1 - \cos 2\theta)$.

The alignment procedure suggested for use with the adapter utilizes the fact that the α misalignment is projected, undistorted, onto a flat-plate camera and can be measured without any need to calibrate the crystal-to-film distance. A 5 to 15° oscillation photograph is taken with one of the arcs of the goniometer head parallel to the X-ray beam. Two super-imposed exposures (with one exposure time twice that of the other), 180° apart on the spindle, are sufficient to locate the equator; the arc correction is then one-half the angle between the two zero-layer lines. Four photographs (alternating the arc that is parallel to the beam) are sufficient to align a crystal that is initially misaligned as much as 10° on each arc (Figure 8–3). The use of this iterative process is practical because the sensitivity of Polaroid film permits short exposure times.

Another application which takes advantage of the sensitivity of Polaroid film is the investigation of systematic absences along the axis of alignment. A crystal grown *in situ* on a Weissenberg goniometer cannot be easily moved to a precession camera or mounted along some other axis. However, when the equiinclination angle (μ) for a particular level is set on the Weissenberg goniometer and the crystal is aligned along an orthogonal axis, the origin of that reciprocal-lattice level is located on the surface of the sphere of reflection and is always in diffracting position. If a series of oscillation photographs (sufficient in angular range to delineate the layer lines on the film) are taken with the μ angle set for each layer in turn, simple inspection of the photographs will reveal the presence or absence of reflections along the axis of alignment (Figure 8–4). This will be sufficient to indicate any systematic absences during a space-group determination.

Single-crystal diffractometer methods for aligning single crystals of known and unknown unit-cell parameters are described by Huffman (1974). A systematic search of reciprocal space (best conducted with an automatic diffractometer), followed by an analysis of all the reflections found, will determine the lattice type and symmetry. Alternatively, if a few reflections are found, an orientation matrix can be calculated and the unit cell determined using standard techniques.

8.3.4. Radiation Damage

Crystals studied at low temperatures are subject to two types of "radiation damage": destruction of the crystal and, for samples maintained in sealed capillary tubes, distillation of the sample from the point where the beam strikes the capillary tube to some other portion of the tube. The latter effect has been observed by the author during a study of the high-temperature phase of methylchloroform on a precession camera and by Burbank (1974) during a study of osmium oxyfluoride.

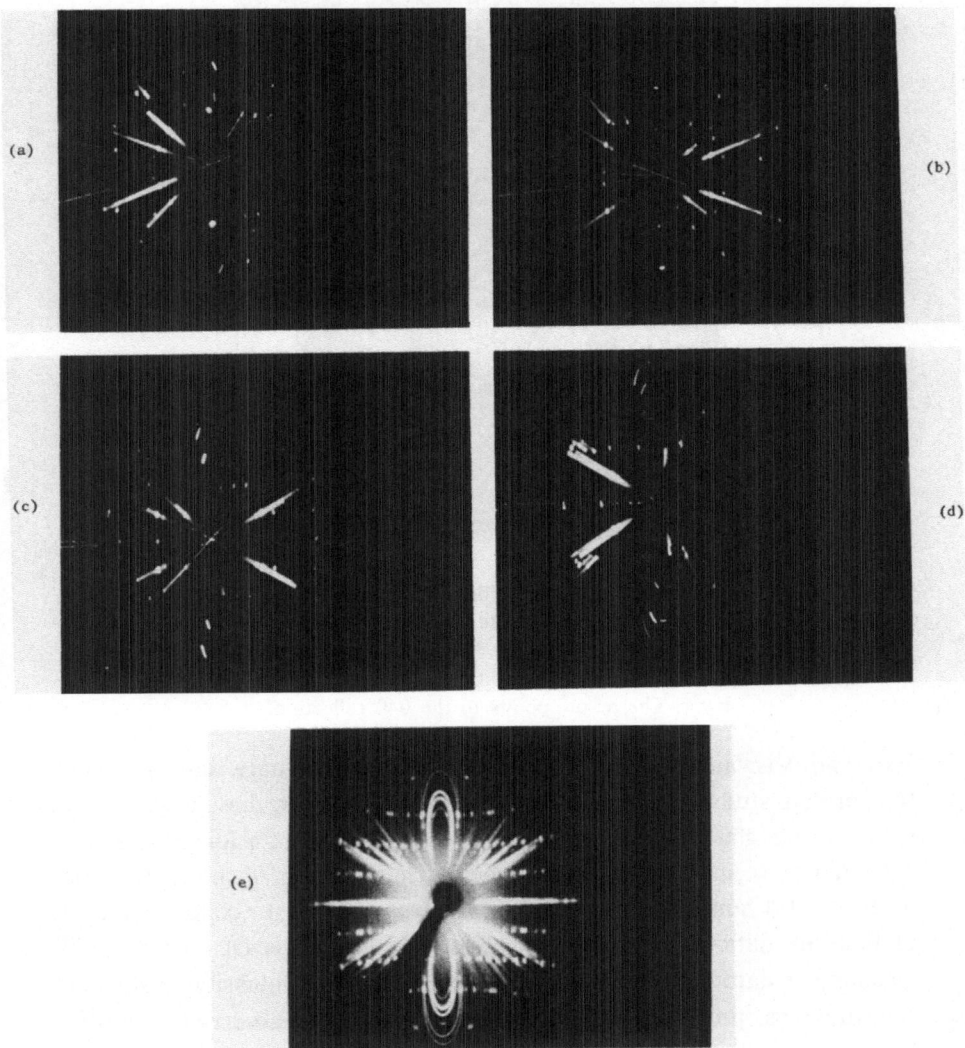

Fig. 8–3. A crystal of tetrachlorophthalic anhydride misset approximately 10° on each arc required four 10° oscillation photographs in order to obtain the rotation photograph shown in (e); arc 1 was parallel to the beam in pictures (a) and (c) and arc 2 in pictures (b) and (d).

If the former, common type of radiation damage occurs, it is usually necessary to select a new sample or to grow a new crystal in the capillary tube. If the same sample is reused, the possible contamination caused by the decay products of the radiation-damaged crystal should be considered.

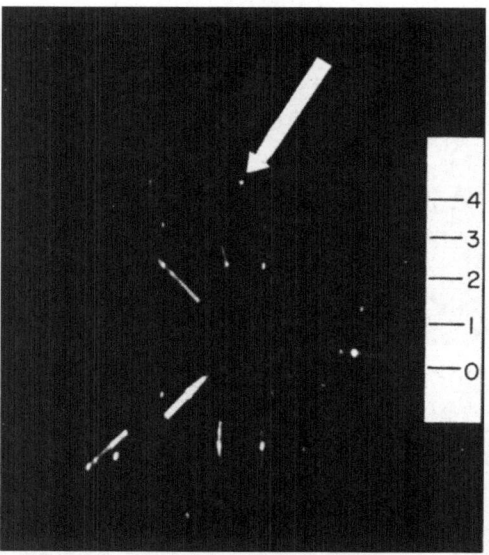

Fig. 8–4. A 10° oscillation photograph of a crystal of tetrachlorophthalic anhydride aligned along *b* taken with the Weissenberg goniometer set at the equiinclination angle for the fourth layer. The arrow points to the 040 reflection.

Rao and Viswamitra (1972) observed radiation damage during a low-temperature study of glycine silver (I) nitrate. However, these crystals were grown inside glass capillary tubes and were nearly 8 mm long. Data were collected by feeding into the X-ray beam fresh regions of the crystal once in every 200 hours of exposure. Thus it was possible to collect a full set of intensity data with a minimum of radiation damage. Of course, if diffractometer data are obtained, a calibration curve of intensity of standard reflections vs. time should be used to correct all the observed intensities. Very often, radiation damage can be reduced or eliminated by lowering the temperature.

The factors that should be considered when data are collected at low temperatures are summarized in Table 8–3.

8.4. Data Reduction

Low-temperature X-ray diffraction data should be analyzed in the same way room-temperature data are treated. However, two special cases are

Table 8–3. Summary of Low-Temperature Data Collection Procedures

---

1. Check quality of crystal after cooling.
2. Collect intense inner sphere first.
3. Check sensitivity of intensity to temperature.
4. Allow for restrictions on setting angles.
5. Allow for absorption by windows and sample holder.

---

encountered often enough in low-temperature studies to deserve special mention: absorption corrections and refinement of disordered structures.

8.4.1. Absorption Corrections

8.4.1.1. Powder Sample

A common method of obtaining low-temperature powder data is to condense the sample on a conduction-cooled copper wire or tube. This may result in a displacement of the powder lines on the Debye–Scherrer photograph. Kogan et al. (1960) discuss the problem of recognizing and correcting for extraneous lines due to the copper and the displacement of the powder lines.

8.4.1.2. Absorption by Windows and Sample Holder

Background and intensity measurements from both single-crystal and powder samples will be affected by any material (e.g., Mylar or beryllium) in the path of the diffracted beam.

If the X-ray beam, in general, passes through the windows obliquely, the resulting absorption is a function of the ω and 2θ values for each reflection. Woodley et al. (1971) obtained a correction curve for their instrument by measuring the window absorption (for six Mylar windows) for the special case of $2\theta = 0$ at various ω settings of the cryostat. Absorption should vary as $\exp(-6\mu t \csc \omega)$. The value of $6\mu t$ was first calculated from measurements made on the direct beam, with and without the cryostat in place. The observed count rate was plotted against ω and good agreement was obtained between the calculated and observed curves.

Honeywell et al. (1964) used a beryllium sample cell for low-temperature, high-pressure diffractometer studies of liquid samples. They present data showing the experimentally observed intensity as a function of the scattering angle for the empty beryllium cell. These results indicate a significant increase in the background, as well as specific peaks, due to sintered beryllium. Similar results have been observed for the beryllium shroud used on the Cryo-Tip goniometer head (see Section 4.7).

However, Coppens (1972) states that scattering by the beryllium windows in the CT-38 single-crystal helium cryostat (see Section 4.4.1) is not a problem, probably because the beryllium is located far from the center of the diffractometer and the receiving beam is finely collimated. If beam size and counter openings are large, resulting errors in the background correction may be appreciable for some reflections. The location of the window off the center of the diffractometer and increased collimation in front of the counter may prevent part of the cryostat-window scattering from entering the counter window.

Coppens also reports that when cylindrical beryllium shields are used the maximum difference in absorption is approximately 3% and that it can be allowed for in a data-processing computer program.

Just et al. (1975) have developed an algorithm for determining the calculated background of a peak (and its standard deviation) in the presence of a systematic variation in the background due to diffraction by the walls of a cryostat.

8.4.1.3. Sample in Capillary Tube and Completely Bathed in the Beam

This situation has been treated by Wells (1960), Smith (1966), Santoro et al. (1968), and Verschoor and Keulen (1971). Absorption corrections for small single crystals held inside glass capillary tubes are given. These are corrections for absorption by the sample tube; corrections for absorption by the crystal must be treated separately using standard methods.

Verschoor and Keulen (1971) report that the mass absorption coefficient (μ) of the capillary tube in which the crystals were placed was 12.5 cm^{-1} for molybdenum radiation while μ for the crystal was only 1.72 cm^{-1}. Corrections for absorption by the capillary tube were calculated using a transmission factor approximated by

$$A = 1 - F_1(\phi, \theta) \cdot F_2(\chi, \theta)$$

where $F_2(\chi, \theta) = (1 - \sin^2 \chi \sin^2 \theta)^{-1/2}$ and $F_1(\phi, \theta)$ was determined graphically.

Smith (1966) based his correction procedure on the assumption of cylindrical symmetry in the glass capillary tube and the aluminum and beryllium windows of the low-temperature apparatus. The expression corrects for the longer path length in the absorbing medium of a reflection making an angle, α, with the plane normal to the axis of symmetry. It is of the form

$$I = I_0 \exp\left[-\mu t(\sec \alpha - 1)\right]$$

where I_0 is the uncorrected intensity, μ is the linear absorption coefficient, and t is the thickness of the material. The sum of the three factors μt (for glass, Al, and Be) is used in the expression.

8.4.1.4. Sample in Capillary Tube and Larger than the Beam

When a crystal is grown from a liquid inside a capillary tube, the crystal is cylindrical and generally longer than the cross section of the beam. The volume of irradiated crystal will vary from reflection to reflection by as much as a factor of two. Parkes and Hughes (1963b) present the correction factors for both Weissenberg and precession techniques, while Coyle and Schroeder (1971) treat the case for Eulerian-cradle geometry. The latter treatment requires that the axis of the cylindrical crystal be coincident with the ϕ axis. The correction factors correct for absorption by the crystal as well as by the sample tube (see Appendix in their paper). The case of general orientation of a crystal is discussed by Coyle (1972).

8.4.1.5. Effect of Capillary Tube on Background Measurement

Krieger et al. (1974) examined capillary-tube effects and background corrections with emphasis on data collection in protein crystallography. Many of their conclusions can be applied to low-temperature studies of crystals inside capillary tubes. Their results can be summarized as follows:

(a) The background will be affected by the amount of capillary tube in the direct beam.

(b) For the normal situation of a crystal mounted, but not centered, in a capillary tube, the crystal is centered in the X-ray beam while the capillary tube itself is eccentric in the beam. The background variation is an approximately sinusoidal function of ϕ, as the capillary tube rotates in and out of the beam.

(c) There is a similar background dependence on χ. The exact nature of this variation depends on the experimental conditions.

(d) The background intensity is essentially uncorrelated with crystal absorption. As a consequence of this, the background intensity should always be subtracted from scan intensity before the absorption corrections are applied.

A number of experimental absorption corrections have been described in the literature. Although these absorption corrections cannot compete with Gaussian integration or analytical methods in accuracy, they are quite useful for routine structure determinations—particularly in cases where the crystal size cannot be measured exactly. As a result, such semiempirical methods are used extensively in protein crystallography where the crystal is located inside a capillary tube in equilibrium with a liquid. They are equally useful in correcting low-temperature data collected from crystals inside capillary tubes.

These methods have been described by North et al. (1968) and Huber and Kopfmann (1969 and earlier papers). The variation in intensity of one or more reflections at or near $\chi = 90°$ as the crystal is rotated about ϕ is used to give a curve of relative transmission vs. azimuthal angle. The transmission coefficient for a general reflection is derived from the transmissions in the directions of the incident and reflected beams for that reflection.

Another method of interest uses the azimuthal scans on a large number of equivalent reflections and expands the transmission factor as a Fourier series in the diffractometer angles (Flack, 1974). However, because this method requires measurement of at least four equivalent reflections, it is not useful for crystals of low symmetry or where equivalent reflections cannot be measured because of instrumental restrictions (e.g., when low-temperature apparatus is used).

Watkin (1975) reports that the same reflections (from a sample in a cappillary tube) that were measured on a four-circle diffractometer using (a) bisecting geometry and (b) fixed-χ geometry differed in intensities by as much as a factor of 10. These reflections had been observed with the incident or reflected beam inclined at a small angle to the capillary-tube axis.

An additional absorption correction factor was applied to the North et al. (1968) correction. This factor falls off sharply as the angle between the capillary-tube axis and the X-ray beam increases. For example: 1°, 92,038; 2°, 308; 4°, 17; 8°, 4; 16°, 2; 32°, 1.5; 64°, 1.25. Further details are given in the paper.

8.4.2. Refinement of Disordered Models

A large number of molecules studied at low temperatures exhibit one or more crystallographic phase transitions as the temperature is lowered. Quite often, the phases closest to the melting point are disordered, resulting in very few observable X-ray reflections. Also, the difficulties of collecting data at low-temperatures occasionally result in poor or incomplete data sets, so that standard crystal-structure analyses cannot be completed. A third problem may arise from the effects of large librational motion on the part of the molecules.

Several techniques and computer programs have been developed to treat these problems. André et al. (1971, 1972) have described two programs (ORION and PYTHIE). The former carries out a least-squares refinement of the parameters of the rigid groups and/or individual atomic coordinates and thermal parameters. The latter also refines the molecules as a rigid group. The molecular parameters, i.e., three rotations and three translations in the general case, are randomly generated by a Monte Carlo method. The packing thus obtained is group-refined, using only a few selected strong reflections, then kept or rejected according to several tests on the weighted reliability index. The best structures are further refined using an increasing number of reflections up to a given limit. The program is fast and automatic.

Press and Hüller (1973) devised a new method of analyzing the orientational structure of molecular solids, based on the expansion of the coherent scattering-length density of a molecule into a complete set of orthonormal functions. It is used to treat systems where no well-defined equilibrium positions exist for the atoms due to the presence of large librational motion or dynamic disorder.

Other recent treatments include: a modification of the King–Lipscomb Hindered-Rotor expression (Bennett et al., 1975); the use of Kubic Harmonics for a plastic crystal in the cubic system (Levy et al., 1975); and an algorithm for refining crystal structures with poor or limited data (Cady and Larson, 1975).

CHAPTER 9

Future Trends in Low-Temperature X-Ray Diffraction

Several trends have become apparent in recently designed LTXRD apparatus and will affect the future development of this type of instrumentation. The most important of these are summarized below.

Miniaturization. Recent technological advances have been used in reducing the size of LTXRD apparatus. The smaller equipment is more efficient, can be used with fewer modifications of existing apparatus, and uses less "cooling power." Examples include the goniometer-head adapters of Renton and Baun (1963) (Figure 4–8a), Petsko (1975), Sharon (1975) (Figure 4–8b), and Air Products and Chemicals (1973) (Figure 4–13), as well as the miniature dewars of Altona (1964a,b) (Figure 3–21) and Thaxton and Jacobson (1970) (Figure 3–22).

Sample Chambers. The availability of sturdy materials with low X-ray absorption characteristics and the development of computer programs for calculating absorption corrections have resulted in the increased use of sample enclosures, even for accurate single-crystal data collection. When the sample is enclosed in a chamber, temperature stability is improved, consumption of cryogen is minimized, frost formation is prevented, and lower temperatures can be reached.

Chambers which have been incorporated in nearly all conduction-cooling cryostats (see Chapter 4) are now becoming more and more common in gas-stream cooling devices, as described in Section 3.4.4.

Outer-Stream Heaters. The gas-stream cooling method became extremely popular after the development of the concentric warm outer stream. However, if the inner and outer gas streams are not carefully balanced, the resulting turbulence tends to draw moisture-laden air into the cold air stream, resulting in deposits of ice on the sample. This problem has been largely solved by using a heater to warm the outer portion of the cold gas stream. Thus, it is no longer necessary to balance two independent gas streams (see Section 3.2.2).

Continuous-Flow System. One of the major limitations on the design of conduction-cooling cryostats has been the restrictions imposed by the presence of a liquid-filled dewar attached to the X-ray apparatus. The availability of efficient flexible transfer lines has resulted in the dewar being located some distance away from the X-ray apparatus. The cryogenic fluid circulates through the cooling chamber, which is relatively light and free of restrictions on its orientation. Details are found in Section 4.2 and Figures 4–3, 4–6, and 4–10.

Mechanical Refrigeration. Closed-cycle refrigerators that reach 1.4 K are available. These devices, as well as mechanically refrigerated dewars and probes, can be used to replace consumable cryogens. They are exceptionally stable, can be used simply by plugging them into an electric outlet, and are in the long run economically competitive with consumable cryogens (see Sections 3.4.1.1 and 4.5, Figure 3–18).

X-Ray Apparatus. Certain modifications to the X-ray diffraction apparatus simplify the design requirements for the low-temperature system. For example, in the θ/θ diffractometer (e.g., Gerard, 1972) the sample is stationary while both the X-ray tube and X-ray detector move symmetrically. All accessories mounted on this powder diffractometer also remain stationary during the course of the investigation. Other examples include the oscillation–rotation cameras employed in macromolecular crystallography (see Section 3.5.2) and the use of an evacuated chamber surrounding an X-ray diffractometer (Figure 4–4).

New Perspectives. Future LTXRD apparatus will no doubt take advantage of new advances in the cryogenic field. Two items which will probably be used in future devices are heat pipes and fiber optics. The heat pipe is an efficient device for transferring heat or cold from one location to another while fiber optics can be used to observe samples inside closed sample chambers (see, e.g., Williams and Packard, 1974).

Over the years, the need for LTXRD apparatus has increased as the applications of LTXRD have been developed. As a result, there are now several excellent LTXRD devices available commercially. Many of these systems require no previous experience in LTXRD and can be used with little or no modifications to existing equipment. Thus, the development of improved, efficient LTXRD equipment and its ready availability will combine to increase the use of low-temperature methods in future X-ray diffraction investigations.

PART IV

Appendices, Bibliography, and Index

Appendices

Appendix 1. Temperature Measurement

The accuracy with which the temperature must be known varies with the needs of the investigation. In some cases, "temperature measurement" is simply a means of monitoring the system in order to know that the low-temperature device has operated continuously and that the sample has not undergone any major temperature fluctuations. In such cases, a temperature sensor with a rapid response time (e.g., a thermocouple) is connected to a continuous-recording device (e.g., a potentiometric recorder). In other cases, it may be necessary to know the temperature to within 0.01° or better (e.g., during thermal expansion measurements).

The theory and practical details of temperature measurement are discussed in many articles and texts. The present discussion is limited to information of greatest interest to workers in low-temperature X-ray diffraction. It is based on information found in the following references: Scott (1959), Gray and Finch (1971), Sparks et al. (1972), ASTM (1972), Sturtevant (1971), and Kinzie (1973). An extensive bibliography is found in Kinzie's book.

Temperature Scale. Modern temperature measurements are based on the temperature scale derived from the second law of thermodynamics. Inasmuch as this scale is not easy to realize in practice, the International Practical Temperature Scale of 1968 (IPTS-68) was chosen in such a way that it approximates the thermodynamic scale. This practical scale assigns values for the temperatures of eleven reproducible equilibrium states, called "fixed points," and standard instruments, calibrated at these fixed points, are used to interpolate between the fixed-point temperatures.

The text of IPTS-68 [published in English in *Metrologia* **5**(2), 35 (1969)] contains supplementary information on the construction and handling procedures recommended as "good practice" for the temperature-measuring instruments required by the scale, instructions on how to maintain the "fixed points," and the temperatures of 27 secondary reference points between 13.956 K and 3387°C.

A note in the IPTS-68 text reminds readers that the 1967 Conférence Générale des Poids et Mesures adopted the name "kelvin," symbol "K," for the unit of thermodynamic temperature, with "degrees Celsius," symbol "°C," as an alternative unit for a temperature interval. Note the omission of the word "degree" and the degree symbol from the kelvin scale. Table A1–1 lists the IPTS-68 values of interest in low-temperature studies.

Thermocouples. The determination of temperature with a thermocouple consists of a measurement of the emf generated in a loop formed by wires of two dissimilar metals joined at both ends. The measuring junction must be at the temperature of the object or environment whose temperature is to be determined. The reference junction is held at a fixed temperature, usually the ice point (0°C). For accurate measurements, the wires must be homogeneous, unstressed, and well insulated from each other except at the junctions.

Iron–constantan (ANSI symbol J) and copper–constantan (ANSI symbol T) thermocouples have been used for temperature measurements as low as 10 K. The latter is recommended since the homogeneity of the component wires can be maintained better than other base metal wires, and

Table A1–1.   IPTS–68   Reference Points

Assigned temperature		Equilibrium state
K	°C	
373.15	100	Boiling point of water
273.16	0.01	Triple point of water
90.188	−182.962	Boiling point of oxygen
54.361	−218.789	Triple point of oxygen
27.102	−246.048	Boiling point of neon
20.28	−252.87	Boiling point of equilibrium hydrogen
17.042	−256.108	Equilibrium between the liquid and vapor phases of equilibrium hydrogen at 33,330.6 N/m² pressure
13.81	−259.34	Triple point of equilibrium hydrogen

errors due to the inhomogeneity of wires are greatly reduced. It is most reliable down to 80 K. These thermocouples have been supplanted to some degree for temperatures below 80 K by gold–iron alloys used in conjunction with normal silver. The newer couples can be used to 4 K and have a greater thermoelectric power at low temperatures. In addition, the high thermal conductivity of copper is sometimes a serious disadvantage because of the possibility of heat leakage (Kinzie, 1973).

Constantan, an alloy of about 40% nickel and 60% copper, is the negative wire and is usually identified by the red insulation surrounding it. Manufacturers of thermocouple wire adjust the composition of constantan to maintain, for example, a copper–constantan output close to the tabulated values when used with a good grade of copper. Therefore, there are some variations in the actual alloy composition. Because of present trends in thermocouple usage, constantan that is acceptable for copper–constantan thermocouples is not acceptable for iron–constantan thermocouples, and *vice versa.*

New thermocouple reference tables have been issued by NBS, updating and expanding the NBS Circular 561. The revisions were made to reflect changes in materials and improvements in data fitting methods, but primarily to take account of changes in the International Practical Temperature Scale. The new data are based on the IPTS-68 for temperatures above 20 K and on the NBS helium-gas acoustical velocity scale from 4 to 20 K. NBS Monograph 125 (SD Cat. No. C13.44:125) is available from the U.S. Government Printing Office, Washington, D.C. 20402.

Extension wires and special connectors, having approximately the same thermoelectric properties as the thermocouple wires with which they are used, can be inserted between the measuring junction and the reference junction. The advantages, proper choice, and limitations of extension wires are discussed in an ASTM (1972) manual and by Kinzie (1973).

The author has found it useful to employ both a standard potentiometer and a potentiometric recorder with a slow chart speed. The former is used for accurate temperature measurements, while the latter is left running when the sample is unattended (e.g., overnight). Any drastic temperature changes, such as might be caused by a power failure of short duration, will be immediately apparent on the chart. This precaution is very important when single crystals grown from liquids are being studied.

It is also possible to monitor the temperature using a thermocouple connected to a pyrometer. Pyrometers, with a variety of temperature ranges (as low as $-200°C$), are available from several manufacturers of electric meters. These relatively inexpensive meters are calibrated in degrees and are useful for situations not requiring a high degree of accuracy.

The thermocouple wires should be located in such a manner that several inches of the wire are maintained in the same environment as the sample. This will reduce any errors due to the conduction of heat along the wires to the measuring junction. Wires of very small diameter (0.08 mm) are also recommended (Young, 1966).

If a large sample is being cooled (e.g., a powder diffractometer specimen), the thermocouple can be embedded in the sample. If a small sample is used (e.g., a single crystal), the thermocouple can be located in the cooling block (conduction cooling) or close to the sample. Care should be taken not to place the thermocouple in the X-ray beam.

A calibration curve correlating the sample temperature with the thermocouple reading should be prepared (see Sections 4.1.3 and 8.1.2).

Resistance Thermometer. The standard thermometer for measuring low temperatures is a compact resistor made of very pure platinum wire contained in a protecting tube. The wire hangs freely, in a strain-free condition, close to the wall of the protecting tube so that heat is easily transferred between the wire and the medium surrounding the tube. Four terminals (for "current" and "potential" leads) allow the resistance of the resistor between branch points to be measured independently of the resistance of the leads. Accuracy to within $0.01°$ can be attained down to 15 K.

Thermometers of this type are rather expensive and fragile. They are recommended for use with conduction-cooling apparatus in which the experimental conditions call for extremely accurate temperature measurements.

Metals and alloys other than platinum may also be used in resistance thermometers. Constantan has been used as a thermometer in the 2–20 K region. Thermistors (thermally sensitive resistors of semiconducting metal oxides) have sufficient reproducibility and stability to be useful for resistance thermometry. Thermistors are also available in a wide variety of forms, including small beads (0.2 mm diameter) with extremely small heat capacities and thermal lags.

Appendix 2. Cryogenic Liquids and Solids

Any substance with a suitably low temperature can be used to cool samples for LTXRD studies. The physical properties of several popularly used cryogens are presented in Table A2–1. Tables of viscosity, thermal conductivity, and thermodynamic properties of these and other cryogens can also be found in Scott (1959), Vance and Duke (1962), Zabetakis (1967), Croft (1970), Haselden (1971), and Rose-Innes (1973).

The choice of cryogen for a particular application will depend on the temperature desired, availability of cryogen, compatibility of cryogen and apparatus, safety of cryogen, and, on a local basis, cost of cryogen. Since many low-temperature systems require regular replenishment of cryogen, the cost factor should not be minimized.

Safety. A number of safety considerations that are common to all lique-fied gases should be emphasized. Safe use of cryogenic liquids requires consideration of two general areas: the properties of the fluids themselves and the low-temperature properties of the materials used in constructing the apparatus. Pressure buildup as a result of vaporization, physical contact with the cryogenic liquid, cold vapor, or cold metal surfaces, replacement of atmospheric oxygen with an asphyxiant, and structural weakening of stress-bearing parts of the apparatus must all be considered. In general, hazards tend to increase as the volume of the system increases.

Table A2–1. Physical Properties of Common Cryogens[a]

Property	N_2	He	H_2	CO_2
Molecular weight	28.013	4.003	2.016	44.01
Freezing point (K)	63.1	2[b]	13.8	—
Boiling point (K)	77.3	4.2	20.3	194.6[c]
Heat of fusion (cal/g)	6.1	—	13.9	47.5[d]
Heat of vaporization (cal/g)	47.7	5.5	106.5	36.2
Specific heat of gas (cal/g at 25° C)	0.249	1.24	3.41	—
Density of liquid at b.p. (g/liter)	808	125	71	1560[e]
Density of vapor at b.p. (g/liter)	4.6	14	1.3	2.8[f]
Density of gas at 300 K (g/liter)	1.14	0.16	0.082	1.98
Liquid-to-gas expansion ratio[g]	710	780	865	790

[a]Data obtained from Zabetakis (1967).
[b]At 25 atm.
[c]Sublimes.
[d]At the triple point.
[e]Of the solid.
[f]At the sublimation point.
[g]Ratio of density of liquid or (solid) at boiling (sublimation) point to density of gas at 300 K.

All cryogenic fluids exist at temperatures low enough to damage tissue. Such cooling does not take place immediately because the blood supply to the tissue acts as a heat source and the contact of a cryogen with warm tissue creates an insulating gas film. If the cryogen should make contact with the skin or clothing, immediately flood the area with water. The large heat capacity of water, its harmlessness, and its ready availability all combine to make water an important safety control factor.

The treatment of cryogenic burns is simply to restore tissue to normal body temperature as rapidly as possible, followed by protection of the injured tissue from further damage and infection. The use of safety glasses and clean, dry, loose-fitting gloves is recommended.

Do not permit the liquefied gas to become trapped in an unvented system (e.g., between two valves). Rather large pressures can be produced by completely vaporizing a cryogenic fluid in a closed chamber. The Clapeyron equation

$$\frac{\triangle P}{\triangle T} = \frac{\triangle H_v}{T(V_g - V_l)}$$

shows that the rate of pressure change with temperature is a function of the molar heat of vaporization, $\triangle H_v$, and the molar volumes of the gas and liquid, V_g and V_l, respectively.

From this equation it can be seen that a container initially filled with liquid helium at 1 atm pressure and 4.2 K would be subject to a pressure of 1000 atm at 270 K. Clearly all systems containing liquefied gases should be equipped with pressure-relief devices.

The properties of most materials change with a decrease in temperature. It is necessary to be familiar with the strength, thermal expansion, thermal conductivity, and heat capacity of the construction materials over the operating temperature range.

For further details, consult Zabetakis, *Safety with Cryogenic Fluids* (1967); and Rose-Innes, *Low Temperature Laboratory Techniques (Using Liquid Helium)* (1973).

Liquid Nitrogen (LN₂). Liquid nitrogen is a clear, waterlike fluid which is only slightly reactive. Both the gas and the liquid are odorless, colorless, nonflammable, nontoxic, and nonexpensive. The gas can act as a simple asphyxiant if it displaces the air needed to support life. At the present time, LN₂ is commercially available at a relatively low price. It appears to be the most suitable cryogen for investigations that must be carried out between 80 and 200 K.

Liquid Helium. Liquid helium is a clear fluid which is nonreactive. The gas, which is inert, nontoxic, colorless, and odorless, can act as a simple asphyxiant if it displaces the air needed to support life. In older systems, the liquid-helium dewar is generally contained within a LN_2 dewar in order to reduce the consumption of liquid helium. Modern dewars, utilizing super-insulation, are constructed without the surrounding LN_2 dewar.

Helium gas has a high specific heat, approximately five times that of nitrogen and exceeded only by that of hydrogen. Helium also conducts heat about five times faster than nitrogen and is an excellent heat-transfer medium. It also exhibits a negative Joule–Thomson effect (i.e., it heats rather than cools upon free expansion at temperatures above 40 K).

If a dewar of liquid helium is evacuated to reduce its temperature, the liquid will boil violently until 2.14 K is reached. At this temperature the liquid undergoes a second-order transformation to liquid helium II. This fluid has properties possessed by no other liquid. Two properties that are of greatest interest to the present discussion are (a) superfluidity, i.e., the viscosity drops to an exceptionally low value, allowing it to flow very rapidly through capillary tubes, and (b) creep, i.e., a thin liquid film forms on the walls of the container and migrates rapidly over the surface (overcoming any gravitational forces) until it comes in contact with a warmer area, where it evaporates and increases the amount of vapor the pump must handle.

For practical purposes, 4.2 K seems to be a practical lower temperature limit, unless the experimenter has considerable experience in dealing with liquid helium.

Liquid Hydrogen. Liquid hydrogen is a flammable, colorless, odorless liquid about 1/14 the density of water. The gas can act as an asphyxiant if it displaces the air needed to support life.

Several serious accidents have been reported over the years, and liquid hydrogen has been largely supplanted by liquid helium. However, in those areas where the use of liquid helium is economically prohibited, liquid hydrogen can be used, providing proper precautions are taken. The hazards of liquid hydrogen are primarily due to the hydrogen gas that forms extremely rapidly from any spills. Its low heat of vaporization, rapid rate of diffusion, and low ignition temperature combine to create dangerous situations in the absence of suitable safety precautions. It is suggested that further study of recommended safety procedures be undertaken prior to the use of liquid hydrogen. However, with proper ventilation and/or recycling of the gas, hydrogen systems are safe.

One final point to consider is that the hydrogen molecule can exist in two forms: *ortho*hydrogen, in which the nuclear spins are parallel, and *para*-hydrogen, in which they are antiparallel. Ordinary gas consists of about 75% *ortho*-H_2 and 25% *para*-H_2. When this is liquefied, the liquid undergoes a slow, natural exothermic conversion from the *ortho* to the *para* form. The heat liberated in this reaction exceeds the heat of vaporization, resulting in boiloff losses of about 1% per hour. A catalyst has been developed to yield better than 95% *para*-H_2 during liquefaction (equilibrium conditions are 99.79% *para*-H_2 and 0.21% *ortho*-H_2). The para form can be stored with nominal losses which are a function of the storage container design.

Solid Carbon Dioxide. For temperatures down to about $-75°C$, solid carbon dioxide, commonly known as dry ice, can be used. This cryogen is used to cool a liquid with a freezing point below $-85°C$ which is in contact with the heat exchanger. Commonly used liquids include acetone (fp $-95°C$), trichloroethylene (fp $-86°C$), and 1-propanol (fp $-127°C$).

The hazards due to the flammability of acetone and 1-propanol are minimized as long as the liquid is cooled. However, for extended investigations it is safest to use trichloroethylene.

The fumes from these substances are toxic. When the dewars containing these substances warm up to room temperature, they should be placed in a well-ventilated area (e.g., under a hood) or stored in sealed containers.

A further word of warning is in order. Although the temperatures of the liquefied gases usually employed in LTXRD are lower than that of dry ice, the potential danger to the skin due to minor spillage is far greater with a dry-ice slurry. This is so because the liquefied gas vaporizes rather quickly while the cold dry-ice slurry remains in contact with the skin for a far longer period. Suitable precautions should be taken with these mixtures to prevent painful blisters and "burns."

Appendix 3. Mechanical Refrigeration

In recent years, closed-cycle mechanical refrigerators capable of cooling a dewar or a flexible probe to temperatures as low as 130 K have become commercially available. Several of these devices have been designed with capacities large enough to cool a stream of flowing gas. They can be used in place of dry ice or liquid nitrogen. Although their initial cost is rather high, this is rapidly paid off in terms of savings of dry ice or liquid nitrogen. There are additional savings in terms of the manpower needed to replenish the cryogen, e.g., moving large dewars of LN_2 or crushing dry ice.

Other devices that can cool to as low as 4 K are also available, but these generally require special design of the X-ray apparatus, with particular care being taken to avoid vibrations of the sample. Details of several devices incorporating mechanical refrigeration are found in Sections 3.4.1 and 4.5.

The cryogenic refrigerators that are best suited for LTXRD studies are those that operate at a single, constant temperature. Control of the specimen temperature can be accomplished by other means.

The most common type of refrigerator consists of a totally enclosed gas-expansion unit with one to three cooling stages. The principles involved in this device are the same as those used in gas liquefaction apparatus. The fact that the gas is totally enclosed and constantly recycled allows the refrigeration unit to operate at a higher level of efficiency than the liquefier.

Three methods are commonly used to liquefy the gas within the refrigerator.

1. *Engine Expansion.* When a compressed gas is made to expand within an engine, work is generated and the gas cools down.
2. *Throttle Expansion.* When a gas expands adiabatically without performing any external work, it will generally experience a temperature change (Joule–Thomson effect). As the starting pressure falls to near atmospheric, the temperature drop becomes vanishingly small. At lower temperatures, large temperature changes can be achieved by throttle expansion.
3. *Vapor Compression.* This cycle is used in most domestic refrigerators. It relies on throttle expansion, but can be differentiated from the normal Joule–Thomson effect. It operates on the fact that the vapor pressure of a fluid varies with its temperature and therefore a boiling liquid can be made to absorb heat at one temperature and reject heat at a second, higher temperature by compressing and condensing the vapor thus generated.

In general, a cryogenic refrigerator incorporates the following general features: compressor, cryogenic fluid (liquid or gas), condenser (to store liquid), expansion device (as described above), and evaporator (exchange of heat between medium to be cooled and fluid at its coldest).

The working parts are mechanical. Difficulties with the apparatus may be due to mechanical problems with the compressor or to a leak of the cryogenic fluid. Devices with flexible probes can develop leaks if the flexible lines are not properly handled. Repairing the leak and recharging the refrigerator with fluid may require returning the device to the manufacturer.

Appendix 4. Joule–Thomson Expansion Cooling

If a gas expands on passing through a porous plug or throttle without doing any external work, it generally exhibits a lowering of its temperature. The cooling is attributed to the work done in overcoming the intermolecular attractions. Hydrogen and helium, however, unless precooled, exhibit an increase in temperature.

If the pressure, volume, and temperature of the gas are represented by P, V, and T, respectively, with subscript 1 representing conditions prior to expansion and the subscript 2 representing conditions after expansion, then, for one mole of gas, $P_2V_2 - P_1V_1$ is the work done by the gas. The process is assumed to be adiabatic, and so the work done must equal the decrease in internal energy (E):

$$P_2V_2 - P_1V_1 = E_1 - E_2$$

and

$$H = E_1 + P_1V_1 = E_2 + P_2V_2$$

This equation shows that the enthalpy (H) remains constant. The laws of thermodynamics show that

$$dH = \left(\frac{\partial H}{\partial P}\right)_T dP + \left(\frac{\partial H}{\partial T}\right)_P dT$$

from which it can be shown that

$$\left(\frac{\partial T}{\partial P}\right)_H = \frac{T(\partial V/\partial T)_P - V}{C_P}$$

The term $(\partial T/\partial P)_H$ is referred to as the Joule–Thomson coefficient. An ideal gas should show no Joule–Thomson effect [since $(\partial V/\partial T)_P = R/P = V/T$]. For real gases, the sign of the Joule–Thomson coefficient will depend on the relative values of $T(\partial V/\partial T)_P$ and V.

From the above equation and the van der Waals equation it can be shown that

$$\frac{\Delta T}{\Delta P} = \frac{2a/RT - b}{C_P}$$

If $2a/RT < b$, there is a decrease in temperature as the gas streams through a porous plug from a higher to a lower pressure. If $2a/RT = b$, the $\triangle T = 0$ and there is no Joule–Thomson effect. When $2a/RT > b$, there is an increase in temperature when the gas passes through a porous plug. The Joule–Thomson inversion temperature (T_I), where $\triangle T = 0$, occurs when $T = 2a/Rb$.

For an ideal gas undergoing an adiabatic change,

$$P_1V_1{}^\gamma = P_2V_2{}^\gamma$$

where $\gamma = C_P/C_V$. From this it is found that

$$(T_1/T_2)^\gamma = (P_1/P_2)^{\gamma-1}$$

The decrease in temperature in an adiabatic expansion is thus greater, the larger the difference between initial and final pressures.

These principles are utilized in many gas liquefiers and cryogenic refrigerators, as discussed in Appendix 3. In addition to these closed systems, there are several devices used in LTXRD which employ the Joule–Thomson effect in an open system (see Section 4.7).

The devices are not often used with helium for several reasons, including the need for precooling the gas (so as to reach the inversion temperature) and the high cost of helium gas (LTXRD apparatus does not usually provide for the recovery of helium gas). However, the cost and inversion temperature of nitrogen made it an ideal gas for use with this apparatus.

When designing apparatus using Joule–Thomson expansion, the following two limitations should be considered:

1. It has been shown above that the temperature differential depends on the pressure drop. Accordingly, in order to attain a temperature drop sufficient to cause liquefaction of the gas, it is necessary to maintain the pressure in the nitrogen tank above 100 kg/cm². Thus the gas cylinders will have to be changed before they are empty. Furthermore, variations in the temperature may occur as the pressure drops. Although this difficulty may be overcome by passing the gas through a compressor prior to its entering the

Joule–Thomson expansion device, the cost involved in this arrangement is rather high.

2. Since the device is attached to a high-pressure source of gas by a flexible line, the weight of this gas transfer line must be compensated for so that the instrument does not become misaligned.

Appendix 5. Thermoelectric Devices

If two conductors are joined together and kept at a constant temperature while a current (J) passes through the junction, heat is generated or absorbed at the junction. This phenomenon is called the Peltier effect. The Peltier coefficient, Π_{12}, is defined so that the heat (q) emitted or absorbed per second at the junction is $\Pi_{12}J$.

If two conductors are joined together at both ends and the two junctions are kept at different temperatures, an electromotive force is set up which is proportional to the temperature difference. This is called the Seebeck effect and is the basis of the thermocouple. The electromotive force per degree Celsius is called the thermoelectric power (Q_{12}).

It can be shown that $\Pi_{12} = Q_{12}T$. The thermoelectric power is easier to determine than the Peltier coefficient and so is often used in design.

A thermoelectric cooling element can be constructed by joining the ends of two semiconductors side by side to a copper plate, which is to act as the cooling plate. The remote ends of the semiconductors are separately placed in thermal contact with a heat sink and are connected to a dc source. If a current, J, flows in the correct direction, the copper plate will start to cool as a quantity of heat is extracted from it.

The effective cooling is reduced by two factors: thermal conduction through the semiconductors to the heat sink and ohmic heating within them.

If the copper plate is thermally isolated from its surroundings, its temperature will drop until the potential cooling effect is balanced by thermal conduction and ohmic heating.

A good cooling element obviously requires a high electrical conductivity and a low thermal conductivity. Various alloys of bismuth make the best material for refrigeration purposes. By properly cascading several cooling elements, large temperature differentials, with respect to the heat sink, can be achieved.

The chief advantage of using such devices is that the only external power source needed to achieve cooling is electricity. However, the cooling capacity of such devices is limited. In order to obtain maximum efficiency, the cooling plate should be isolated from its surroundings, e.g., by placing it in an evacuated chamber. This results in difficulties of design for LTXRD applications. A further problem is that each cooling stage used increases the height of the cooling device.

Several systems designed for use with X-ray apparatus have been built and are described in Section 4.6. Because of the limitations discussed above, the lowest operating temperature for these devices seems to be about $-50°C$, although thermoelectric devices that will reach 77 K have been built successfully for other applications.

Appendix 6. Dewars

Cryogenic fluids are usually stored in special insulated containers such as that devised by Sir James Dewar in 1898. Most small dewars are made of glass or aluminum, while the larger units are made of stainless steel or copper. These vessels consist of two containers, one within the other, separated by an evacuated space and with very few points of contact. Dewars up to 250 liters in size are normally used for the storage and transport of liquid gases, while smaller ones (0.25–40 liters) are used for experimental purposes.

Storage Dewars. Dewars up to 50 liters in size, used for the transportation of liquefied gases, generally consist of a spherical inner container within an outer casing and have a relatively high evaporation-loss rate. Furthermore, material purchased in these small quantities are proportionately more expensive. Therefore, it is more economical to purchase LN_2 in 110- to 175-liter cylindrical dewars, which, when pressurized, can be used to deliver a steady stream of LN_2. Pressurization is normally attained by letting the dewar stand 1–2 days. However, if a freshly filled container has to be used, it may be necessary to attach it to a cylinder of *dry,* pressurized nitrogen gas. Some containers can be obtained with a built-in vaporizer that allows one to obtain a steady flow of nitrogen gas. For example, the Linde PGS-45 unit has a 175-liter capacity and can deliver both gas and liquid. It is supplied with a pressure-building coil and can deliver gas at constant pressure as long as there is LN_2 in the tank. At a flow rate of 12 ft^3/h, it will deliver dry nitrogen gas for twelve days.

At this low flow rate it can also be refilled with LN_2 without significant change in gas pressure.

Laboratory Dewars. Dewars used for laboratory purposes vary considerably in design, construction material, and volume. Vessels as small as 0.25 liter may be used for thermocouple reference junction baths, while vessels as large as 40 liters may be used as evaporators for gas-stream cooling systems. Materials of construction include glass, stainless steel, plastic, and aluminum, with the primary insulation consisting of an evacuated space and superinsulation. The dewar openings may be either wide-mouthed or narrow.

Glass dewars made for laboratory use are usually quite strong; those made for household use should not ordinarily be used for low-temperature service, as they break easily when mishandled. In any case, most workers wrap the outer exposed glass surface with tape to furnish a surface that can be handled easily and to prevent fragmentation of the glass should a break occur.

Some of the LTXRD apparatus described in Part II utilizes cryostat dewars with windows able to transmit X-rays. Several adhesives which have been used for sealing thin, fragile X-ray-transparent windows to the metal parts of the apparatus are discussed in Appendix 11.

Safety. Special precautions must be observed when handling liquefied gases. If the liquid absorbs heat and vaporizes in a closed container, pressures great enough to rupture the vessel may be reached. The explosive force of this rupture can cause severe damage to people and apparatus alike (see Appendix 2). The large storage dewars are normally designed with safety valves and rupture discs. However, the smaller long-necked dewars can become a potential hazard if the neck should become clogged. The covers of these dewars are purposely loose-fitting. A frequent cause of potentially dangerous blockages of neck tubes is the formation of solid plugs of substances which are fluids at room temperature. A common cause of a plug in the neck tube of a liquid-helium vessel is the condensed air which may drip down when a transfer line is removed. This may solidify on parts of the neck which are below the melting point. Subsequent accumulations may result in the formation of a solid plug. In addition, the diffusion of atmospheric air into a vessel left open to the atmosphere can occur. To remove the plug, a stream of room-temperature helium or nitrogen gas should be directed at the plug. A simple type of relief valve is the Bunsen valve: a longitudinal slit in a piece of rubber tubing sealed at one end.

Appendix 7. Transfer Lines

Specially designed tubes are used to transfer cryogenic materials from one location to another. Two applications are of immediate interest: the transfer of liquid nitrogen or helium from a storage dewar to the cooling apparatus and the transfer of cold gas from the apparatus to the sample. The devices discussed in this book cover a wide range of temperatures. The insulating requirements for a given transfer line depend on the temperature of the material being transferred. The basic transfer line consists of an insulated tube used to transport cold fluids between two points.

Some of the problems associated with transfer lines include proper insulation, flexibility, structural stability, and ease of handling. Both stock and custom-made transfer lines are available commercially (see Appendix 12). For ease of handling and greatest adaptability, at least a short length of line should be flexible.

Descriptions of some easily made, reasonably efficient transfer lines have been presented in Section 3.3. However, the most efficient transfer lines are constructed from flexible or rigid stainless steel tubing or glass tubing and are evacuated before use. The glass transfer lines should be silvered before evacuation. Detailed instructions for silvering glass are found in several older experimental physics books or in the "Arts and Recipes" section of some older editions of the *Handbook of Chemistry and Physics* (e.g., pages 2989 and 2996–2998 of the 36th edition). The surface of the glass tubing that is to be silvered should be cleaned properly prior to the fabrication of the dewar tube or nozzle and again afterward to prevent blistering of the deposited silver.

Two solutions are needed. Solution A is prepared by dissolving 5 g of silver nitrate in 300 ml of distilled water and adding dilute aqueous ammonia until the precipitate that forms is nearly, but not entirely, redissolved. The solution is then filtered and water is added to a total volume of 500 ml. Solution B is made by dissolving 1 g of silver nitrate in a small quantity of water and pouring it into about 500 ml of boiling water. 0.83 g of Rochelle salt is dissolved in a small quantity of water and added to the boiling solution. Continue the boiling for half an hour, until the gray precipitate collects as a powder in the bottom of the flask. Filter while hot and add water to a total volume of 500 ml. These solutions may be stored in the dark for a month or two.

For silvering, equal volumes of the two solutions are mixed. Best results are obtained if the two solutions are mixed together inside the dewar tube and thoroughly shaken.

Appendix 8. Insulation

A good insulator minimizes heat transfer by both conduction and radiation. The use of a hard vacuum will nearly eliminate conduction but will accentuate the effect of radiation. On the other hand, reducing radiation is not effective if at the same time conduction is allowed to increase excessively.

Solid-Phase Conduction. The presence of breaks in a solid phase is a source of high resistance to heat flow. The main factors affecting heat flow across the breaks are the effective contact area between particles, the composition and pressure of gas between the particles, and the temperature.

At atmospheric pressure the increase in thermal conductivity with bulk density is linear; under vacuum conditions it decreases with increasing bulk density. Thus, when selecting an insulant for use at atmospheric pressures, a material of lighter weight should be chosen, but for vacuum conditions denser materials may be advantageous.

Gas-Phase Conduction. This can be divided into true conduction and convective heat transfer. The former depends upon momentum transfer by collisions between the gas molecules and the container surfaces and between adjacent gas molecules. The latter allows heat transfer by successive collision with other molecules of gas before reaching the cold wall. The factor that determines which mechanism operates is the mean free path of the molecules.

Radiation. Radiation is an electromagnetic phenomenon. Its contribution to heat flow becomes significant at cryogenic temperatures whenever vacuum is used. High vacuum alone without a good reflecting surface facing the vacuum space is quite inadequate. The use of several radiation barriers reduces radiation effects even more.

There are several categories of insulation, each of which is available in several forms.

Powder Insulation. Certain minerals, consisting mainly of silica, are crushed, ground, and then heated. The result is a white powder, each particle of which consists of a mass of cavities with a pore size of approximately 10^{-6} cm. When the pressure surrounding the powders is of the order of 10^{-3} mm Hg, the mean free path of the gas molecules becomes large compared to the pore size, and heat flow occurs mainly by conduction through the particles and by radiation.

Solid Foam. Materials such as expanded polystyrene, expanded polyurethane, and glass foam have very low thermal conductivities. Included in this category are the do-it-yourself polyurethane foams that can be utilized in the construction of low-temperature equipment.

Superinsulation. A number of reflecting screens thermally insulated from each other are interposed between a warm outer surface and the cold inner surface being insulated. The original form of multiple-layer insulation consisted of alternating layers of aluminum foil and glass-fiber paper. Current techniques utilize polyester film (e.g., Melinex or Mylar) coated on one or both sides with vacuum-deposited aluminum. This is crinkled or dimpled and wrapped around the dewar. This insulation is extremely efficient and lightweight.

Further details have been given by Haselden (1971) (in Chapter 4 of his book) and by Tien and Cunnington (1972).

Kaufman and Bullitt [*Rev. Sci. Instrum.* **45,** 127 (1974)] describe a method for preparing molds and insulated dewars using a commercially available resin and catalyst called Eccofoam FPA manufactured by Emerson and Cuming, Inc., Dielectric Materials Division, Canton, Massachusetts, U.S.A.

Appendix 9. Frost Prevention

The need to keep the cold sample free from condensed vapors cannot be overemphasized. The presence of frost (i.e., water, carbon dioxide, or other materials) presents several sources of potential difficulty: (1) the crystal may be misaligned due to the weight of the condensate; (2) in the case of liquids contained in capillary tubes, the presence of a film over the tube may prevent observation of the growing crystal; (3) during intensity data collection, uneven absorption of the X-ray beam may result in incorrect relative intensities; and (4) diffraction from the condensate may be incorrectly interpreted. In addition, frost forming on movable parts of the instrument may cause rusting or immobility of the apparatus.

Several methods have been developed for preventing frost formation. Generally speaking, each of the systems discussed in Part II incorporates at least one of these methods. They may be categorized as follows.

Evacuated Chamber. If the sample is maintained in a vacuum and the chamber walls are properly insulated, no frost can form on any critical part of the apparatus. This method is most often used when liquid-helium temperatures are approached. If the sample has a low vapor pressure, it is necessary to seal it in a sample holder. The common use of small beryllium or Mylar windows often makes it difficult to see the sample during the alignment process.

Dry-Box Arrangement. Several systems have been set up with the apparatus enclosed in a specially designed dry box. Dry nitrogen or dry air (depending on the operating temperature) is used to flush out the chamber, and a steady stream is maintained. This system is cheaper to build and easier to design than the vacuum chamber. However, it also requires an essentially permanent setup and presents some difficulties in reaching the crystal for adjustment. Excellent results have been obtained with it (Reed and Lipscomb, 1953; Drenth and Wiebenga, 1955; Viswamitra, 1962; Allen et al., 1963; Prakash, 1966; Abowitz and Ladell, 1968).

E. N. Maslen (as reported by Coppens, 1972) recycled the dry-box atmosphere through a drying agent and attained a relative humidity of $<1\%$. Prakash (1966) kept P_2O_5 desiccant inside the dry box.

Plastic Tent. If a framework (e.g., aluminum rods and connectors) is set up around the apparatus and a plastic film draped over it, a simpler version of the dry box can be cheaply and quickly set up. The edges of the plastic are taped down and the tent is constantly flushed with dry gas. The advantage of this arrangement is that it does not require the expenditure of large sums, is adaptable to the physical requirements of particular experimental conditions, and can be stored and modified as needed. Large glove bags, with entrance ports and built-in gloves, can also be used (Haas and Rossman, 1970).

Outer Stream. The use of a dry stream of room-temperature gas surrounding the cold stream of gas has been described in Chapter 3. This is probably the most versatile cooling device. It can be used without any other frost-prevention devices. If the temperature is above $-100°C$ or if the time during which the sample is cooled is not too long, this arrangement can be used quite successfully. For extended use at lower temperatures, it is generally more convenient to use an auxiliary device such as the plastic tent. Crystals have been successfully maintained at 120 K for periods up to two months using a combination of the outer stream and plastic tent.

Local conditions are extremely important in determining the efficiency of the outer stream. The relative humidity and the presence of drafts or air currents near the sample must be controlled.

In air-conditioned buildings, air currents are determined by the location of vents. Occasionally, it is necessary to simply shield the apparatus from the air current; at other times a tent must be employed. The degree of moisture in the air will vary from day to day and so will the effectiveness of the outer stream. The use of a relative-humidity indicator or dew-point measuring device in the vicinity of the apparatus is helpful in controlling the appearance of frost on the sample.

Heating Mantles and Tapes. The parts of the apparatus that are cooled by the cold stream can be kept at room temperature by wrapping them with insulated heating tapes or by insulating them with glass wool shields. The use of these devices has been discussed in Part II.

Appendix 10. Miscellaneous Gas-Flow Accessories

Flowmeters. The use of flowmeters in the various gas lines of a gas-stream cooling apparatus facilitates quantitative reproduction of the experimental conditions. A flowmeter indicates the rate of flow of the fluid. It does not control the rate of flow unless equipped with a control valve or flow controller.

Two types of flowmeters are available. One kind, *mass flowmeters,* consist of an electrically heated tube and an arrangement of thermocouples to measure the differential cooling caused by a gas passing through the tube. Thermoelectric elements generate a dc voltage proportional to the rate of mass flow gas through the tube. This design depends only on the mass flow and the heat capacity of the gas and is, therefore, almost insensitive to pressure and temperature change.

The other type of flowmeters are *rotameters,* which operate on the variable-area principle. They consist of a tapered tube oriented with its axis vertical and the smaller end at the bottom. The fluid is conveyed in an upward direction, and a restriction called the float is free to move vertically within the tube. At any flow rate, within the range of the meter, the float will rise vertically, increasing the flow area, until it is stabilized at a point where the upward hydraulic forces are just balanced by the net weight of the buoyed float. Flow rate, therefore, is indicated by the position of the float with reference to a scale on the tube. Spherical floats of different densities are utilized to provide versatility in the flow-measurement range of any tube. The position of the spherical float should always be read at a point corresponding to the center of the float.

Rotameters, unlike mass flowmeters, are affected by pressure and temperature variations. For this reason, it is always advisable to control the pressure of a gas stream containing a rotameter with a good pressure regulator.

In a rotameter, the flow-control valve constitutes the separation between the inlet and outlet pressures. When the meter is equipped with a valve at the inlet, the readings on the tube are correct providing the outlet pressure is close to atmospheric. When there is a valve at the outlet, the readings on the tube are correct if the gas pressure is equal to the pressure for which the tube was calibrated.

Heat Exchangers. A study of systems in which heat is transferred from an entering warm gas stream is presented by White (1968) in Chapter III of his book. Theoretical considerations and methods of constructing heat exchangers are discussed.

Gas Dryers. Several models of commercial gas dryers are available. One that is particularly useful is the Self-Regenerating Dry-Pak, Model 4103–3, manufactured by the Wilkerson Corporation (Appendix 12). This is a relatively inexpensive model in which the air is dried in one drying tower while a small portion of dry air is used to purge a second drying tower. After approximately one minute, the situation is reversed automatically. The unit can operate months on end without interruption provided the incoming compressed air has passed through suitable water and oil (coarse) filters.

A drying tube containing a drying agent which turns color as it absorbs moisture is a useful check on the operation of the gas dryer. It should be placed in the line between the dryer and the heat exchanger. In a properly operating system, no color change will be observed even after several months of use.

Appendix 11. Low-Temperature Adhesives and Greases

The following substances have been recommended for use as low-temperature adhesives or for improving thermal contact at low temperatures.

Mylar. The aluminum backing on aluminized Mylar can be removed prior to applying the epoxy by using 0.5 M NaOH solution, washing in distilled water, and roughening with No. 600 emery cloth [*Rev. Sci. Instrum.* **41,** 1107–1108 (1970)]. Hysol adhesive R8–2038 and Hysol hardener H8L–3466   (Hysol Division, The Dexter Corporation, Olean, New York) and Narmco base resin 3135 with Narmco curing agent 7111 (Narmco Materials Division of Wittaker Corporation, 600 Victoria Street, Costa Mesa, California) are two epoxies that are recommended for holding Mylar windows in place at low temperatures under vacuum conditions (*ibid.;* Peterson and Simmons, 1965). Boiko et al. (1972b) suggest an adhesive with a composition of "20 parts of dibutylphthalate and 8 parts polymethylpolyamine to 100 parts ED–6 epoxy resin." The film was pressed onto the components with a force of approximately 5 kg/cm^2, and the adhesive was polymerized at 180°C for 2 hours.

Other suggested adhesives are as follows: *Araldite* (epoxy resin MY753, hardener HY951) is good for cementing windows and holds a vacuum to liquid-helium temperatures (Thomas, 1972). *G. E. Adhesive No. 7031* remains adhesive at low temperatures and is also a satisfactory thermal conductor (Abrahams, 1960; Morosin, 1966). It has also been used as a binder for holding a powder sample in a cryostat (Etourneau et al., 1975). *"Goo,"* a rubber-base adhesive which is removable with acetone and is usable to liquid-helium temperatures, is sold by William K. Walthers, Inc., 4050 North 34, Milwaukee, Wisconsin [*Rev. Sci. Instrum.* **41,** 1189 1970)].

Teflon. Standard rubber cement is recommended (Reichert, 1972), although its ability to form a vacuum seal has not been tested. "Goo" might also be satisfactory.

Abell and King (1970) repeat the warning of Barrett et al. (1967) that the use of Mylar film limits the lower range of Bragg angles, because this material gives a broad diffraction profile at $d \approx 3.1$ Å. The latter suggest using a Teflon film to measure angles in this region.

Thermal Contact. Good thermal contact must be maintained between the sample and the cooling block when conduction cooling is employed. Smith and Leider (1968) used Apiezon 'N' grease mixed with fine copper powder. A similar mixture called "Cry-Con" conductive grease is available from Air Products and Chemicals, Inc. (Advanced Products Department, Allentown, Pennsylvania 18105). Also, see M. M. Kreitman, "Low-temperature thermal conductivity of several greases," *Rev. Sci. Instrum.,* **40,** 1562 (1969). For a discussion of the strain effects caused by using a sample binder, see Section 4.1.3.

Appendix 12. List of Manufacturers of Low-Temperature Equipment and Accessories

The following is a list of manufacturers of equipment that is useful in conducting LTXRD investigations. This list is not meant to be exhaustive; rather it is meant to be representative of the type of apparatus that is available. The last part of the list contains the names of companies manufacturing apparatus that is compatible with the X-ray diffraction instruments currently in use. Other names are found in Appendices 8 and 11.

Air Dryers

Wilkerson Corporation
Englewood, Colorado 80110, USA

Capillary Tubes

Karl Hackl
211 Sumatra Road
London NW 6
  United Kingdom
  (see Nelmes, 1970)

Paul Raebiger
Flankenschanze 30/36
100 Berlin 20, West Germany

Unimex-Caine Corporation
1829 North Arbogast Avenue
Griffith, Indiana 46319, USA

Cryogenic Components (Valves, bayonet connectors, phase separators, etc.)

Linde Division
Union Carbide Corp.
270 Park Avenue
New York, New York 10017, USA

Pennwalt Corp.
P.O. Box 2138
Columbus, Ohio 43216, USA

Dewars (Glass, stainless steel)

Linde Division of Union Carbide
  (see Cryogenic Components)

Minnesota Valley Engineering
New Prague, Minnesota 56071, USA

Flowmeters and Rotameters

Aalborg Instruments and
  Controls, Inc.
57 Regina Road
Monsey, New York 10952, USA

Datametrics
340 Fordham Road
Wilmington, Massachusetts 01887, USA

Insulation

Armstrong Armaflex
Armstrong Cork Co.
Lancaster, Pennsylvania, USA

Norton Company
37 East Street
Winchester, Massachusetts 01890,
USA
(Superinsulation)

Low-Temperature Pyrometers and Meters

Omega Engineering, Inc.
Box 4047, Springdale Station
Stamford, Connecticut 06907, USA

Simpson Electric Co.
853 Dundee Ave.
Elgin, Illinois 60120, USA

Mechanical Refrigerators (Probes, refrigerated dewars, closed-cycle refrigerators)

FTS Systems, Inc.
P.O. Box 158
Stone Ridge, New York 12484, USA

Grant Instruments (Cambridge) Ltd.
Barrington
Cambridge CB2 5QZ, United Kingdon

Malaker Laboratories, Inc.
High Bridge, New Jersey 08829, USA

Neslab Instruments, Inc.
871 Islington Street
Portsmouth, New Hampshire 03801,
USA

Polycold System, Inc.
178 Paul Drive
San Rafael, California 94903, USA

Thermocouples and Accessories

Leeds and Northrup
4907 Stenton Ave.
Philadelphia, Pennsylvania 19144,
USA

Omega Engineering, Inc.
Box 4047, Springdale Station
Stamford, Connecticut 06907, USA

Thermoelectric Coolers

Borg-Warner Thermoelectrics
Des Plaines, Illinois 60018, USA

Cambion Thermoelectrics
Cambridge Thermionic Corp.
445 Concord Ave.
Cambridge, Massachusetts 02138,
USA

Jepson Thermoelectric, Inc.
6001 North Keystone
Chicago, Illinois, USA

X-Ray, Low-Temperature Instruments

a. Low-Temperature Apparatus Available from Manufacturers of X-Ray
   Equipment

Anton Paar K.G.

Kärntnerstrasse 322
A-8054 Graz, Postfach 17, Austria
(Conduction, LN_2 powder
    diffractometer)

Enraf-Nonius

Röntgenweg 1, P.O. Box 483
Delft, The Netherlands
    (Gas stream, LN_2, Weissenberg
    goniometer, precession camera,
    Guinier–Lenné camera,
    Guinier–Simon camera,
    single-crystal diffractometer)

Philips X-Ray Instruments

Eindhoven, The Netherlands
    (Gas stream, LN_2, single-crystal
    diffractometer)

Rich, Seifert and Co.

Bogenstrasse 41
2070 Ahrensburg, West Germany
    (Gas stream, LN_2, Weissenberg
    goniometer, precession camera,
    single-crystal diffractometer)

Rigaku Denki Co., Ltd.

9-8 2-chome
Sotokanda, Chiyoda-Ku
Tokyo, Japan
    (Conduction, LN_2, powder
    diffractometer)

Stoe Instrumente

Hilperstrasse 10
61 Darmstadt, Postfach 4162
West Germany
    (Gas stream, LN_2, Weissenberg
    goniometer, precession camera,
    reciprocal-lattice explorer)

Syntex Analytical Instruments

10040 Bubb Road
Cupertino, California 95014, USA
    (Gas stream, LN_2, single-crystal
    diffractometer)

Techsnabexport

Moscow, USSR
    (Gas stream, LN_2, single-crystal
    diffractometer)

b. Low-Temperature Apparatus Available from Manufacturers of Cryogenic Equipment: (Most of these manufacturers carry a full line of
   cryogenic apparatus and will custom-build apparatus.)

Air Products and Chemicals, Inc.

Advanced Products Department
P.O. Box 538
Allentown, Pennsylvania 18105, USA
    (Joule–Thomson expansion,
    LN_2 or LHe, continuous-flow
    cryostat, powder or single-crystal
    systems)

Cincinnati Sub-Zero Products, Inc.

2612 Gilbert Avenue
Cincinnati, Ohio 45206, USA
    (Refrigerated cold box, to $-50°$ C)

Cryogenic Associates

1718 North Luett Avenue
Indianapolis, Indiana 46222, USA
    (Conduction, LHe, diffractometers;
    transfer lines)

M.E.R.I.C.

91290 Arpajan, France
    (Conduction, powder, topography)

Oxford Instrument Co., Ltd.

Osney Mead
Oxford, OX2 OXD, United Kingdom
    (Conduction, LHe, powder or
    single-crystal systems,
    continuous-flow cryostat)

Appendix 13. Recommended Procedure for Describing Low-Temperature Apparatus in Publications

During the course of preparing this book, it was noted that many LTXRD devices are not fully described in the literature. Several pertinent pieces of information should be reported in order to assist the reader in assessing the instrument. These include the six basic points utilized in the code number (see Bibliography):

1. Type of sample that can be studied
2. Type of cooling system used
3. Method of preventing frost formation
4. Minimum temperature attainable
5. Type of X-ray instrument used
6. Any special characteristics

A brief description of the system and a diagram are helpful, particularly if unique features are incorporated in the apparatus. The manufacturer, his address, and the model number of any special components should be given. If several components are used, they can be listed in an appendix to the paper.

Other information that is useful to the reader contemplating the construction of a particular device includes (where applicable):

1. Rate of cryogen consumption and frequency of refill
2. Accuracy of temperature measurement
3. Long- and short-range temperature stability
4. Temperature gradient in the vicinity of the crystal

Most important, in order to facilitate the retrieval of this information, the *abstract* should mention that a LTXRD device is described in the article.

Bibliography

B.1. Introduction

This Bibliography serves simultaneously as a list of references cited in the book, as an author index, and as a general bibliography on LTXRD apparatus and techniques. The citations are meant to be representative of earlier papers in LTXRD (prior to 1945) and comprehensive in the coverage of more recent low-temperature articles. Papers dealing solely with results of LTXRD studies are not included, except as examples of LTXRD applications.

Secondary sources have not been relied upon. The original (or a copy of the original) of every citation in this Bibliography has been checked by the author. It is hoped that no errors have been introduced during the key-punching of the data.

A unique feature of this Bibliography is the assignment of a six-digit code number, characterizing the apparatus involved, to every reference. In addition, LTXRD techniques are assigned separate code numbers. The key to the code numbers is found in Table B–1. This table is also repeated on a foldout page at the back of the book which can be kept in view simultaneously with the reference list.

Thus, for example, the code number 111312 following a reference means that the LTXRD device described in that article can be used with any type of sample, cools the sample with a cold gas stream, prevents ice condensation by means of a concentric warm outer gas stream, can reach a minimum temperature between 78 and 200 K, and was designed for use with a Debye–Scherrer camera with a vertical sample axis.

Immediately following the list of citations (Section B.2), all the six-

Table B–1. Code Numbers Used in Classifying Low-Temperature X-Ray
Diffraction Apparatus and Techniques

a. Type of Apparatus

First digit: Type of sample that can be Studied

1. Any type
2. Single crystal
3. Powder
4. Metal
5. Protein
6. Neutron diffraction

Second digit: Type of cooling used

1. Cold gas stream
2. Conduction (cryogenic fluid as coolant)
3. Conduction (thermoelectric cooling)
4. Conduction (mechanical refrigeration)
5. Joule–Thomson expansion
6. Immersion of sample
7. Immersion of camera
8. Use of cold room

Third digit: Method of frost prevention

1. Dry gas stream
2. Dry chamber
3. Evacuated chamber
4. Not given
5. None

Fourth digit: Minimum temperature attainable (Kelvin)

1. Less than 20
2. 20–78
3. 78–200
4. 200–260
5. Greater than 260
6. Not available

Fifth digit: Type of X-ray instrument mentioned

0. Any type
1. Debye–Scherrer camera (includes back-reflection)
2. Flat-cassette and Laue cameras
3. Guinier camera
4. Oscillation–rotation camera

5. Weissenberg goniometer
6. Precession camera
7. Diffractometer
8. Small-angle instrument
9. Topographic studies

Sixth digit: Special characteristics

0. None
1. Horizontal
2. Vertical
3. Back-reflection
4. High-temperature capability also
5. High-pressure capability also
6. Weissenberg goniometer accessories
7. Cold-working at low temperatures

b. Techniques and Applications

1. Sample preparation if solid at room temperature
2. Sample preparation if liquid at room temperature
3. Sample preparation if gas at room temperature
4. Crystal growth if liquid or gas at room temperature
5. Techniques for randomly aligning powder sample
6. Reactive or radioactive sample
7. Applications
11. Purpose of low-temperature method
12. Choice of temperature
13. Choice of cooling method
14. Temperature calibration
15. Characteristics of X-ray film at low temperatures
16. Alignment of crystal
17. Special programming of automatic diffractometer
18. Correction for absorption due to sample holder
19. Refinement of disordered models
20. Review article
21. General low-temperature techniques

digit apparatus code numbers are listed in order of increasing magnitude together with the accession number of the reference(s) in which that device is described (Section B.3). This list is followed by other listings in which the references are sorted on the basis of the *individual* digits of the code number. Thus, if you wish to know how many low-temperature devices have been built using Joule–Thomson expansion as the cooling mechanism, look under category 5 in the list where the references are sorted on the basis of the *second* digit of the apparatus code number. The last part of the Bibliography (Section B.4) is a sorting on the basis of the technique code number.

The italic numbers following each citation in Section B.2 refer to the pages in the text on which the reference is mentioned.

It is thus possible to "design" an instrument by constructing a suitable six-digit code number and to check the list to see if such an apparatus has been described previously in the literature. Also, one can easily obtain access to all the references, e.g., for growing single crystals from materials that are liquids at room temperature, by looking at code number 4 in the technique code-number sorting. [Checking the main listing of each citation will give the page(s) in the book on which that citation is discussed. All references in the Bibliography are not necessarily discussed in the text.]

Most of the categories listed in Table B–1 are self-explanatory. A few require further comments.

First Digit. Some devices have been designed for recording data from a large single-crystal plate that is held in only one orientation (e.g., during thermal expansion measurements). These are generally listed under category 3 (powders) rather than category 2 (single crystals) since they are not three- or four-circle devices and are similar to powder diffractometers.

Categories 4 and 5 (metals and proteins) are included because of the special interest in these materials.

Although a comprehensive description of low-temperature neutron diffraction instrumentation is not included in this text, a few devices (category 6) have been mentioned, primarily because they have been or can be adapted easily for X-ray instruments.

Fifth Digit. Some early low-temperature instruments were constructed for use with the ionization chamber counters of the early 1930s. These are included in category 7 (diffractometers) even though they are somewhat impractical for use with modern diffractometers.

Sixth Digit. Categories 1 and 2 (horizontal and vertical) are meant to specify further the type of instrument indicated by the fifth digit of the code number. Thus, XXXX71 describes a horizontal diffractometer. The terms "horizontal" and "vertical" are used in their popular sense: for the Debye–Scherrer and oscillation–rotation cameras and Weissenberg goniometer they refer to the orientation of the rotation axis, while for the diffractometer (both single-crystal and powder) they describe the orientation of the $\theta/2\theta$ circle.

A number of instruments have been designed to serve more than one purpose during X-ray diffraction investigations. This includes simultaneous low-temperature and high-pressure investigations, cold-working of metals at low temperatures, or a single apparatus with high-temperature as well as low-temperature capabilities. These special features have been indicated by the use of categories 4, 5, and 7.

Techniques and Applications. Category 7 is used to designate those references which have been referred to in Chapter 2 as examples of the applications of LTXRD investigations. New apparatus or techniques are not necessarily described in these articles.

Category 20 (review articles) refer to reviews of LTXRD apparatus and/or applications. A separate apparatus code number is not used for a review article unless new, previously unreported apparatus is also described in the article.

The last category, 21, refers to books and articles dealing with general low-temperature techniques and apparatus. Most of these references are used in the Appendices.

Russian articles are cited in their English translations, where these exist. Readers who wish to consult the originals will find references in the translations.

B.2. Alphabetical Listing

1   ABELL,J.S. AND KING, H.W. (1970)
    CRYOGENICS 10, 119-122
    AN ATTACHMENT FOR STRAINING X-RAY DIFFRACTOMETER SPECIMENS AT
    CRYOGENIC TEMPERATURES
            323177
                                                    113, 117, 129, 136, 137, 249

2   ABOWITZ,G. AND LADELL,J. (1968)
    J. PHYS. E., SCI. INSTRUM. 1, 113-117
    A LOW TEMPERATURE SYSTEM FOR AUTOMATIC SINGLE-CRYSTAL
    DIFFRACTOMETRY
            112300   14
                                              13, 36, 60, 77, 78, 199, 200, 246

3   ABRAHAMS,S.C. (1960)
    REV. SCI. INSTRUM. 31, 174-176
    LIQUID-HELIUM GONIOMETER-MOUNTED CRYOSTAT FOR A SINGLE CRYSTAL
    AUTOMATIC NEUTRON DIFFRACTOMETER
            623171   21
                                                          108, 123, 249

    ABRAHAMS,S.C.   SEE CALHOUN AND ABRAHAMS (1953)

4   ABRAHAMS,S.C., COLLIN, R.L., LIPSCOMB,W.N., AND REED,T.B (1950)
    REV. SCI. INSTRUM. 21, 396-397
    FURTHER TECHNIQUES IN SINGLE-CRYSTAL X-RAY DIFFRACTION STUDIES
    AT LOW TEMPERATURES
            112300   2   3   4   31, 46, 62, 67, 170, 172, 185, 191, 204

5   ABRAHAMS,S.C. AND KALNAJS,J. (1954)
    J. CHEM. PHYS. 22, 434-436
    THE LATTICE CONSTANTS OF THE ALKALI BOROHYDRIDES AND
    THE LOW-TEMPERATURE PHASE OF NABH4
            322372
                                                          117, 131, 139

6   ABRAHAMS,S.C. AND LIPSCOMB,W.N. (1952)
    ACTA CRYST. 5, 93-99
    THE CRYSTAL STRUCTURE OF THIOPHENE AT -55 D-C
            112356
                                                                  88

    ADAM,M.F.   SEE HEATON ET AL. (1970)

7   AGRON,P.A., LEVY,H.A., AND BOGARDUS,B.J. (1972)
    J. APPL. CRYST. 5, 432-433
    THERMOELECTRIC COOLING DEVICE FOR A SINGLE-CRYSTAL
    NEUTRON DIFFRACTOMETER
            633470   17
                                                              116, 145

8   AIR PRODUCTS AND CHEMICALS (1973)
    MANUAL FOR CRYO-TIP MODEL AC-1-101A
            153200
                                                    108, 113, 145, 147, 223

    AKMECHET,H.N.   SEE SANDLER AND AKMECHET (1958)

9   AKNAZAROV,S.K., SHABELNIKOV,L.G., AND SHEKHTMAN,V.S. (1974)
    INSTRUM. EXPTL. TECH. (USSR) 17, 1774-1775
    (ALSO PUBLISHED IN CRYOGENICS 15, 613-614 (1975))
    A DEVICE FOR DIVERGENT-BEAM X-RAY RECORDING AT
    LOW TEMPERATURES IN ELECTRIC FIELDS
            112390
                                                              84, 117

10 ALBRACHT,S.P.J. (1974)
    J. MAG. RES. 13, 299-303
    A LOW-COST COOLING DEVICE FOR EPR MEASUREMENTS AT 35 GHZ
    DOWN TO 4.8 K
                            7
                                                                  26

11   ALIEV,N.A. AND IBRAGIMOV,N.I. (1959)
     DOKLADY AKAD. NAUK AZERBAIDZHAN S.S.R. 15, 289-292
     LOW AND HIGH TEMPERATURE VACUUM X-RAY CAMERA
                 323314

     ALIEV,N.A.   SEE MAMEDOV AND ALIEV (1955)

12   ALLEN,K.W., JEFFREY,G.A., AND MCMULLAN,R.K. (1963)
     REV. SCI. INSTRUM. 34, 300-301
     COLD BOX FOR USE WITH A WEISSENBERG X-RAY DIFFRACTION CAMERA
                 113600
                                                                              246

13   ALTONA,C. (1964A)
     THESIS, UNIVERSITY OF LEIDEN
     MOLECULAR STRUCTURE AND CONFORMATION OF SOME
     HALOGENO-1,4-DIOXANES
                 212350
                                                            35, 39, 40, 70, 81, 223

14   ALTONA,C. (1964B)
     ACTA CRYST. 17, 1282-1286
     A VERSATILE COOLING TECHNIQUE FOR X-RAY DIFFRACTION BY SINGLE
     CRYSTALS AT TEMPERATURES BELOW 90 D°K
                 212350
                                                            35, 39, 40, 70, 81, 223

     AMMA,E.L.   SEE TANAKA AND AMMA (1964)

15   AMOROS,J.L., CARBONELL,A., AND CANUT,M.L. (1962)
     REV. CIENC. APL. 16, 385-396
     TWO-CIRCLE X-RAY DIFFRACTOMETER FOR HIGH AND LOW TEMPERATURES
     AND ITS APPLICATION TO ABSOLUTE MEASUREMENTS OF
     DIFFUSE DIFFRACTION
                 211374

16   AMOROS,J.L., GUIBERT,M., CANUT,M.L., AND ARRESE,F. (1961)
     REV. CIENC. APL. 15, 289-297
     A VERTICAL WEISSENBERG GONIOMETER FOR HIGH AND LOW TEMPERATURES
                 212354

     ANDERSON,D.L.   SEE JOHNSON AND ANDERSON (1966)

17   ANDO,M. (1973)
     J. APPL. CRYST. 6, 418-419
     A CRYOSTAT FOR X-RAY TOPOGRAPHY
                 223390
                                                                          25, 117

18   ANDRÉ,D., FOURME,R., AND RENAUD,M. (1971)
     ACTA CRYST. B27, 2371-2380
     REFINEMENT BY RIGID GROUPS
                 19
                                                                          191, 221

19   ANDRÉ,D., FOURME,R., AND RENAUD,M. (1972)
     ACTA CRYST. A28, 458-463
     APPLICATION OF RIGID GROUP CONCEPT TO THE DETERMINATION OF SIMPLE
     CRYSTALLINE STRUCTURES
                 19
                                                                              221

20   ANDRÉ,D., FOURME,R., AND ZECHMEISTER,K. (1972)
     ACTA CRYST. B28, 2389-2395
     CRYSTAL AND MOLECULAR STRUCTURE OF DIETHYL ETHER AT 128 D°K
                 4
                                                                              191

     ANDRÉ,D.   SEE KAHN ET AL. (1973)

APPEL,A.   SEE INTRATER AND APPEL (1961)

21 ARONOVA,P.N., GEGUSIN,Y.E., AND OVCHARENKO,N.N. (1959)
    INDUSTRIAL LAB. USSR 25, 646
    X-RAY PHOTOGRAPHY AT LOW TEMPERATURES
                     365311   157

    ARRESE,F.   SEE AMOROS ET AL. (1961)

    ASADA,E.   SEE GOTO ET AL. (1969)

22 ASHBY,E.C. AND SCHWARTZ,R.D. (1974)
    J. CHEM. EDUC. 51, 65-68
    A GLOVE BOX SYSTEM FOR THE MANIPULATION OF AIR
    SENSITIVE COMPOUNDS
                        1   21

23 ASTM (1972)
    MANUAL ON THE USE OF THERMOCOUPLES IN TEMPERATURE MEASUREMENT,
    ASTM SPECIAL PUBLICATION 470, PHILADELPHIA, PA. 19103
                        21   229, 231

24 ATOJI,M. (1965)
    NUCL. INSTRUM. METHODS 35, 13-33
    MULTIPURPOSE NEUTRON DIFFRACTION INSTRUMENTATIONS
                     623170   27, 108, 112

25 ATOJI,M. AND LIPSCOMB,W.N. (1954)
    ACTA CRYST. 7, 173-175
    THE CRYSTAL STRUCTURE OF HYDROGEN FLUORIDE
                        3   6   170, 177

26 ATOJI,M., SCHIRBER,J.E., AND SWENSON,C.A. (1959)
    J. CHEM. PHYS. 31, 1628-1629
    CRYSTAL STRUCTURE OF BETA-MERCURY
                     323110   130

27 ATTARD,A.E. AND AZAROFF,L.V. (1960)
    J. SCI. INSTRUM. 37, 238-239
    LIQUID NITROGEN CRYOSTAT FOR SINGLE-CRYSTAL DIFFRACTION
                     212370   68

28 AXON,H.,HELLAWELL,A., POOLE,D., AND HUME-ROTHERY,W. (1953)
    BRIT. J. APPL. PHYS. 4, 188-189
    A NOTE ON TEMPERATURE CONTROL DURING LATTICE SPACING MEASUREMENTS
                        14   24, 199

    AZAROFF,L.V.   SEE ATTARD AND AZAROFF (1960)

    BABIC,E.   SEE GIRT ET AL. (1974)

29 BACON,G.E., CURRY,N.A., AND WILSON,S.A. (1964)
    PROC. ROY. SOC. (LONDON) A279, 98-110
    A CRYSTALLOGRAPHIC STUDY OF SOLID BENZENE BY NEUTRON DIFFRACTION
                     282520   158

30 BAGARYATSKII,Y.A., KOLONTSOVA,E.V., AND
    RUSAKOVA-LUKOVSKAYA,N.Y. (1951)
    ZH. TEKH. FIZ. 21, 658-662
    X-RAY STUDIES OF AGING OF ALUMINUM ALLOYS
                     423320

    BALAKINA,L.M.   SEE SHEVELEV AND BALAKINA (1960)

31   BALZER,R. AND SIMMONS,R.O. (1974)
     PROC. INTL. CONF. ON LOW-TEMPERATURE PHYSICS AND CHEMISTRY, LT13
     ED., K.D. TIMMERHAUS, ET AL.,
     PLENUM PRESS, PPS. 115-119
     THERMAL DEFECTS IN BCC HELIUM-3 CRYSTALS
     DETERMINED BY X-RAY DIFFRACTION
                 222175   21, 23, 129

     BARBERICH,G.S.   SEE MCDONALD ET AL. (1966)

32   BARNES,W.H. (1929)
     PROC. ROY. SOC. (LONDON) A125, 670-693
     THE CRYSTAL STRUCTURE OF ICE BETWEEN 0 D-C AND -183 D-C
                 222340   12, 115, 127

33   IBID.
                 322310

34   IBID.
                 362310

35   BARNES,W.H. AND HAMPTON,W.F. (1935)
     REV. SCI. INSTRUM. 6, 342-344
     A VARIABLE TEMPERATURE X-RAY POWDER CAMERA
                 322410   115, 134

36   BARRETT,C.S. (1956)
     ACTA CRYST. 9, 671-677
     X-RAY STUDY OF THE ALKALI METALS AT LOW TEMPERATURES
                 423177   1   108, 129, 137, 166

37   BARRETT,C.S. (1957)
     TRANS. AM. SOC. METALS 49, 53-117
     METALLURGY AT LOW TEMPERATURES
                 7   20   11, 23

38   BARRETT,C.S. (1962)
     ADV. X-RAY ANALYSIS 5, 33-47
     X-RAY DIFFRACTION STUDIES AT LOW TEMPERATURES
                 7   20

39   BARRETT,C.S. (1967)
     HANDBOOK OF X-RAYS
     MCGRAW-HILL, NEW YORK
     ED., E.F. KAELBLE
     CHAPTER 15 - LOW TEMPERATURE DIFFRACTION
                 20

     BARRETT,C.S.   SEE MEHL AND BARRETT (1930)

     BARRETT,C.S.   SEE BATTERMAN AND BARRETT (1966)

40   BARRETT,C.S. AND MEYER,L. (1964)
     J. CHEM. PHYS. 41, 1078-1081
     X-RAY DIFFRACTION STUDY OF SOLID ARGON
                 323177   3   129, 137, 175

41   BARRETT,C.S., MEYER,L., AND WASSERMAN,J. (1967)
     J. CHEM. PHYS. 47, 740-743
     ARGON-FLUORINE PHASE DIAGRAM
                 323177   6   114, 129, 137, 177, 249

42   BARRETT, C.S. AND TRAUTZ,O.R. (1948)
     TRANS. MET. SOC. AIME 175, 579-605
     LOW TEMPERATURE TRANSFORMATIONS IN LITHIUM
                 362372

43 IBID.
                312372

44 BATTERMAN,B.W. AND BARRETT,C.S. (1966)
    PHYS. REV. 145, 296-301
    LOW-TEMPERATURE STRUCTURAL TRANSFORMATION IN V3SI
                223190   117

45 BAUN,W.L. (1959)
    APPL. SPECT. 13, 79-80
    SIMPLE LOW-TEMPERATURE DIFFRACTOMETER SPECIMEN MOUNT
                325372   131

46 BAUN,W.L. (1961)
    ADV. X-RAY ANALYSIS 4, 201-211
    DESIGN AND APPLICATIONS OF A VARIABLE-TEMPERATURE DIFFRACTOMETER
    SPECIMEN MOUNT
                323374   131

    BAUN,W.L.   SEE RENTON AND BAUN (1963)

47 BAUN,W.L. AND RENTON,J.J. (1963)
    WRIGHT-PATTERSON AIR FORCE BASE REPORT NO. ASD-TDR-63-278
    LOW TEMPERATURE X-RAY DIFFRACTION TECHNIQUES
                123371   20   26, 132, 139

48 BAUN,W.L. AND RENTON,J.J. (1964)
    ADV. X-RAY ANALYSIS 7, 302-313
    THE DESIGN AND USE OF SPECIAL-PURPOSE ATTACHMENTS FOR THE
    HORIZONTAL DIFFRACTOMETER
                123371   126, 132, 139

    BAXTER,R.   SEE CLARK AND BAXTER (1966)

49 BECK,A., CHIEH,P.C., PEARSON,W.B., AND STOKHUYZEN,R. (1973)
    ACA MEETING ABSTRACTS, SER.2, 1, 156
    A CAMERA FOR LOW-TEMPERATURE CRYSTALLOGRAPHY
                272150   155

50 BECKMAN,O. AND KNOX,K. (1961)
    PHYS. REV. 121, 376-380
    MAGNETIC PROPERTIES OF K(MN)F3, I.
                213141   82

    BEEMAN,W.W.   SEE NEYNABER ET AL. (1959)

51 BENNETT,M.J., HUTCHEON,W.L., AND FOXMAN,B.M. (1975)
    ACTA CRYST. A31, 488-494
    APPLICATIONS OF THE KING AND LIPSCOMB EXPRESSION FOR THE
    X-RAY SCATTERING OF A HINDERED ROTOR
                           19   221

    BENSEY,F.N.   SEE BURBANK AND BENSEY (1951)

    BENSEY,F.N.   SEE BURBANK AND BENSEY (1953)

    BERAR,J.F.   SEE CALVARIN AND BERAR (1975)

    BEREZNYAK,N.   SEE BOGOYAVLENSKII AND BEREZNYAK (1967)

    BERGER,J.   SEE LOW ET AL. (1966)

52 BERTIE,J.E., CALVERT,L.D., AND WHALLEY,E. (1963)
    J. CHEM. PHYS. 38, 840-846
    TRANSFORMATIONS OF ICE II, ICE III AND ICE IV
    AT ATMOSPHERIC PRESSURE
                        2   5   178

53   BERTOLUCCI,M.D. AND MARSH,R.E.   (1974)
     J. APPL. CRYST. 7, 87-88
     LATTICE PARAMETERS OF C6F6 AND C6H3F3 AT -17 D-C
                     185450   2   4   6   158, 168

54   BILDERBACK,D.H. AND COLELLA,R.   (1975)
     PHYS. REV. B11, 793-797
     VALENCE CHARGE DENSITY IN GREY TIN
                            1   162

55   BISWAS,S.G.   (1958)
     INDIAN J. PHYS. 32, 13-18
     X-RAY ANALYSIS OF FROZEN PYRIDINE
                     363311   152, 154

56   BLACK,I.A., BOLZ,L.H., BROOKS,F.P., MAUER,F.A.,
     AND PEISER,H.S.   (1958)
     J. RESEARCH NAT. BUR. STAND. 61, 367-371
     A LIQUID HELIUM COLD CELL FOR USE WITH AN X-RAY DIFFRACTOMETER
                     323171   129

     BLACK,R.E.   SEE GRUBER ET AL. (1969)

     BLESSING,R.H.   SEE COPPENS ET AL. (1974)

     BLESSING,R.H.   SEE WANG ET AL. (1976)

     BLEWITT,T.H.   SEE GRUBER ET AL. (1969)

     BLOCK,S. (1975)   PERSONAL COMMUNICATION   23

     BLOCK,S.   SEE SANTORO ET AL. (1968)

57   BOCHKAREV,V.F. AND EGOROV,V.A.   (1971)
     INSTRUM. EXPTL. TECH. (USSR) 14, 1584-1585
     VACUUM ATTACHMENT TO THE URS-501M DIFFRACTOMETER
     FOR INVESTIGATING THIN FILMS
                     323374   132

     BOGARDUS,B.J.   SEE AGRON ET AL. (1972)

58   BOGOYAVLENSKII,I. AND BEREZNYAK,N.   (1967)
     APP. METODY RENTGENOVSK. ANAL. 1, 176-178
     CRYOSTAT FOR X-RAY DIFFRACTION ANALYSIS OF SOLID HELIUM
                     323125   130

59   BOIKO,A.A.   (1966)
     ACTA CRYST. 21, A215
     LOW-TEMPERATURE X-RAY DIFFRACTION APPARATUS
                     212350

60   IBID.
                     324370

61   BOIKO,A.A., ERMAKOV,V.M., MEDVEDEV,V.S. AND PODOLICH,V.B.   (1973A)
     INSTRUM. EXPTL. TECH. (USSR) 16, 1571-1572
     AN X-RAY CRYOGENIC CHAMBER
                     313171   84, 101

62   BOIKO,A.A., GORBATII,L.Y., KORYASHKIN,V.I., KUDRYASHOV,I.H.,
     MIRENSKII,A.V., MICHALENKO,V.H., AND KHEIKER,D.M.   (1972A)
     APP. METODY RENTGENOVSK. ANAL. 10, 29-33
     LOW TEMPERATURE APPARATUS FOR DIFFRACTOMETERS
     WITH INCLINATION GEOMETRY
                     211371   50, 52, 129

63 BOIKO,A.A., KUCHERYAVII,V.A., AND PROKHVATILOV,A.I. (1972B)
INSTRUM. EXPTL. TECH. (USSR) 15, 929-931
(ALSO PUBLISHED IN CRYOGENICS 13, 237-239 (1973))
A HELIUM CRYOSTAT FOR AN X-RAY DIFFRACTOMETER
FOR STRUCTURE STUDIES
313171   137, 249

64 BOIKO,A.A. AND OVCHINNIKOV,A.S. (1968)
INDUSTRIAL LAB. USSR 34, 1527-1528
LOW-TEMPERATURE ATTACHMENT TO THE RKVT-400 CAMERA
212342   127

65 BOIKO,A.A. AND SHMYTKO,I.M. (1974)
INSTRUM. EXPTL. TECH. (USSR) 17, 873-874
(ALSO PUBLISHED IN CRYOGENICS 15, 35-36 (1975))
CRYOSTAT FOR X-RAY REFLECTION PHOTOGRAPHY
IN A WIDELY DIVERGING BEAM
112390   84, 117

66 BOIKO,A.A., SUVOROV,E.V., AND MUKHIN,K.Y. (1973B)
INSTRUM. EXPTL. TECH. (USSR) 16, 648-650
LOW TEMPERATURE ATTACHMENT FOR THE KFOR-4 X-RAY GONIOMETER
212360   84, 92, 116

67 BOIKO,A.A. AND UMANSKII,M.M. (1967)
APP. METODY RENTGENOVSK. ANAL. 1, 154-172
APPARATUS FOR LOW-TEMPERATURE X-RAY STUDIES
20

68 BOIKO,A.A., ZUBENKO,V.V., MARKINA,L.I., MYASNIKOV,Y.G.
SOLOVEICHIK,M.B., TATKIN,L.Z., UMANSKII,M.M.,
AND FINKELSHTEIN,Y.N. (1972C)
APP. METODY RENTGENOVSK. ANAL. 10, 22-28
APPARATUS FOR X-RAY DIFFRACTION AT LOW TEMPERATURES
312371

69 BOLHUIS, F., VAN (1971)
J. APPL. CRYST. 4, 263-264
COOLING APPARATUS FOR A SINGLE-CRYSTAL X-RAY DIFFRACTOMETER
111300   36, 56, 71, 96, 204

BOLLING,G.F.   SEE ROESSLER AND BOLLING (1964)

70 BOL'SHUTKIN,D.N., BOSIN,M.E., GASAN,V.M., ZINOVIEV,M.V.,
MARKOV,A.S., AND MOKRII,N.I. (1969)
APP. METODY RENTGENOVSK. ANAL. 4, 20-25
CRYOSTAT FOR X-RAY STUDIES IN THE TEMPERATURE RANGE 4.2-300 D-K
323171

71 BOL'SHUTKIN,D.N., GASAN,V.M., KUCHERYAVII,V.A.,
MERONOV-KOPISOV,V.S., MOKRII,N.I., PROKHVATILOV,A.I.,
AND ERENBERG,A.I. (1970)
APP. METODY RENTGENOVSK. ANAL. 6, 12-15
HELIUM CRYOSTAT FOR X-RAY DIFFRACTOMETER WITH TEMPERATURE REGULATOR
323171   175

72 BOL'SHUTKIN,D.N., GASAN,V.M., PROKHVATILOV,A.I.,
AND ERENBERG,A.I. (1972)
ACTA CRYST. B28, 3542-3548
THE CRYSTAL STRUCTURE OF ALPHA-CF4
5   178

BOL'SHUTKIN,D.N.   SEE GRUSHKO ET AL. (1970)

73 BOLZ,L.H., BROIDA,H.P., AND PEISER,H.S.   (1962)
    ACTA CRYST. 15, 810-811
    SOME OBSERVATIONS ON GROWING CRYSTALS OF ARGON
                            4

74 BOLZ,L.H. AND MAUER,F.A. (1963)
    ADV. X-RAY ANALYSIS 6, 242-249
    MEASUREMENT OF THE LATTICE CONSTANTS OF NEON ISOTOPES
    IN THE TEMPERATURE RANGE 4 TO 24 D-K
                        323171   3   129

    BOLZ,L.H.   SEE BLACK ET AL. (1958)

    BOLZ,L.H.   SEE MAUER AND BOLZ (1961)

75 BONFIGLIOLI,A.F. AND TESTARD,O. (1964)
    ACTA CRYST. 17, 668-670
    APPARATUS FOR MEASURING X-RAY DIFFRACTION AT LOW TEMPERATURE
                        313371   101

76 BONILLA,A., GARLAND,C.W., AND SCHUMAKER,N.E.   (1970)
    ACTA CRYST. A26, 156-158
    LOW TEMPERATURE INVESTIGATION OF NH4(BR)
                        153100   149

    BOSIN,M.E.   SEE BOL'SHUTKIN ET AL. (1969)

77 BOUTTIER,L. (1949)
    C.R. ACAD. SCI. 228, 1419-1421
    APPARATUS FOR OBTAINING AND TRANSPORTING SINGLE CRYSTALS
    AT LOW TEMPERATURE
                            4   186

78 BOUTTIER,L. AND DUNOYER,J.M. (1953)
    BULL. SOC. FR. MINERAL. CRISTALLOGR. 76, 79-85
    DESCRIPTION OF A PHOTOGRAPHIC CHAMBER FOR TAKING DEBYE-SCHERRER
    DIAGRAMS OF SPECIMENS AT VERY LOW TEMPERATURES
                        313312   46, 99

    BOWLES,B.B.   SEE DUMBLETON AND BOWLES (1966)

79 BRADY,J.H. AND REUTH,E.C.,VAN (1966)
    J. SCI. INSTRUM. 43, 833-834
    A SIMPLE SPECIMEN HOLDER FOR LOW-TEMPERATURE X-RAY INTENSITY
    STUDIES USING A G.E. DIFFRACTOMETER
                        323371   112, 132

    BRAMMER,W.G.   SEE NEYNABER ET AL. (1959)

    BROIDA,H.P.   SEE BOLZ ET AL. (1962)

80 BRONSVELD,P.M., KUMRA,S.K., AND STRYLAND,J.C.   (1973)
    CAN. J. PHYS. 51, 25-35
    X-RAY DIFFRACTION BY COMPRESSED SOLID ARGON
                        123345   4   23, 128, 134, 193

    BROOKS,F.P.   SEE BLACK ET AL. (1958)

81 BROOMÉ,B. (1923)
    PHYSIK. Z. 24, 124-130
    X-RAY INVESTIGATION OF SOLID BENZENE
                        385510   157

82   BROWN,D.S. AND WALLWORK,S.C.   (1962)
     J. SCI. INSTRUM.   39, 319-320
     A METHOD OF OBTAINING EQUI-INCLINATION WEISSENBERG PHOTOGRAPHS
     AT LOW TEMPERATURES
                    212350
                                                                    86

83   BRUEHL,H.G. AND SCHMIDT,W.   (1971)
     KRIST. TECH.   6, K17-K19
     SIMPLE VAPOR CRYOSTAT FOR X-RAY DIFFRACTION STUDIES
     BETWEEN 77 AND 370 D-K
                    212350

84   BUERGER,M.J.   (1964)
     THE PRECESSION METHOD
     WILEY, NY   PAGES 257-260
                         20

     BULATOV,A.S.   SEE KOGAN ET AL. (1964)

     BULATOVA,R.F.   SEE KOGAN ET AL. (1960)

     BULATOVA,R.F.   SEE KOGAN AND BULATOVA (1969)

85   BURBANK,R.D.   (1953)
     ACTA CRYST.   6, 55-56
     THE USE OF LOW TEMPERATURES IN ACCURATE STRUCTURE ANALYSIS
                         11
                                                                    12

86   BURBANK,R.D.   (1973)
     J. APPL. CRYST.   6, 437-441
     LOW-TEMPERATURE X-RAY SYSTEM FOR MATERIALS OF HIGH CHEMICAL
     REACTIVITY, HIGH VAPOR PRESSURE OR LOW MELTING POINT
                    111300   2   4   6   13   36, 39, 42, 46, 49, 54, 57, 77, 79, 91,
                                                                    93, 177, 181, 182, 183, 185, 186, 187, 195, 200

87   BURBANK,R.D.   (1974)
     J. APPL. CRYST.   7, 41-44
     X-RAY STUDY OF AN OSMIUM OXYFLUORIDE OF UNKNOWN COMPOSITION
                         4
                                                                    214

     BURBANK,R.D.   SEE JONES ET AL. (1970)

88   BURBANK,R.D. AND BENSEY,F.N.   (1951)
     REPORT NO. K-841, CARBIDE AND CARBON CHEMICALS CO.,
                              K-25 PLANT, OAK RIDGE,TENN.
     THE APPLICATION OF THE BUERGER PRECESSION CAMERA
     TO LOW TEMPERATURE STUDIES
                    211360
                                                                    49, 69, 91

89   BURBANK,R.D. AND BENSEY,F.N.   (1953)
     J. CHEM. PHYS.   21, 602-608
     THE STRUCTURE OF THE INTERHALOGEN COMPOUNDS, I.
                    211360
                                                                    49

90   BURBANK,R.D. AND JONES,G.R.   (1974)
     INORG. CHEM.   13, 1071-1074
     CRYSTAL STRUCTURE OF IODINE PENTAFLUORIDE AT -80 D-C
                         4

91   BURTON,E.F. AND OLIVER,W.F.   (1936)
     PROC. ROY. SOC. (LONDON)   153A, 166-172
     THE CRYSTAL STRUCTURE OF ICE AT LOW TEMPERATURES
                    323310
                                                                    110, 133, 135

92   BUTTERS,R.G. AND MYERS, H.P. (1955)
     CAN. J. TECHNOLOGY 33, 356-359
     A LOW-TEMPERATURE ATTACHMENT FOR AN X-RAY GONIOMETER
                     323372   131

     BUTUZOV,V.P.   SEE FLINT AND BUTUZOV (1937)

     CABELKA,D.   SEE KING ET AL. (1969)

93   CADY,H.H. AND LARSON,A.C. (1975)
     ACTA CRYST. B31, 1864-1869
     AN ALGORITHM FOR REFINEMENT OF CRYSTAL STRUCTURES WITH POOR DATA
                      19   221

94   CALHOUN,B.A. AND ABRAHAMS,S.C. (1953)
     REV. SCI. INSTRUM. 24, 397
     A LOW-TEMPERATURE ADAPTOR FOR THE NORELCO HIGH ANGLE SPECTROMETER
                     322372   131, 139, 140

95   CALVARIN,G. AND BERAR,J.F. (1975)
     J. APPL. CRYST. 8, 380-385
     X-RAY POWDER DIFFRACTION STUDY OF THE ORDER-DISORDER
     TRANSITION OF FERROCENE
                     323171   130

     CALVERT,L.D.   SEE BERTIE ET AL. (1963)

96   CAMPBELL,J.A. AND HILDEBRAND,J.H. (1943)
     J. CHEM. PHYS. 11, 334-337
     THE STRUCTURE OF LIQUID XENON
                     312315   43, 46, 99

     CANUT,M.L.   SEE AMOROS ET AL. (1961)

     CANUT,M.L.   SEE AMOROS ET AL. (1962)

     CARBONELL,A.   SEE AMOROS ET AL. (1962)

97   CARPENTER,G.B. AND RICHARDS,S.M. (1962)
     ACTA CRYST. 15, 360-364
     THE CRYSTAL STRUCTURE OF BETA-IODINE MONOCHLORIDE
                      4   179, 191

     CARPENTER,G.B.   SEE DOHLEN AND CARPENTER (1955)

     CARPENTER,G.B.   SEE SHALLCROSS AND CARPENTER (1958)

     CARTZ,L.   SEE GOPALAKRISHNA AND CARTZ (1972)

     CASPAR,D.   SEE COHEN ET AL. (1971)

     CHAMBERS,J.L.   SEE KRIEGER ET AL. (1974)

     CHAPMAN,B.F.   SEE UNDERWOOD AND CHAPMAN (1976)

98   CHARBIT,P. AND DUCROS,P. (1968)
     BULL.SOC. FR. MINERAL. CRISTALLOGR. 91, 116-125
     APPARATUS FOR X-RAY DIFFRACTOMETRY AT VERY LOW TEMPERATURES
                     323271   16

99   CHAWDHURY,S.A. (1968)
     PAK. J. SCI. IND. RES. 11, 151-153
     EFFICIENT METHOD FOR RECORDING LOW-TEMPERATURE XRAY
     DIFFRACTION PHOTOGRAPHS
                     212350   54, 63

100 CHEESMAN,G.H. AND SOANE,C.M. (1957)
     PROC. PHYS. SOC. (LONDON) B70, 700-702
     THE LATTICE CONSTANTS OF THE INERT ELEMENTS
                 323310   3   5
                                                            130, 135, 175

     CHEN,C.   SEE LOW ET AL. (1966)

     CHEVALIER,B.   SEE ETOURNEAU ET AL. (1975)

     CHIEM,P.C.   SEE BECK, ET AL. (1973)

101 CHOPRA,K.L. (1962)
     CRYOGENICS 2, 167-169
     LIQUID HELIUM CRYOSTAT FOR X-RAY DIFFRACTION STUDIES
                 123171
                                                            108, 123

102 CHRISTOE,C.W. AND DRICKAMER,H.G. (1969)
     REV. SCI. INSTRUM. 40, 169-170
     CLAMP CELL FOR HIGH PRESSURE - LOW TEMPERATURE X-RAY AND MOSSBAUER
     RESONANCE STUDIES
                 324625

     CHRISTOPH,G.G.   SEE KRIEGER ET AL. (1974)

103 CIMINO,A., PARRY,G.S., AND UBBELOHDE,A.R. (1959)
     PROC. ROY. SOC. (LONDON) A252, 445-456
     X-RAY STUDIES OF THERMAL CYCLES IN THE TRANSFORMATION OF KCN
                 111343
                                                            74

104 CIOFFI,P.P. AND TAYLOR,L.S. (1922)
     J. OPT. SOC. AM. 6, 906-909
     A METHOD OF MAINTAINING SMALL OBJECTS AT ANY TEMPERATURE
     BETWEEN -180 D-C AND +20 D-C
                 115300
                                                31, 43, 44, 46, 55, 67, 86

     CIOFFI,P.P.   SEE MCKEEHAN AND CIOFFI (1922)

105 CLARK,N.H. AND BAXTER,R. (1966)
     J. SCI. INSTRUM. 43, 757
     SPLIT WEISSENBERG SCREENS FOR LOW-TEMPERATURE SINGLE CRYSTAL
     X-RAY WORK
                 214356
                                                            88

106 CLAUDET,G.M., TIPPE,A., AND YELON,W.B. (1976)
     J. PHYS. E., SCI. INSTRUM. 9, 259-261
     A CRYOSTAT FOR FOUR-CIRCLE NEUTRON DIFFRACTION
                 613171

107 CLAUS,W.D. (1931)
     PHYS. REV. 38, 604-617
     EFFECT OF TEMPERATURE ON THE INTENSITY OF X-RAYS
     DIFFUSELY SCATTERED FROM ROCK SALT
                 222370
                                                            112, 131

108 CLIFTON,D.F. (1950)
     REV. SCI. INSTRUM. 21, 339-342
     LOW-TEMPERATURE X-RAY DIFFRACTION APPARATUS
                 165300
                                                            46, 67, 134, 153

109 IBID.
                 312371

110 IBID.
                 322371

COCKETT,A.H.   SEE DIN AND COCKETT (1960)

111   COCKS,F.H., PREECE,C.M., AND KING,M.W. (1966)
      PHYS. LETTERS 22, 287-288
      X-RAY PEAK SHIFTS ON COOLING A COMPOSITE OF GOLD PARTICLES
      IN A COLLODION-AMYL ACETATE MATRIX
                        1   5
                                                              110, 179

      COCKS,F.H.   SEE COGAN AND COCKS (1975)

112   COGAN,S.F. AND COCKS,F.H. (1975)
      J. APPL. CRYST. 8, 571-572
      A METHOD FOR MOUNTING POWDER SAMPLES THAT AVOIDS ADVERSE
      STRAIN EFFECTS AT LOW TEMPERATURES
                        1   5
                                                                  179

113   COHEN,C., CASPAR,D., PARRY,D. AND LUCAS,R. (1971)
      COLD SPRING HARBOR SYMP., QUANT. BIOL., 36, 205-216
      TRUPOMYOSIN CRYSTAL DYNAMICS
                      531560
                                                                  145

114   COLAPIETRO,M. (1975)
      ACTA CRYST. A31, S239
      A LOW-TEMPERATURE APPARATUS FOR SINGLE-CRYSTAL DIFFRACTOMETRY
                      111300
                                                              96, 98

115   COLE,E.A. AND HOLMES,D.R. (1960)
      J. POLYMER SCI. 46, 245-256
      CRYSTAL LATTICE PARAMETERS OF HYDROCARBONS
                      312374
                                                              46, 101

      COLELLA,R.   SEE BILDERBACK AND COLELLA (1975)

116   COLLIN,R.L. (1952)
      ACTA CRYST. 5, 431-432
      THE CRYSTAL STRUCTURE OF SOLID CHLORINE
                        3
                                                              12, 175

117   COLLIN,R.L. AND LIPSCOMB,W.N. (1951)
      ACTA CRYST. 4, 10-14
      THE CRYSTAL STRUCTURE OF HYDRAZINE
                      112306   2   4
                                                    88, 90, 175, 186, 210

      COLLIN,R.L.   SEE ABRAHAMS, ET AL. (1950)

      COLOONY,P.C.   SEE STAMMLER ET AL. (1963)

      COLSON,S.D.   SEE WHEELER AND COLSON (1975)

      COOPER,W.F.   SEE COPPENS ET AL. (1974)

118   COPPENS,P. (1972)
      PROCEEDINGS OF THE ADVANCED STUDY INSTITUTE ON EXPERIMENTAL
      ASPECTS OF X-RAY AND NEUTRON SINGLE CRYSTAL DIFFRACTION METHODS,
      AARHUS (UNPUBLISHED)
      INSTRUMENTATION FOR LOW TEMPERATURES AND THE ADVANTAGES OF LOW
      TEMPERATURE DATA IN ACCURATE CRYSTALLOGRAPHY
                      11   12   13   14   20   14, 197, 200, 203, 206,
                                                                  218, 246

119   COPPENS,P. (1975)
      MTP INTERNATIONAL REVIEW OF SCIENCE
      SERIES 2 - PHYSICAL CHEMISTRY, VOLUME 11,
      CHEMICAL CRYSTALLOGRAPHY, J.M. ROBERTSON, ED.
      BUTTERWORTHS, LONDON   PPS. 21-56
                        7
                                                                  15

COPPENS,P.   SEE REES AND COPPENS (1973)

COPPENS,P.   SEE WANG ET AL. (1976)

COPPENS,P.   SEE WANG AND COPPENS (1976)

120 COPPENS,P., GODEL,J., AND SABINE,T.M. (1967)
     REV. SCI. INSTRUM. $\underline{38}$, 1333-1334
     USE OF A JOULE-THOMSON REFRIGERATOR FOR NEUTRON DIFFRACTION
     AT LIQUID NITROGEN TEMPERATURES
                653371   108, 145, 146, 149

121 COPPENS,P., ROSS,F.K., BLESSING,R.H., COOPER,W.F., LARSEN,F.K.,
     LEIPOLDT,J.G., REES,B., AND LEONARD,R. (1974)
     J. APPL. CRYST. $\underline{7}$, 315-319
     A CRYOSTAT FOR COLLECTION OF THREE-DIMENSIONAL DIFFRACTOMETER DATA
     AT LIQUID HELIUM TEMPERATURES
                223171   17   108, 113, 115, 119, 121, 200

122 COPPENS,P. AND VOS,A. (1971)
     ACTA CRYST. B$\underline{27}$, 146-158
     ELECTRON DISTRIBUTION IN CYANURIC ACID, II.
                          11   15

123 COULON,G., LECOQ,J., AND ESCAIG,B. (1974)
     J. PHYSIQUE $\underline{35}$, 557-569
     X-RAY DISLOCATION SUBSTRUCTURE OBSERVATIONS IN ALPHA-IRON
     SINGLE CRYSTAL BETWEEN ROOM TEMPERATURE AND 123 D°K
                224390   25, 117

124 COX,E.G. (1932)
     PROC. ROY. SOC. (LONDON) A$\underline{135}$, 491-498
     THE CRYSTALLINE STRUCTURE OF BENZENE
                222440   115, 134

125 COYLE,B.A. (1972)
     ACTA CRYST. A$\underline{28}$, 231-233
     ABSORPTION AND VOLUME CORRECTIONS FOR A CYLINDRICAL SPECIMEN,
     LARGER THAN THE BEAM, AND IN GENERAL ORIENTATION
                          18   219

126 COYLE,B.A. AND SCHROEDER,L.W. (1971)
     ACTA CRYST. A$\underline{27}$, 291-295
     ABSORPTION AND VOLUME CORRECTIONS FOR A CYLINDRICAL SAMPLE
     LARGER THAN THE X-RAY BEAM, EMPLOYED IN EULERIAN GEOMETRY
                          18   219

127 CRANDALL,P.B. (1969)
     REV. SCI. INSTRUM. $\underline{40}$, 954-955
     LOW TEMPERATURE ADAPTOR FOR THE DEBYE-SCHERRER POWDER CAMERA
                311310   100

128 CROFT, A.J. (1970)
     CRYOGENIC LABORATORY EQUIPMENT
     PLENUM PRESS, N.Y.
                          21   233

CROFT,W.J.   SEE HORNE ET AL. (1959)

129 CRUICKSHANK,D.W.J. (1956)
     ACTA CRYST. $\underline{9}$, 1005-1009
     THE VARIATION OF VIBRATION AMPLITUDES WITH TEMPERATURE
     IN SOME MOLECULAR CRYSTALS
                          12   12, 197, 198

130   CRUICKSHANK,D.W.J. (1960)
      ACTA CRYST. 13, 774-777
      THE REQUIRED PRECISION OF INTENSITY MEASUREMENTS FOR
      SINGLE-CRYSTAL ANALYSIS
                              11   12

131   CRUICKSHANK,D.W.J., MENDOZA,E. AND ROBERTSON,J.H. (1956)
      BRIT. J. APPL. PHYS. 7, 425-435
      SUMMARIZED PROCEEDINGS OF A CONFERENCE ON LOW TEMPERATURE
      CRYSTALLOGRAPHY
                              20   37, 206

132   CUCKA,P., SINGMAN,L., LOVELL,F.M., AND LOW,B.W. (1970)
      ACTA CRYST. B26, 1756-1760
      STUDIES OF INSULIN CRYSTALS AT LOW TEMPERATURES. EFFECTS ON LATTICE
      DIMENSIONS, TEMPERATURE PARAMETERS, AND STRUCTURE
                              215571   95, 96

      CUNICELLA,V.T.   SEE MCDONALD ET AL. (1966)

      CUNNINGTON,G.R.   SEE TIEN AND CUNNINGTON (1972)

      CURRY,N.A.   SEE BACON, ET AL. (1964)

133   DANIELSSON,S., GRENTHE,I., AND OSKARSSON,A. (1976)
      J. APPL. CRYST., 9, 14-17
      A LOW-TEMPERATURE APPARATUS FOR SINGLE-CRYSTAL DIFFRACTOMETRY
                              111300   72

134   DARLINGTON,C.N.W. AND MEGAW,H.D. (1973)
      ACTA CRYST. B29, 2171-2185
      THE LOW-TEMPERATURE PHASE TRANSITION OF SODIUM NIOBATE
                              212350   38, 54

      DAVEY,A.R.   SEE THEWLIS AND DAVEY (1955)

135   DAVIES,G.R., JARVIS,J.A.J., KILBOURN,B.T., MAIS,R.H.B.,
      AND OWSTON,P.G. (1970)
      J. CHEM. SOC. (A) 1970, 1275-1283
      CRYSTAL AND MOLECULAR STRUCTURE OF BIS-(NN-DIMETHYLDITHIOCARBAMATO)-
      NITROSYLIRON (AT -80 D-C)
                              212371   68, 96

      DAVIS,B.L.   SEE KAMB AND DAVIS (1964)

136   DAVIS,G.T. AND EBY,R.K. (1975)
      REV. SCI. INSTRUM. 46, 1285-1286
      LOW-TEMPERATURE X-RAY ATTACHMENT
                              323372   132

137   DEISEROTH,H.J., WESTERBECK,E., AND SIMON,A. (1975)
      ACTA CRYST. A31, S93
      X-RAY INVESTIGATION OF AIR SENSITIVE AND LOW MELTING COMPOUNDS-
      THE CRYSTAL STRUCTURE OF (CS)4-O
                              4   188

      DELÉMOUZÉE,L.   SEE GHISLAIN ET AL. (1965)

138   DEL PRA,A., VALLE,G., AND MAMMI,M. (1967)
      CHIM. IND. (MILAN) 49, 1183-1186
      COOLING SYSTEM FOR LOW TEMPERATURE X-RAY ANALYSES
                              212356

139   DENNISON,D.M.   (1921)
      PHYS. REV.   17, 20-22
      THE CRYSTAL STRUCTURE OF ICE
                  324310   5   *133, 178*

140   DERNIER,P.D.   (1972)
      REV. SCI. INSTRUM.   43, 931-933
      A LOW TEMPERATURE BAFFLE FOR SINGLE CRYSTAL X-RAY DIFFRACTION
                  212171   *82, 83*

      DERUYTERRE,A.   SEE GHISLAIN ET AL. (1965)

141   DE SMEDT,J. AND KEESOM,W.H. (1924)
      PROC. FOURTH INTERNATIONAL CONF. REFRIGERATION   4, 107A-115A
      THE STRUCTURE OF SOLID NITROUS OXIDE AND CARBON DIOXIDE
                  322312   *133, 135*

142   DE SMEDT,J., KEESOM,W.H., AND MOOY,H.H.   (1930)
      PROC. ACAD. SCI. AMSTERDAM   33, 255-257
      ON THE CRYSTAL STRUCTURE OF NEON
                  322110   3   *12, 130, 175*

      DE SMEDT,J.   SEE KEESOM AND DE SMEDT (1922)

      DE SMEDT,J.   SEE KEESOM ET AL. (1930)

143   DICKENS,B.   (1966)
      U.S. NAVAL PROPELLANT PLANT, REPORT NPP/RR 66-9, 21 PPS.
      TECHNIQUES FOR THE LOW-TEMPERATURE X-RAY STUDIES OF
      SENSITIVE MATERIALS
                  211350   1   2   6   *71, 87, 88, 89, 162*

      DIERKS,H.   SEE DIETRICH AND DIERKS (1970)

144   DIETRICH,H. (1968)
      MESSTECHNIK   76, 303-308
      LOW TEMPERATURE EQUIPMENT FOR SINGLE-CRYSTAL DIFFRACTOMETER
      AND WEISSENBERG CAMERA
                  111306   *50, 53, 55*

145   DIETRICH,H. AND DIERKS,H.   (1970)
      MESSTECHNIK   78, 184
      IMPROVEMENT IN LOW-TEMPERATURE APPARATUS FOR AUTOMATIC FOUR-CIRCLE
      SINGLE-CRYSTAL DIFFRACTOMETER
                  111300

146   DIETZE,M.   (1958)
      EXP. TECH. DER PHYS. 6, 120-124
      DESCRIPTION OF A LOW-TEMPERATURE X-RAY CAMERA
                  323312

147   DIN,F. AND COCKETT,A.H. (1960)
      LOW-TEMPERATURE TECHNIQUES
      GEORGE NEWNES,LMTD., LONDON
                                   21

148   DOBBS,E.R. (1961)
      EXPERIMENTAL CRYOPHYSICS
      EDS., F.E. HOARE, L.C. JACKSON, AND N. KURTI
      BUTTERWORTHS, LONDON   PAGES 336-343
      X-RAY METHODS AT LOW TEMPERATURES. A REVIEW
                                   20

149   DOHLEN,W.C.,VON, AND CARPENTER,G.B.   (1955)
      ACTA CRYST.   8, 646-651
      THE CRYSTAL STRUCTURE OF ISOCYANIC ACID
                   4
                                                                                                  190

150   DONDE,A., FOMENKO,N., AND KHOTKEVICH,V.I.   (1967)
      APP. METODY RENTGENOVSK. ANAL.   1, 173-175
      CRYOSTAT FOR X-RAY DIFFRACTION STUDIES OF DEFORMED METALS
      AT LOW TEMPERATURES
                   46310⁷
                                                                                                  157

151   DOWELL,L.G. AND RINFRET,A.P.   (1960)
      NATURE   188, 1144-1148
      LOW-TEMPERATURE FORMS OF ICE AS STUDIED BY X-RAY DIFFRACTION
                   323372
                                                                                                  131

152   DRENTH,W. AND WIEBENGA,E.H.   (1955)
      ACTA CRYST.   8, 755-760
      STRUCTURE OF ALPHA,OMEGA-DIPHENYL-POLYENES, IV.
                   212350   11
                                                                                                  246

153   DREW,R.E. AND EINSTEIN,F.W.B.   (1973)
      INORG. CHEM.   12, 829-835
      CRYSTAL STRUCTURE AT -100 D°C OF VC7O8N2H9•X H2O
                   253371
                                                                                                  145

      DRICKAMER,H.G.   SEE CHRISTOE AND DRICKAMER (1969)

      DUCROS,P.   SEE CHARBIT AND DUCROS (1968)

      DUFOUR,L.   SEE HARTOULARI AND DUFOUR (1970)

      DUKE,W.M.   SEE VANCE AND DUKE (1962)

154   DULMAGE,W.J. AND LIPSCOMB,W.N.   (1951)
      ACTA CRYST.   4, 330-334
      THE CRYSTAL STRUCTURES OF HYDROGEN CYANIDE
                   4
                                                                                             195, 209

155   DULMAGE,W.J. AND LIPSCOMB,W.N.   (1952)
      ACTA CRYST.   5, 260-264
      THE CRYSTAL AND MOLECULAR STRUCTURE OF PENTABORANE
                   4
                                                                                              12, 196

156   DUMBLETON,J.H. AND BOWLES,B.B.   (1966)
      REV. SCI. INSTRUM.   37, 1613-1614
      HEATING AND COOLING UNIT FOR USE WITH POLAROID X-RAY CAMERA
                   321320
                                                                                                  134

157   DUNOYER,J.M.   (1952)
      HELV. CHIM. ACTA   35, 840-845
      DEBYE-SCHERRER DIAGRAMS OF COMPOUNDS CRYSTALLIZING AT
      LOW TEMPERATURES
                   313312
                                                                                               46, 99

      DUNOYER,J.M.   SEE BOUTTIER AND DUNOYER (1953)

158   EASTMAN,E.D.   (1924)
      J. AM. CHEM. SOC.   46, 917-923
      X-RAY DIFFRACTION PATTERNS FROM CRYSTALLINE AND LIQUID BENZENE
                   112300   31, 43, 46, 67, 97, 178

159   EBDON,F.R. AND WHEELER,D.A.   (1971)
      J. APPL. CRYST.   4, 254-255
      A SELF-REGULATING MINI-COOLER FOR THREE-DIMENSIONAL SINGLE-CRYSTAL
      NEUTRON-DIFFRACTION MEASUREMENTS
                   653371
                                                                                             145, 149

EBY,R.K.   SEE DAVIS AND EBY (1975)

160 EELES,E.G. (1960)
ADV. CRYOGENIC ENGINEERING 3, 248-253
A PRECISION LOW-TEMPERATURE X-RAY CAMERA
            323112
                                                    130

EGOROV,V.A.   SEE BOCHKAREV AND EGOROV (1971)

161 EINSTEIN,F.W.B., GILBERT,M.M., AND TUCK,D.G. (1972)
INORG. CHEM. 11, 2832-2836
CRYSTAL STRUCTURE OF (IN)(C5H5)3 AT -100 D-C
            253371
                                                    145

EINSTEIN,F.W.B. SEE DREW AND EINSTEIN (1973)

ENEMARK,J.   SEE FRENZ ET AL. (1969)

162 ENRAF-NONIUS (1971)
UNIVERSAL LOW TEMPERATURE DEVICE- INSTRUCTION MANUAL
            111300
                                          35, 49, 71, 97

ERENBERG,A.I.   SEE BOL'SHUTKIN ET AL. (1970)

ERENBERG,A.I.   SEE BOL'SHUTKIN ET AL. (1972)

ERICKSON,R.A.   SEE KHAN AND ERICKSON (1970)

ERMAKOV,V.M.   SEE BOIKO ET AL. (1973)

ESCAIG,B.   SEE COULON ET AL. (1974)

163 ETOURNEAU,J., CHEVALIER,B., AND RABARDEL,L. (1975)
J. PHYS. E., SCI. INSTRUM. 8, 930-934
X-RAY DIFFRACTOMETRY ON POLYCRYSTALLINE MATERIALS BETWEEN
300 AND 4.2 K - PROBLEMS CONCERNING THE METHOD OF
COOLING THE SAMPLE
            323172
                                    109, 130, 179, 202, 249

164 EUBIG,C. AND TOMIZUKA,C.T. (1972)
REV. SCI. INSTRUM. 43, 1804-1810
A METHOD FOR PERFORMING HIGH PRECISION LATTICE PARAMETER
CHANGE MEASUREMENTS ON QUENCHED ALUMINUM
            422371   1   14
                                               24, 110, 202

FALCONER,W.E.   SEE JONES ET AL. (1970)

FANKUCHEN,I.   SEE KAUFMAN AND FANKUCHEN (1949)

FANKUCHEN,I.   SEE POST ET AL. (1951)

FANKUCHEN,I.   SEE POST ET AL. (1952)

FANKUCHEN,I.   SEE POST AND FANKUCHEN (1953)

FANKUCHEN,I.   SEE STEINFINK ET AL. (1953)

165 FÉHER,F. AND KLOTZER,F. (1935)
Z. ELEKTROCHEM. 41, 850-851
THE CRYSTAL STRUCTURE OF THE HYDROGEN SUPEROXIDES
            312310

166 FÉHER,F. AND KLOTZER,F. (1937)
Z. ELEKTROCHEM. 43, 822-826
THE CRYSTAL STRUCTURE OF THE HYDROGEN SUPEROXIDES
            212340
                                                    43

167   FIGGINS,B.F., JONES,G.O., AND RILEY,D.P. (1956)
      PHIL. MAG. 1, 747-758
      THERMAL EXPANSION OF AL AT LOW TEMPERATURES
                   323113   130

168   FILIPPINI,G., GRAMACCIOLI,C.M., SIMONETTA,M., AND
      SUFFRITTI,G.B. (1974)
      ACTA CRYST. A30, 189-196
      ON SOME PROBLEMS CONNECTED WITH THERMAL MOTION IN MOLECULAR
      CRYSTALS AND A LATTICE-DYNAMICAL INTERPRETATION
                   7   13

      FINCH,D.I.   SEE GRAY AND FINCH (1971)

169   FINKEL,V.A. (1971)
      LOW-TEMPERATURE X-RAY EXAMINATION OF METALS
      METALLURGIYA, MOSCOW   256 PPS.
                   400000   20

      FINKEL,V.A.   SEE SMIRNOV AND FINKEL (1965)

      FINKELSHTEIN,Y.N.   SEE BOIKO ET AL. (1972)

      FIRTH,E.M.   SEE JAMES AND FIRTH (1928)

170   FLACK,H.D. (1974)
      ACTA CRYST. A30, 569-573
      AUTOMATIC ABSORPTION CORRECTION USING INTENSITY MEASUREMENTS
      FROM AZIMUTHAL SCANS
                   18   220

      FLEISCHMANN,K.   SEE GIEREN ET AL. (1973)

171   FLINN,P.A., MCMANUS,G.M, AND RAYNE,J.A. (1961)
      PHYS. REV. 123, 809-812
      EFFECTIVE X-RAY AND CALORIMETRIC DEBYE TEMPERATURE FOR COPPER
                   323174   129

172   FLINT,E.E. AND BUTUZOV,V.P. (1937)
      ZAVOD. LAB. 6, 91-95
      CAMERAS FOR ROENTGENOGRAPHY AT LOW TEMPERATURES
                   20

      FOMENKO,N.   SEE DONDE ET AL. (1967)

173   FORD,P.T. AND POWELL,H.M. (1954)
      ACTA CRYST. 7, 604-605
      THE UNIT CELL OF KBH4 AT 90 D-K
                   6   176

      FORMANEK,H.   SEE THOMANEK ET AL. (1973)

174   FORRESTOR,J.D. (1961)
      J. SCI. INSTRUM. 38, 153-156
      X-RAY DIFFRACTION DATA FROM SINGLE CRYSTALS AT 20 D-K
                   212140   82

175   FOURME,R. (1970)
      THESIS, FACULTE DES SCIENCES DE PARIS
      X-RAY TECHNIQUES FOR VARIABLE PRESSURE AND TEMPERATURE
                   211300   2   4

176   FOURME,R. (1972)
      ACTA CRYST. B28, 2984-2991
      X-RAY DIFFRACTION STUDY OF CRYSTALLINE FURAN
      AT ATMOSPHERIC PRESSURE
                   4   195

177   FOURME,R. (1975)
      PERSONAL COMMUNICATION
                  111300
                                                    35, 50, 54, 72, 96, 204

      FOURME,R.   SEE RENAUD AND FOURME (1966)

      FOURME,R.   SEE RENAUD AND FOURME (1967)

      FOURME,R.   SEE ANDRE ET AL. (1971)

      FOURME,R.   SEE ANDRE ET AL. (1972)

      FOURME,R.   SEE KAHN ET AL. (1973)

      FOXMAN,B.M.   SEE BENNETT ET AL. (1975)

178   FRANCOMBE,M.H. (1957)
      J. SCI. INSTRUM. 34, 35
      GAS-COOLING DEVICE FOR USE WITH A 19 CM X-RAY POWDER CAMERA
                  312310
                                                              99, 152

179   FRAZER,B.C. AND PEPINSKY,R. (1950)
      PHYS. REV. 80, 124
      LOW-TEMPERATURE X-RAY DIFFRACTION GONIOMETER
                  223300
                                                              133

      FRAZER,B.C.   SEE KEELING ET AL. (1953)

180   FRENZ,B., ENEMARK,J., SCHROEDER,L.W., HODGSON,D., ROBINSON,W.
      LOYD,R., AND IBERS,J.A. (1969)
      J. APPL. CRYST. 2, 112-115
      A SIMPLE COLD GAS FLOW APPARATUS
                  111300
                                                              36, 55, 56, 69

181   FRENZ,B.A. AND IBERS,J.A. (1972)
      INORG. CHEM. 11, 1109-1116
      THE CRYSTAL AND MOLECULAR STRUCTURE OF (NI(PHEN)3)(MN(CO)5)2
                  253370
                                                              145

182   FRIDRICHSONS,J. AND MATHIESON,A.MCL. (1958)
      REV. SCI. INSTRUM. 29, 784-785
      LOW-TEMPERATURE ATTACHMENT FOR A SINGLE-CRYSTAL EQUI-INCLINATION
      WEISSENBERG GONIOMETER
                  212350
                                                              46, 86

      FURUSETH,S.   SEE HOPE ET AL. (1973)

      GALY,J.   SEE TRUT ET AL. (1973)

      GARLAND,C.W.   SEE BONILLA ET AL. (1970)

      GASAN,V.M.   SEE BOL'SHUTKIN ET AL. (1969)

      GASAN,V.M.   SEE BOL'SHUTKIN ET AL. (1970)

      GASAN,V.M.   SEE BOL'SHUTKIN ET AL. (1972)

      GATALO,Z.   SEE GIRT ET AL. (1974)

      GAWLIK,D.   SEE GOBRECHT ET AL. (1971)

      GEGUSIN,Y.E.   SEE ARONOVA ET AL.(1959)

183 GERARD,N. (1972)
     J. PHYS. E., SCI. INSTRUM. 5, 524-526
     A SYMMETRICALLY MOVING THETA-THETA X-RAY DIFFRACTION GONIOMETER
                    300070   224

184 GERARD,N. AND PERNOLET,R. (1973)
     J. PHYS. E., SCI. INSTRUM. 6, 512
     LOW TEMPERATURE, HIGH PRESSURE CELL FOR X-RAY DIFFRACTION
                    322475   23, 134

     GEROLD,V.K.   SEE GOULD AND GEROLD (1965)

185 GERVAIS,M. SELLA,C., AND SPRITZER,C. (1966)
     BULL. SOC. FR. MINERAL. CRISTOLLAGR. 89, 237-242
     X-RAY DIFFRACTION CAMERAS AT LOW-TEMPERATURES
                    321300   86, 133

186 GHISLAIN,R., DELÉHOUZÉE,L., AND DERUYTERRE,A. (1965)
     J. SCI. INSTRUM. 42, 502-503
     A VERY SIMPLE LOW TEMPERATURE ATTACHMENT
     FOR THE NORELCO X-RAY DIFFRACTOMETER
                    321372   132, 140, 141

187 GIBBONS,D.F. (1958)
     PHYS. REV. 112, 136-140
     THERMAL EXPANSION OF SOME CRYSTALS WITH THE DIAMOND STRUCTURE
                    14   16

188 GIEREN,A., HOPPE,W., AND FLEISCHMANN,K. (1973)
     ANGEW. CHEM. INTERNAT. ED. 12, 322
     LOW TEMPERATURE X-RAY STRUCTURE ANALYSIS OF AMMONIUM CARBAMATE
                    182400   157

     GILBERT,M.M.   SEE EINSTEIN ET AL.(1972)

189 GIRT,E., BABIĆ,E., GATALO,Z., KURŠUMOVIĆ,A.,
     AND LEONTIĆ,B. (1974)
     J. PHYS. E., SCI. INSTRUM. 7, 354-355
     MODIFICATION OF SEEMAN-BOHLIN X-RAY CAMERA OF METASTABLE
     STATES OF ULTRARAPIDLY QUENCHED SAMPLES
                    323110   7   23, 107, 130, 202

190 GIRT,E., LEONTIĆ,B., AND KURŠUMOVIĆ,A. (1975)
     J. PHYS. E., SCI. INSTRUM. 8, 59-61
     X-RAY SCATTERING CAMERA FOR HIGH PRECISION MEASUREMENTS OF THE
     LATTICE PARAMETER AT THREE TEMPERATURES
                    323110   23, 130, 202

191 GOBRECHT,H., GAWLIK,D., AND GROSSE,R. (1971)
     J. PHYS. E., SCI. INSTRUM. 4, 913-914
     LIQUID AIR CRYOSTAT FOR THE BUERGER PRECESSION CAMERA
                    211364   68, 92, 95, 116

     GODEL,J.   SEE COPPENS ET AL. (1967)

     GODEL,J.   SEE RUDMAN AND GODEL (1969)

192 GOETZ,A. AND HERGENROTHER,R.C. (1932)
     PHYS. REV. 40, 643-661
     X-RAY STUDIES OF THE THERMAL EXPANSION OF
     BISMUTH SINGLE CRYSTALS
                    165300   152

193   GOLDAK,J.A.   (1964)
      J. SCI. INSTRUM.   41, 722-726
      A 77-600 D-K VACUUM PATH X-RAY DIFFRACTION APPARATUS
                        323374   117, 118, 119, 131

194   GOLDSCHMIDT,V.M.   (1912)
      Z. F. KRISTALLOGR.   51, 1-27
      CHANGES OF ANGLES IN MINERALS AT LOW TEMPERATURES   3

195   GOLIK,V.R.   (1960)
      INDUSTRIAL LAB. USSR   26, 391-392
      A LOW TEMPERATURE X-RAY CAMERA
                        465323   157

196   GOLOB,P. AND HORN,M.   (1973)
      J. APPL. CRYST.   6, 410-411
      AN X-RAY CAMERA FOR SPECIAL LOW-TEMPERATURE STUDIES
                        323310   133

197   GOPALAKRISHNA,E.M. AND CARTZ,L.   (1972)
      ACTA CRYST.   B28, 2917-2924
      CRYSTAL STRUCTURE OF FROZEN O-ETHOXYBENZOIC ACID
                        111400   54, 89, 95, 96, 167

      GORBATII,L.Y.   SEE BOIKO ET AL. (1972)

198   GOULD,R.W. AND GEROLD,V.K.   (1965)
      NORELCO REPORTER   12, 12-13
      INEXPENSIVE CONTROLLED TEMPERATURE SAMPLE HOLDER FOR LOW ANGLE
      SCATTERING STUDIES OF QUENCHED METAL FOILS
                        325472   131, 140

199   GOTO,M., ASADA,E., UCHIDA,T., AND ONO,K.   (1969)
      YUKAGAKU   18, 299-305
      SIMULTANEOUS MEASUREMENTS OF X-RAY DIFFRACTION AND DTA
                        322374   25

      GRAMACCIOLI,C.M.   SEE FILIPPINI ET AL. (1974)

200   GRÄNICHER,H., HELG,U., AND SCHAR,R.   (1959)
      HELV. PHYS. ACTA   32, 474-476
      AN X-RAY CAMERA FOR VERY LOW TEMPERATURES
                        313111   82, 130

201   GRAY,W.T. AND FINCH,D.I.   (1971)
      PHYSICS TODAY   24 (SEPTEMBER) PPS. 32-40
      HOW ACCURATELY CAN TEMPERATURE BE MEASURED.
                        21   229

      GREGORY,E.   SEE MCDONALD ET AL. (1966)

      GRENTHE,I.   SEE DANIELSSON ET AL. (1976)

      GRENVILLE-WELLS,H.J.   SEE LONSDALE AND GRENVILLE-WELLS (1956)

      GROSSE,R.   SEE GOBRECHT ET AL. (1971)

202   GRUBER,E.E., BLEWITT,T.H., TESK,J.A., SHARMA,B.D.,
      AND BLACK,R.E.   (1969)
      REV. SCI. INSTRUM.   40, 1429-1432
      CRYOSTAT-SPECTROMETER SYSTEM FOR PRECISION MEASUREMENT
      OF LATTICE PARAMETERS OF IRRADIATED MATERIALS
                        223170   129

203   GRUSHKO,V.I., BOL'SHUTKIN,D.N., SHCHERBAKOV,G.N.,
      AND SHEVCHENKO,N.F.   (1970)
      SOVIET PHYSICS - CRYSTALLOGRAPHY   15, 536-538
      GROWING OF LARGE ARGON CRYSTALS
                  263320   4
                                                              154, 157

204   GUENGANT,L.   (1958)
      J. RECH. DU C.N.R.S.   9, 371-373
      CRYOSTAT FOR X-RAY DIFFRACTION STUDIES OF ICE AT HIGH PRESSURES
                  322415
                                                              134

      GUIBERT,M.   SEE AMOROS ET AL. (1961)

205   GÜNTHER,P., HOLM,K., AND STRUNZ,H.   (1939)
      Z. PHYSIK. CHEM.   B43, 229-239
      THE STRUCTURE OF SOLID HYDROGEN FLUORIDE
                  322310   3
                                                              130, 175

206   GUTTORMSON,R.J. AND ROBERTSON,B.E.   (1973)
      ACTA CRYST.   B29, 173-179
      THE CRYSTAL STRUCTURE OF C8H6CL4 AT -65 D-C
                  253370   17
                                                  18, 145, 148, 163

      GUTTZEIT,R.   SEE ZICKERT AND GUTTZEIT (1966)

207   HAAS,D.J.   (1968)
      ACTA CRYST.   B24, 604
      X-RAY STUDIES ON LYSOZYME CRYSTALS AT -50 D-C
                  500000   1
                                                              164

208   HAAS,D.J. AND ROSSMANN,M.G.   (1970)
      ACTA CRYST.   B26, 998-1004
      CRYSTALLOGRAPHIC STUDIES ON LACTATE DEHYDROGENASE AT -75 D-C
                  500000   1
                                                              165, 246

      HAENDLER,H.M.   SEE LANGE AND HAENDLER (1972)

      HALL,W.J.   SEE SPARKS ET AL. (1972)

      HALVERSON,G.   SEE HANSON AND HALVERSON (1948)

      HAMPTON,W.F.   SEE BARNES AND HAMPTON (1935)

209   HANSON,A.W.   (1965A)
      ACTA CRYST.   18, 599-604
      THE CRYSTAL STRUCTURE OF C2204H20
                  12
                                                              198, 208

210   HANSON,A.W.   (1965B)
      ACTA CRYST.   19, 19-26
      THE CRYSTAL STRUCTURE OF THE AZULENE,S-TRINITROBENZENE COMPLEX
                  12
                                                          198, 199, 208

211   HANSON,E.E. AND HALVERSON,G.   (1948)
      J. AM. CHEM. SOC.   70, 779-783
      X-RAY DIFFRACTION STUDY OF SOME SYNTHETIC RUBBERS
      AT LOW TEMPERATURES
                  322420   7
                                                              115, 134

212   HARDING,T.T.   (1956)
      CAN. J. CHEM.   34, 371-375
      LOW-TEMPERATURE APPARATUS FOR SINGLE CRYSTAL X-RAY GONIOMETERS
                  111300
                                                  38, 64, 66, 67, 72

HARDING,T.T.   SEE WALLWORK AND HARDING (1954)

213 HARDY,J.A. (1968)
     J. PHYS. E., SCI. INSTRUM. 1, 580-582
     THE COOLING OF CRYSTALS IN X-RAY CRYSTALLOGRAPHY
            114300
                                                            35

     HARDY,L.H.   SEE MUELLER ET AL. (1958)

214 HARTOULARI,R. AND DUFOUR,L. (1970)
     BULL. SOC. CHIM. FR. 1970, 4274-4275
     X-RAY DIFFRACTION GONIOMETER SAMPLE HOLDER FOR THE LOW TEMPERATURE
     STUDY OF   A POWDER SAMPLE IN AN AMMONIUM ATMOSPHERE
            332470   6
                                                            145

215 HARVEY,G.G. (1933)
     PHYS. REV. 43, 707-710
     DIFFUSE SCATTERING OF X-RAYS FROM SYLVINE. III.
            223370
                                                            131

216 HASELDEN,G.G. (1971)
     CRYOGENIC FUNDAMENTALS
     ACADEMIC PRESS, N.Y.
                            21
                                                         233, 245

217 HAWES,L.L. (1959)
     ACTA CRYST. 12, 34-35 (ALSO IBID. 142-143, 477-478)
     THE THERMAL EXPANSION OF SOLID BROMINE
            372310   5
                                                            155

     HEADY,H.H.   SEE SMITH AND HEADY (1955)

218 HEATON,L., MUELLER,M.H., ADAM,M.F. AND HITTERMAN,R.L. (1970)
     J. APPL. CRYST. 3, 289-294
     NEUTRON DIFFRACTION CRYO-ORIENTER
            623171
                                                         108, 122

     HEISKANEN,K.   SEE HOVI ET AL. (1964)

     HELG,U.   SEE GRANICHER ET AL. (1959)

     HELLAWELL,A.   SEE AXON ET AL. (1953)

219 HELMHOLZ,L. (1935)
     J. CHEM. PHYS. 3, 740-747
     THE CRYSTAL STRUCTURE OF HEXAGONAL SILVER IODIDE
            223340
                                                            133

220 HENGSTENBERG,J. AND MARK,H. (1928)
     Z. F. KRISTALLOGR. 69, 271-284
     THE FORM AND SIZE OF THE MICELLES OF CELLULOSE AND RUBBER
            362320
                                                            152

221 HENSHAW,D.E. (1957)
     J. SCI. INSTRUM. 34, 270-271
     A METHOD OF STABILIZING THE SPECIMEN TEMPERATURE
     FOR SINGLE-CRYSTAL X-RAY CRYSTALLOGRAPHY
            111300   4
                                                          38, 187

222 HERBSTEIN, F.H. (1963)
     ACTA CRYST. 16, 255-263
     ACCURATE DETERMINATION OF CELL DIMENSIONS FROM
     SINGLE-CRYSTAL X-RAY PHOTOGRAPHS
                            11

HERGENROTHER,R.C. SEE GOETZ AND HERGENROTHER (1932)

HIDAKA,M. SEE SUGINO ET AL. (1973)

HILDEBRAND,J.H. SEE CAMPBELL AND HILDEBRAND (1943)

223 HIMMLER,U., PEISL,H., SEPP,A., AND WAIDELICH,W. (1969)
Z. ANGEW. PHYS. 28, 104-109
EVAPORATION CRYOSTAT FOR HIGH PRECISION X-RAY DIFFRACTION
MEASUREMENTS BETWEEN 4 AND 350 D-K
323174   108, 129

224 HINE,R., RICHARDS,J.P.G., AND TICHY,K. (1975)
J. APPL. CRYST. 8, 37-41
THE AVOIDANCE OF MULTIPLE DIFFRACTION ERRORS IN SINGLE-CRYSTAL
INTENSITY MEASUREMENTS AT LOW TEMPERATURES
16   209

HINE,R. SEE WOODLEY ET AL. (1971)

225 HIRSHFELD,F.L. AND SCHMIDT,G.M.J. (1956)
ACTA CRYST. 9, 233-236
THREE EXAMPLES OF ACCURATE CRYSTAL-STRUCTURE ANALYSIS BY
LOW-TEMPERATURE X-RAY PHOTOGRAPHY
162350   11   12, 154

HITTERMAN,R.L. SEE HEATON ET AL. (1970)

HODGSON,D. SEE FRENZ ET AL. (1969)

226 HÖHNE,G., KLIPPING,G., LUDEWIG,J., TIPPE,A., AND WALTER,H. (1968)
Z. ANGEW. PHYS. 25, 250-255
LOW-TEMPERATURE X-RAY CAMERAS
212100   82

227 IBID.
312110

228 IBID.
312130

HOLM,K. SEE GUNTHER ET AL. (1939)

HOLMES,D.R. SEE COLE AND HOLMES (1960)

229 HONEYWELL,W.I., KNOBLER,C.M., SMITH,B.L., AND PINGS,C.J. (1964)
REV. SCI. INSTRUM. 35, 1216-1219
APPARATUS FOR X-RAY DIFFRACTION STUDIES OF CONFINED FLUIDS
323372   16   132, 217

230 HOPE,H. (1969)
J. APPL. CRYST. 2, 308-309
FLAT FILM WEISSENBERG ATTACHMENT FOR THE
POLAROID XR-7 CASSETTE
16   90, 210

HOPE,H.   193

231 HOPE,H., FURUSETH,S., SELTE,K. AND KJEKSHUS,A. (1973)
ACA WINTER MEETING ABSTRACTS, PAPER B6
STRUCTURE OF (IO)2SO4
111300   4   50, 51

HOPPE,W.   SEE GIEREN ET AL. (1973)

HOPPE,W.   SEE THOMANEK ET AL. (1973)

232 HORI,T. AND MATSUNO,K. (1972)
ACTA CRYST. A28, S247
LOW-TEMPERATURE EQUIPMENT FOR AUTOMATIC FOUR-CIRCLE DIFFRACTOMETER
          211371   36, 77

HORN,H.   SEE GOLOB AND HORN (1973)

233 HORNE,R.A., CROFT,W.J., AND SMITH, L.B. (1959)
REV. SCI. INSTRUM. 30, 1132-1134
THERMOELECTRIC THERMOSTAT FOR X-RAY DIFFRACTION
          335572   144

234 HOSPITAL,M. (1968)
THESIS - UNIVERSITY OF BORDEAUX
CRYSTAL STRUCTURES OF N-ALIPHATIC DIAMIDES
          211300   50, 71

235 HOVI,V. HEISKANEN,K. AND VARTEVA,M. (1964)
ANN. ACAD. SCI. FENNICAE, SER. A., VI. PHYSICA 144, 3-13
(ALSO PUBLISHED IN LOW TEMPERATURE PHYSICS,LT9(B),1179-1183(1965))
X-RAY INVESTIGATION OF THE MODIFICATIONS II AND III OF NH4(BR)
AT TEMPERATURES BETWEEN 22 AND -125 D-C
          311310   46, 99, 127

236 HOVI,V., MÄNTYSALO,E., AND TIUSANEN,K. (1966)
ACTA MET. 14, 67-69
DETERMINATION OF THE MARTENSITIC CRITICAL TEMPERATURE OF THE
LITHIUM SINGLE CRYSTAL BY USING THE LAUE METHOD
          123320

HUBER,H.   SEE IHRINGER ET AL. (1975)

237 HUBER,J.C. (1969)
J.PHYS. E., SCI. INSTRUM. 2, 294-296
GAS-FLOW TEMPERATURE CONTROLLER
          115300   27, 37

238 HUBER,R. AND KOPFMANN,G. (1969)
ACTA CRYST. A25, 143-152
EXPERIMENTAL ABSORPTION CORRECTIONS - RESULTS
          18   220

239 HUBER-BUSER,E. (1971)
Z.F.KRISTALLOGR. 133, 150-167
REFINEMENT OF LIGHT ATOMS IN HEAVY ATOM STRUCTURES
          11   12   198

240 HUFFMAN,J.C. (1974)
THESIS, INDIANA UNIVERSITY
A CRYSTALLOGRAPHIC STUDY OF ELECTRON DEFICIENT COMPOUNDS
          111300   2   3   4   16   20   50, 53, 58, 175, 185, 189,
                                                     190, 210, 214

241 HUFFMAN,J.C., STREIB,W.E., AND MUELLER,J.M. (1973)
ACA MEETING ABSTR. SER.2, 1, 156
LOW TEMPERATURE TECHNIQUES AND EQUIPMENT FOR
SINGLE CRYSTAL STUDIES
          211300   50, 53

HUGHES,R.E.   SEE PARKES AND HUGHES (1963)

HUGGINS,F.   SEE KING ET AL. (1969)

HÜLLER,A.   SEE PRESS AND HULLER (1973)

242 HUME-ROTHERY,W. AND STRAWBRIDGE,D.J. (1947)
J. SCI. INSTRUM. 24, 89-91
A GENERAL PURPOSE DEBYE-SCHERRER CAMERA AND ITS APPLICATION TO
WORK AT LOW TEMPERATURES
          312310
                                                        46, 99

HUME-ROTHERY,W. SEE AXON ET AL. (1953)

243 HURWITT,S.D. (1967)
RESEARCH/DEVELOPMENT (R/D) 1967 (JANUARY) 24-27
OPEN-AIR FROST-FREE CRYOGENIC COOLING
          111300

HUTCHEON,W.L.   SEE BENNETT ET AL. (1975)

244 HUTCHINSON,T.E. AND MILLER,J.E. (1958)
BULL. AMER. PHYS. SOC. 3, 304
A LOW-TEMPERATURE X-RAY DIFFRACTION CAMERA
          323310
                                                        131

IBERS,J.A.   SEE FRENZ ET AL. (1969)

IBERS,J.A.   SEE FRENZ AND IBERS (1972)

IBRAGIMOV,N.I.   SEE ALIEV AND IBRAGIMOV (1959)

245 ICHIKAWA,M. (1974)
ACTA CRYST. B30, 651-655
THE CRYSTAL STRUCTURE AND PHASE TRANSITION OF
AMMONIUM HYDROGEN BIS-CHLOROACETATE
          212350
                                                        86

246 IHRINGER,J., HUBER,H., WEINER,K.L., AND PRANDL,W. (1975)
ACTA CRYST. A31, S240
A LOW TEMPERATURE GUINIER CAMERA AND DIFFRACTOMETER
          323130

IIDA,T.   SEE ODA ET AL. (1943)

247 INTRATER,J. AND APPEL,A. (1961)
REV. SCI. INSTRUM. 32, 1065-1066
SYSTEM FOR LOW-TEMPERATURE X-RAY DIFFRACTION STUDIES
          323370
                                                        132

248 ITO,T. AND SAKURAI,T. (1973)
ACTA CRYST. B29, 1594-1603
THE STRUCTURE AND ELECTRON DENSITY OF ETHYLENEIMINE QUINONE
          253370   14
                                                        145, 147

249 IWASAKI,F. AND IWASAKI,H. (1972)
ACTA CRYST. B28, 3370-3376
THE CRYSTAL AND MOLECULAR STRUCTURE OF TRIBENZYLAMINE AT -70 D-C
          253370
                                                        145, 148

250 JACCARD,C. KÄNZIG,W. AND PETER,M. (1953)
HELV. PHYS. ACTA 26, 521-544
THE BEHAVIOR OF COLLOIDAL KH2PO4
          312311
                                                        84

JACOBSON,R.A.   SEE THAXTON AND JACOBSON (1970)

251   JAMES,R.W. AND FIRTH,E.M.   (1928)
PROC. ROY. SOC. (LONDON) A117, 62-87
AN X-RAY STUDY OF THE HEAT MOTIONS OF THE ATOMS IN A
ROCK-SALT CRYSTAL
            121370   45

JAMES,W.J.   SEE JOHNSON ET AL. (1974)

JARVIS,J.A.J.   SEE DAVIES ET AL.(1970)

JAULIN,M.   SEE NAUDON AND JAULIN (1968)

252   JAUNCEY,G.E.M. AND RICHARDSON,H.W.   (1934)
J. OPT. SOC. AM. 24, 125-126
SENSITIVITY OF PHOTOGRAPHIC FILMS TO X-RAYS AT
LOW TEMPERATURES
                    15   205

JEFFREY,G.A.   SEE ALLEN ET AL. (1963)

253   JETTER,L.K., MCHARGUE,C.J., WILLIAMS,R.O.,
AND YAKEL,H.L.,JR.   (1957)
REV. SCI. INSTRUM. 28, 1087-1088
LOW-TEMPERATURE CAMERA FOR X-RAY DIFFRACTOMETER
            323377   112, 131

254   JOHNSON,A.D. AND ANDERSON,D.L.   (1966)
NORELCO REPORTER 13, 64
GLOVE BOX HANDLING OF GLASS CAPILLARIES
                    6   176

255   JOHNSON,L.R. (1975)
J. APPL. CRYST. 8, 507-514
THE PHASES OF THE AGI-KI-H2O SYSTEM
            332470   145

256   JOHNSON,P.E., SHAH,J.S., VORA,P.M., AND JAMES,W.J.   (1974)
ACA MEETING ABSTR. SER. 2, 2, 52
A LOW TEMPERATURE X-RAY POWDER CAMERA
            343111   143

JONES,G.O.   SEE FIGGINS ET AL. (1956)

257   JONES,G.R., BURBANK,R.D., AND FALCONER,W.E.   (1970)
J. CHEM. PHYS. 53, 1605-1606
CRYSTALLINE MODIFICATIONS OF XENON HEXAFLUORIDE
                    4

JONES,G.R.   SEE BURBANK AND JONES (1974)

258   JURNAK,F.A. AND RAYMOND,K.N. (1974)
INORG. CHEM. 13, 2387-2397
EEFFECT OF PACKING FORCES ON THE GEOMETRY OF THE (NI(CN)5)3-ION
                    7   20

259   JUST,A., REES,B., AND YELON,W.B. (1975)
ACTA CRYST. B31, 2649-2658
ELECTRONIC STRUCTURE OF CHROMIUM HEXACARBONYL AT 78 K
                    18   218

KABALKINA,S.S.   SEE TRIOTSKAYA ET AL. (1969)

KABALKINA,S.S.　　SEE MALYUSHITSKAYA ET AL. (1975)

260　KAHN,R., FOURME,R., ANDRÉ,D., AND RENAUD,M.　(1973)
ACTA CRYST. B29, 131-138
CRYSTAL STRUCTURES OF CYCLOHEXANE I AND II
　　　　4

261　KAGAN,A.S. AND UMANSKII,Y.S.　(1960)
INDUSTRIAL LAB. USSR　26, 112-113
CAMERAS FOR OBTAINING X-RAY DIFFRACTION AT HIGH AND
LOW TEMPERATURES WITH THE URS-501 DIFFRACTOMETER
　　　　　　323370　　　　　　　　　　　　　　　　　　　　　　131

KALNAJS,J.　　SEE ABRAHAMS AND KALNAJS (1954)

262　KAMB,B. AND DAVIS,B.L.　(1964)
PROC. NAT. ACAD. SCI. (U.S.)　52, 1433-1439
ICE VII, THE DENSEST FORM OF ICE
　　　　　　324475　　　　　　　　　　　　　　　　　　　　　　134

263　KAN,L.S. AND LAZAREV,B.G.　(1951)
ZH. TEKH. FIZ.　21, 1542-1543
A LOW-TEMPERATURE X-RAY CAMERA
　　　　　　323310　　　　　　　　　　　　　　　　　　　　　　133

KANNAN,K.K.　　SEE VISWAMITRA AND KANNAN (1962)

KÄNZIG,W.　　SEE JACCARD, ET AL. (1953)

KARPUKHIN,V.I.　　SEE NIKOLAENKO AND KARPUKHIN (1967)

KATO,T.　　SEE MAETA ET AL. (1975)

KATSENEL'SON,A.A.　　SEE ZHURAVLEV AND KATSENEL'SON (1959)

264　KAUFMAN,H.S. AND FANKUCHEN,I.　(1949)
REV. SCI. INSTRUM.　20, 733-734
A LOW-TEMPERATURE SINGLE CRYSTAL X-RAY DIFFRACTION TECHNIQUE
　　　　　　112340　　　　　4　　　　　　　　　　　　　31, 46, 67, 186

KAWAMINAMI,M　　SEE OKAZAKI AND KAWAMINAMI (1973)

265　KAY,H.F. AND VOUSDEN,P.　(1949)
PHIL. MAG.　40, 1019-1040
SYMMETRY CHANGES IN BARIUM TITANATE AT LOW TEMPERATURES
　　　　　　111343　　　　　　　　　　　　　　　　　　　　　74

266　KEELING,R., FRAZER,B.C., AND PEPINSKY,R.　(1953)
REV. SCI. INSTRUM.　24, 1087-1095
LOW-TEMPERATURE X-RAY GONIOMETER FOR STRUCTURAL STUDIES
OF CRYSTAL TRANSITIONS
　　　　　　223352　　　　　　　　　　　　　　　　　　　　116, 126

267　IBID.
　　　　　　323311

268　IBID.
　　　　　　223371

269　KEESOM,W.H.　(1924)
PROC. FOURTH INTERNATIONAL CONF. REFRIGERATION　4, 117A-120A
REPORT ON THE X-RAY INVESTIGATION OF THE CONSTITUTION IN THE
LIQUID AND SOLID STATES OF SUBSTANCES AT LOW TEMPERATURES
　　　　　　　7　　20　　　　　　　　　　　　　　　　　　　11

270 KEESOM,W.H. (1929)
     COMM. PHYS. LAB. UNIV. LEIDEN, SUPPL. NO. 64, 39-45
     REPORT ON RESEARCHES CONCERNING THE STRUCTURE OF SUBSTANCES
     IN THE SOLID AND LIQUID STATES AT LOW TEMPERATURES
                          20

     KEESOM,W.H.   SEE DE SMEDT AND KEESOM (1924)

     KEESOM,W.H.   SEE DE SMEDT ET AL. (1930)

271 KEESOM,W.H. AND DE SMEDT,J. (1922)
     PROC. ACAD. SCI. AMSTERDAM 25, 118-124
     ON THE DIFFRACTION OF ROENTGEN RAYS IN LIQUIDS
                    363312   135

272 KEESOM,W.H., DE SMEDT,J., AND MOOY,H.H. (1930)
     PROC. ACAD. SCI. AMSTERDAM 33, 814-819
     THE CRYSTAL STRUCTURE OF PARA-H2 AT LIQUID HELIUM TEMPERATURE
                         7   4

273 KEESOM,W.H. AND KOHLER,J.W.L. (1934)
     PHYSICA 1, 655-658
     THE LATTICE CONSTANT AND EXPANSION COEFFICIENT OF
     SOLID CARBON DIOXIDE
                    322110   130

274 KEESOM,W.H. AND TACONIS,K.W. (1935)
     PHYSICA 2, 463-471 (SEE ALSO IBID. 3, 238 (1936))
     AN X-RAY GONIOMETER FOR INVESTIGATION OF THE CRYSTAL STRUCTURE
     OF SOLIDIFIED GASES
                    223350   3   125, 175

275 KEESOM,W.H. AND TACONIS,K.W. (1938)
     PHYSICA 5, 161-169
     STRUCTURE OF SOLID HELIUM
                    223145   130

276 KELLETT,E.A. AND STEWARD,E.G. (1962)
     J. SCI. INSTRUM. 39, 306-308
     HEATING AND COOLING ATTACHMENTS FOR X-RAY POWDER DIFFRACTOMETRY
                    315372   101

     KEULEN,E.   SEE VERSCHOOR AND KEULEN (1971)

277 KHAN,D.C. AND ERICKSON,R.A. (1970)
     REV. SCI. INSTRUM. 41, 107-110
     AN APPARATUS FOR NEUTRON DIFFRACTION FROM SINGLE CRYSTALS
                    633471   144

     KHEIKER,D.M.   SEE BOIKO ET AL. (1972)

     KHEIKER,D.M.   SEE ZEVIN AND KHEIKER (1958)

278 KHOTKEVICH,V.I. (1952)
     ZH. TEKH. FIZ. 22, 474-479
     THE STRUCTURE OF METALS PLASTICALLY DEFORMED AT LOW TEMPERATURES
                    423217

     KHOTKEVICH,V.I.   SEE DONDE ET AL. (1967)

     KILBOURN,B.T.   SEE DAVIES ET AL.(1970)

     KINDER,H.   SEE SEGMULLER ET AL. (1974)

279   KING,H.W. (1975)
      PERSONAL COMMUNICATION
                  323171   *115, 135*

      KING,H.W.   SEE COCKS ET AL. (1966)

      KING,H.W.   SEE ABELL AND KING (1970)

280   KING,H.W., CABELKA,D. AND HUGGINS,F. (1969)
      ACTA CRYST A25, S70
      1.5 D°K CRYOSTAT FOR POWDER DIFFRACTOMETRY IN HIGH MAGNETIC FIELDS
                  323170   *129, 137*

281   KING,H.W. AND PREECE,C.M. (1967)
      ADV. X-RAY ANALYSIS 10, 354-365
      PRECISION LATTICE PARAMETER DETERMINATION AT LIQUID HELIUM
      TEMPERATURES BY DOUBLE-SCANNING DIFFRACTOMETRY
                  323171   14   *24, 108, 129, 137, 201, 206*

282   KING,H.W. AND PREECE,C.M. (1968)
      SIEMENS REVIEW 35, 1-7
      X-RAY DIFFRACTOMETRY AT LIQUID HELIUM TEMPERATURES
                  323171   14   16   *138, 206*

283   KING,M.V. (1954)
      ACTA CRYST. 7, 601-602
      AN EFFICIENT METHOD FOR MOUNTING WET PROTEIN CRYSTALS
      FOR X-RAY STUDIES
                  500000   1   2   *170, 171, 173*

284   KINZIE,P.A. (1973)
      THERMOCOUPLE TEMPERATURE MEASUREMENT
      J. WILEY AND SONS, NEW YORK
                                  21   *229, 231*

285   KISELEVA,K.V. AND MIKHALENKO,V.N. (1962)
      INDUSTRIAL LAB. USSR 28, 1476
      X-RAY CAMERA FOR OBTAINING DEBYE-SCHERRER DIAGRAMS AT
      LOW TEMPERATURES
                  362311   *157*

      KJEKSHUS,A.   SEE HOPE ET AL. (1973)

      KLIPPING,G.   SEE MOHNE ET AL. (1968)

      KLOTZER,F.   SEE FEHER AND KLOTZER (1935)

      KLOTZER,F.   SEE FEHER AND KLOTZER (1937)

      KLUG,H.P.   SEE SEARS AND KLUG (1962)

      KNOBLER,C.M.   SEE HONEYWELL ET AL. (1964)

      KNOX,K.   SEE BECKMAN AND KNOX (1961)

286   KOGAN,V.S., BULATOV,A.S., AND YAKIMENKO,L.F. (1964)
      SOVIET PHYS. JETP 19, 107-109
      TEXTURE IN LAYERS OF HYDROGEN ISOTOPES CONDENSED
      ON A COOLED SUBSTRATE
                  323110   *130, 134*

287   KOGAN,V.S. AND BULATOVA,R.F. (1969)
      APP. METODY RENTGENOVSK. ANAL. 5, 15-23
      APPARATUS FOR X-RAY STRUCTURAL ANALYSIS AT LOW TEMPERATURES
                  322110   3   *130, 154, 175*

288 KOGAN,V.S., LAZAREV,B.G., AND BULATOVA,R.F. (1960)
     SOVIET PHYS. JETP 10, 485-488
     X-RAY DIFFRACTION BY POLYCRYSTALLINE SAMPLES OF HYDROGEN ISOTOPES
              323110   3   6   18
                                                        130, 135, 175, 177, 217

289 KOGAN,V.S. AND OMAROV,T.G. (1965)
     IZV. AKAD. NAUK AZERB. SSR, SER. FIZ.-TEKH. I MAT. NAUK 1965, 87-89
     VACUUM AND LOW-TEMPERATURE X-RAY CAMERA
              323310   6
                                                        133, 177

     KOHLER,J.W.L.   SEE KEESOM AND KOHLER (1934)

     KOLONTSOVA, E.V. SEE BAGARYATSKII ET AL. (1951)

     KOPFMANN,G.   SEE HUBER AND KOPFMANN (1969)

290 KOROLEVA,G.M. (1973)
     INSTRUM. EXPTL. TECH. (USSR) 16, 661
     THE URNT-180 LOW TEMPERATURE X-RAY INSTALLATION
              312370

     KORYASHKIN,V.I. SEE BOIKO ET AL. (1972)

291 KRAMER,W.E., VENTURINO,A.S., AND MAZELSKY,R. (1963)
     REV. SCI. INSTRUM. 34, 933-934
     LOW TEMPERATURE CONVERSION UNIT FOR THE PHILLIPS POWDER CAMERA
              311310
                                                        46, 100

292 KREUGER,A. (1955)
     ACTA CRYST. 8, 348-349
     A WEISSENBERG CAMERA FOR USE AT CONSTANT TEMPERATURES BETWEEN
     ABOUT -150 D°C AND 300 D°C
              212356   4
                                                        38, 55, 86, 191

293 KRIEGER,M., CHAMBERS,J.L., CHRISTOPH,G.G., STROUD,R.M.,
     AND TRUS,B.L. (1974)
     ACTA CRYST. A30, 740-748
     DATA COLLECTION IN PROTEIN CRYSTALLOGRAPHY -
     CAPILLARY EFFECTS AND BACKGROUND CORRECTIONS
              500000   18
                                                        219

294 KRISHNA MURTI,G.S.R. AND SEN,S.N. (1956)
     INDIAN J. PHYS. 30, 242-249
     ON THE CRYSTAL STRUCTURE OF P-DICHLOROBENZENE
     AT DIFFERENT TEMPERATURES
              362312

     KRIVY,I.   SEE SKABA AND KRIVY (1970)

295 KRUNER,H. (1926)
     Z. F. KRISTALLOGR. 63, 275-283
     THE CRYSTAL STRUCTURE OF SOLID CARBON DIOXIDE
              323310
                                                        134

     KUCHERYAVII,V.A. SEE BOL'SHUTKIN ET AL. (1970)

     KUCHERYAVII,V.A. SEE BOIKO ET AL. (1972)

     KUDRYASHOV,I.H. SEE BOIKO ET AL. (1972)

     KUMRA,S.K.   SEE BRONSVELD ET AL.(1973)

     KURŠUMOVIĆ,A.   SEE GIRT ET AL. (1974)

KURŠUMOVIĆ,A.   SEE GIRT ET AL. (1975)

296   KUZNETSOV,V.G. (1956)
RUSSIAN J. INORG. CHEM. 1 (NO. 7) 95-114
X-RAY ANALYSIS AT HIGH AND LOW TEMPERATURES
            114300
                                                                              153, 154

297   LA COUR,T. (1974)
ACTA CRYST. B30, 1642-1643
(AND PERSONAL COMMUNICATION)
1,3,4-THIADIAZOLE AT 220 K
            111300
                                                                              39, 77

LADELL,J.   SEE STEINFINK ET AL. (1953)

LADELL,J.   SEE ABOWITZ AND LADELL (1968)

LAING,M.   SEE NOLTE ET AL. (1975)

298   LANGE,B.A. AND HAENDLER,H.M. (1972)
J. APPL. CRYST. 5, 310
A CAPILLARY SUPPORT APPARATUS FOR USE IN GLOVE BAGS AND DRY BOXES
            1   6
                                                                              176

LA PLACA,S. (1965)   PERSONAL COMMUNICATION
                                                                              56, 185

LARSON,A.C.   SEE CADY AND LARSON (1975)

LARSEN,F.K.   SEE COPPENS ET AL. (1974)

LAZAREV,B.G.   SEE KAN AND LAZAREV (1951)

LAZAREV,B.G.   SEE KOGAN ET AL. (1960)

LECOQ,J.   SEE COULON ET AL. (1974)

LEIDER,H.R.   SEE SMITH AND LEIDER (1968)

LEIPOLDT,J.G.   SEE COPPENS ET AL. (1974)

299   LENHERT,P.G. AND TAKAGI,S. (1974)
ACA MEETING ABSTR. SER. 2, 2, 214
THE USE OF THE CT-38 CRYOSTAT WITH LIQUID NITROGEN COOLANT
            223171   17
                                                                              122

LEONARD,R.   SEE COPPENS ET AL. (1974)

LEONTIĆ,B.   SEE GIRT ET AL. (1974)

LEONTIĆ,B.   SEE GIRT ET AL. (1975)

300   LÉVY,F. (1969)
PHYS. COND. MAT. 10, 71-84
SPONTANEOUS MAGNETOSTRICTION EFFECTS IN SOME RARE-EARTH COMPOUNDS
            312111
                                                                              99, 202

LEVY,H.A.   SEE AGRON ET AL. (1972)

301   LEVY,J.H., SANGER,P.L., TAYLOR,J.C., AND WILSON,P.W. (1975)
ACTA CRYST. B31, 1065-1067
KUBIC HARMONIC ANALYSIS OF THE NEUTRON DIFFRACTION PATTERN
OF THE BODY-CENTERED CUBIC PHASE OF (MO)F6 AT 266 K
            19
                                                                              221

302 LIEBLING,G. AND MARSH,R.E. (1965)
     ACTA CRYST. 19, 202-205
     THE CRYSTAL AND MOLECULAR STRUCTURE OF CYCLOPENTADIENE
                165300   4
                                                              152, 183

303 LING,D. AND WAGENFELD,H. (1965)
     PHYS. LETTERS 15, 8-10
     ANOMALOUS TRANSMISSION OF X-RAYS IN PERFECT SINGLE GERMANIUM
     CRYSTALS AT LIQUID NITROGEN TEMPERATURE
                165320
                                                              157

304 LINKOAHO,M. (1968)
     ANN. ACAD. FENNICAE, SER. A., VI, PHYSICA 284, 5-31
     PARAMETERS OF NACL MEASURED AT 300, 80, AND 4 D-K
                323171   14
                                                              129

305 LIPPMAN,R. AND RUDMAN,R. (1976)
     J. APPL. CRYST. 9, 220-222
     A MECHANICALLY REFRIGERATED GAS STREAM (TO -120 D-C)
     AND SOME USEFUL ACCESSORIES
                111300
                                          64, 65, 91, 92, 187, 200, 209, 210

306 LIPSCOMB,W.N. (1957)
     NORELCO REPORTER 4, 54,75
     LOW-TEMPERATURE CRYSTALLOGRAPHY
                20

     LIPSCOMB,W.N.   SEE ABRAHAMS, ET AL. (1950)

     LIPSCOMB,W.N.   SEE COLLIN AND LIPSCOMB (1951)

     LIPSCOMB,W.N.   SEE DULMAGE AND LIPSCOMB (1951)

     LIPSCOMB,W.N.   SEE DULMAGE AND LIPSCOMB (1952)

     LIPSCOMB,W.N.   SEE ABRAHAMS AND LIPSCOMB (1952)

     LIPSCOMB,W.N.   SEE REED AND LIPSCOMB (1953)

     LIPSCOMB,W.N.   SEE ATOJI AND LIPSCOMB (1954)

     LIPSCOMB,W.N.   SEE STREIB AND LIPSCOMB (1962)

     LIPSCOMB,W.N.   SEE SMITH AND LIPSCOMB (1965)

307 LISOIVAN,V.I. (1975)
     INSTRUM. EXPTL. TECH. (USSR) 18, 1609-1610
     A LOW-TEMPERATURE ATTACHMENT TO AN X-RAY CAMERA
                111300

308 LONSDALE,K. AND GRENVILLE-WELLS,H.J. (1956)
     BRIT. J. APPL. PHYS. 7, 380
     LARGE INCREASE OF LIGHT SENSITIVITY AT LOW TEMPERATURES
     FOR TYPES OF X-RAY FILM
                15
                                                              205

309 LONSDALE,K. AND SMITH,H. (1941)
     J. SCI. INSTRUM. 18, 133-135
     X-RAY CRYSTAL PHOTOGRAPHY AT LOW TEMPERATURES
                165300   11
                                                              152, 153

310 LOSEE,D.L. AND SIMMONS,R.O. (1968)
     PHYS. REV. 172, 934-943
     EQUILIBRIUM VACANCY CONCENTRATION MEASUREMENTS ON SOLID KRYPTON
                323125   3   4   21, 107, 129, 136, 193, 201

LOVELL,F.M.   SEE CUCKA ET AL. (1970)

311   LOW,B.W., CHEN,C., BERGER,J., SINGMAN,L., AND PLETCHER,J.   (1966)
      PROC. NATL. ACAD. SCI. (U.S.)   56, 1746-1750
      STUDIES OF INSULIN CRYSTALS AT LOW TEMPERATURES
                  262320

      LOW,B.W.   SEE CUCKA ET AL. (1970)

      LOYD,R.   SEE FRENZ ET AL. (1969)

      LUCAS,R.   SEE COHEN ET AL. (1971)

      LUDEWIG,J.   SEE HOHNE ET AL. (1968)

312   LUNDGREN,J. (1970)
      ACTA CRYST. B26, 1893-1899
      HYDROGEN BOND STUDIES. XL.
                  4
                                                                            179, 194

      LUNDGREN,J.   SEE SPENCER AND LUNDGREN (1973)

313   LUZZATI,V. (1951)
      ACTA CRYST. 4, 120-131
      CRYSTAL STRUCTURE OF ANHYDROUS NITRIC ACID
                  4
                                                                            186

314   LUZZATI,V. (1953)
      ACTA CRYST. 6, 152-157
      CRYSTAL STRUCTURE OF NITRIC ACID TRIHYDRATE
                  2   4
                                                                            184, 186

315   LYNCH,R.W. AND MOROSIN,B. (1971)
      J. APPL. CRYST. 4, 352-356
      A HEMISPHERICAL FURNACE
                  21
                                                                            112

316   LYTLE,F.W. (1964)
      ADV. X-RAY ANALYSIS 7, 136-145
      X-RAY DIFFRACTOMETRIC EXAMINATION OF LOW-TEMPERATURE PHASE
      TRANSFORMATIONS IN SINGLE-CRYSTAL STRONTIUM NITRATE
                  323171
                                                                      109, 129, 137

317   MAETA,H., KATO,T., AND OKUDA,S. (1975)
      J. PHYS. E., SCI. INSTRUM. 8, 577-578
      A SIMPLE METHOD TO TRANSFER SPECIMENS
      AT LIQUID HELIUM TEMPERATURE
                  21
                                                                            184

      MAIS,R.H.B.   SEE DAVIES ET AL.(1970)

318   MALYUSHITSKAYA,Z.V., MALYUSHITSKII,G.B., NOVICHKOV,V.P.,
      AND KABALKINA,S.S. (1975)
      INSTRUM. EXPTL. TECH. (USSR) 18, 256-258
      X-RAY MEASUREMENTS AT HIGH PRESSURES AND LOW TEMPERATURES
                  322315
                                                                            134

      MALYUSHITSKII,G.B.   SEE MALYUSHITSKAYA ET AL. (1975)

319   MAMEDOV,K.P. AND ALIEV,N.A. (1955)
      TRUD. INST. FIZ. I MAT. AKAD. NAUK. AZERB. SSR, SER. FIZ. 7, 61-67
      AN X-RAY CAMERA FOR STUDIES OF VARIOUS TEMPERATURES
                  300310

MAMMI,M.   SEE DEL PRA ET AL. (1967)

MÄNTYSALO,E.   SEE HOVI ET AL. (1966)

MARGRAVES,J.L.   SEE TALLMAN ET AL. (1961)

MARK,H.   SEE HENGSTENBERG AND MARK (1928)

MARKINA,L.I.   SEE BOIKO ET AL. (1972)

MARKOV,A.S.   SEE BOL'SHUTKIN ET AL. (1969)

320 MARSH,D.J. AND PETSKO,G.A. (1973)
J. APPL. CRYST. 6, 76-80
A LOW-TEMPERATURE DEVICE FOR PROTEIN CRYSTALLOGRAPHY
        515400   54, 60, 65

MARSH,R.E.   SEE LIEBLING AND MARSH (1965)

MARSH, R.E.   SEE BERTOLUCCI AND MARSH (1974)

321 MASCARENHAS,Y. AND MASCARENHAS,S. (1967)
REV. SCI. INTRUM. 38, 141-142
DIFFRACTION CAMERA FOR PRECISION MEASUREMENT OF LATTICE
PARAMETERS AT LOW TEMPERATURES
        323313   116, 134, 135

MASLEN, E.N.   246

322 MASSON,D.B. (1960)
TRANS. MET. SOC. AIME 218, 94-97
COMPOSITION-TEMPERATURE BEHAVIOR OF THE MARTENSITIC
TRANSFORMATION IN BETA-AGCL
        322371   132, 142

MATHEWS,F.S.   SEE NORTH ET AL. (1968)

MATHIESON, A.MCL.   SEE FRIDRICHSONS AND MATHIESON (1958)

323 MATHIOT,A. AND PETROFF,J.F. (1975)
ACTA CRYST. A31, S149
LOW-TEMPERATURE X-RAY TOPOGRAPHY -
APPLICATION TO FERRIMAGNETIC GARNETS
        123190   117

MATSUNO,K.   SEE HORI AND MATSUNO (1972)

324 MAUER,F.A. AND BOLZ,L.H. (1961)
J. RES. NAT. BUR. STAND. C 65, 225-229
AN X-RAY DIFFRACTOMETER CRYOSTAT PROVIDING TEMPERATURE CONTROL
IN THE RANGE OF 4 TO 300 D-K
        323171   3   6   129, 175, 177

MAUER,F.A.   SEE BLACK ET AL. (1958)

MAUER,F.A.   SEE BOLZ AND MAUER (1963)

MAZELSKY,R.   SEE KRAMER ET AL. (1963)

325 MCDONALD,T.R.R., BARBERICH,G.S., CUNICELLA,V.T.,
AND GREGORY,E. (1966)
REV. SCI. INTRUM. 37, 1071-1076
APPARATUS AND TECHNIQUES FOR RETENTION AND X-RAY EXAMINATION
OF METASTABLE HIGH PRESSURE PHASES AT 4.2 D-K
        373120   23, 109, 112, 154, 157

326   MCFARLAN,R.L.   (1936)
       REV. SCI. INSTRUM. 7, 82-85
       APPARATUS FOR X-RAY PATTERNS OF THE HIGH-PRESSURE
       MODIFICATIONS OF ICE
                    322310   1
                                                                          116, 133, 168, 178

       MCHARGUE,C.J.   SEE JETTER ET AL. (1957)

327   MCKEEHAN,L.W. AND CIOFFI,P.P. (1922)
       PHYS. REV. 19, 444-446
       THE CRYSTAL STRUCTURE OF MERCURY
                    112300
                                                                          4, 12, 43, 97, 98

328   MCLENNAN,J.C. AND WILHELM,J.O. (1925)
       TRANS. ROY. SOC. CANADA 19, 51-56
       THE CRYSTAL STRUCTURE OF CARBON DIOXIDE
                    322312   5
                                                                          133

       MCMANUS,G.M.   SEE FLINN ET AL. (1961)

       MCMULLAN,R.K.   SEE ALLEN ET AL. (1963)

       MEDVEDEV,V.S.   SEE BOIKO ET AL. (1973)

       MEGAW,H.D.   SEE DARLINGTON AND MEGAW (1973)

329   MEHL,R.F. AND BARRETT,C.S.   (1930)
       TRANS. MET. SOC. AIME 89, 575-589
       THE SYSTEM CADMIUM-MERCURY
                    312310
                                                                          63

       MELCHER,R.L.   SEE SEGMULLER ET AL. (1974)

       MENDOZA,E.   SEE CRUICKSHANK ET AL. (1956)

       MERONOV-KOPISOV,V.S.   SEE BOL'SHUTKIN ET AL. (1970)

330   MEYER,E. (1973)
       J. APPL. CRYST. 6, 45
       FLAMELESS MOUNTING OF CRYSTALS IN CAPILLARIES
                              1   2   6
                                                                          174. 176

331   MEYER,L. (1963)
       Z. ANGEW. PHYS. 15, 438-440
       A LOW-TEMPERATURE CAMERA FOR AN X-RAY SPECTROGRAPH
        WITH HORIZONTAL AXIS OF ROTATION
                    323177

       MEYER,L.   SEE BARRETT AND MEYER (1964)

       MEYER,L.   SEE BARRETT ET AL. (1967)

       MICHALENKO,V.H.   SEE BOIKO ET AL. (1972)

       MIKHALENKO,V.N.   SEE KISELEVA AND MIKHALENKO (1962)

332   MIKSIC,M.G., SEGERMAN,E., AND POST,B. (1959)
       ACTA CRYST. 12, 390-393
       THE SOLID PHASE TRANSFORMATION IN DIMETHYLACETYLENE AT -119 D-C
                    311372   2
                                                                          62, 101, 102, 168

333   MILLEDGE,H.J. (1969)
       ACTA CRYST. A25, 173-180
       REAL CRYSTALS AS A SOURCE OF ERROR
                              1
                                                                          162, 163

MILLER,J.E.   SEE HUTCHINSON AND MILLER (1958)

MILLER,K.T.   SEE MUELLER ET AL. (1958)

334 MILLS,R.L. AND SCHUCH,A.F. (1974)
     J. LOW TEMP. PHYS. 16, 305-308
     X-RAY STUDY OF HELIUM-4 CLOSE-PACKED STRUCTURES AT HIGH PRESSURE
                 323125   23, 115, 130

     MILLS,R.L.   SEE SCHUCH AND MILLS (1962)

335 MINDUKSHEEV,V.F. AND TERMINASOV,Y.S. (1958)
     INDUSTRIAL LAB. USSR 24, 721
     EQUIPMENT FOR X-RAY PHOTOGRPAHY OF SAMPLES AT LOW TEMPERATURES
                 362310

     MIRENSKII,A.V.   SEE BOIKO ET AL. (1972)

     MOKRII,N.I.   SEE BOL'SHUTKIN ET AL. (1969)

     MOKRII,N.I.   SEE BOL'SHUTKIN ET AL. (1970)

     MOOY,H.H.   SEE DE SMEDT ET AL. (1930)

     MOOY,H.H.   SEE KEESOM ET AL. (1930)

336 MOROSIN,B. (1966)
     J. CHEM. PHYS. 44, 252-257
     COBALT CHLORIDE DIHYDRATE FROM 5 TO 298 D-K
                 223171   108, 123, 249

     MORE,M.   SEE ODOU AND MORE (1975)

     MOROSIN,B.   SEE LYNCH AND MOROSIN (1971)

337 MOROSIN,B. AND SCHIRBER,J.E. (1974)
     J. APPL. CRYST. 7, 295-296
     LOW-TEMPERATURE, HIGH-PRESSURE X-RAY CELL
                 212365   23, 24, 116, 154

     MÖSSBAUER,R.L.   SEE THOMANEK ET AL. (1973)

338 MUELLER,M.H., HARDY,L.H., AND MILLER,K.T. (1958)
     REV. SCI. INSTRUM. 29, 253-254
     CAPILLARY SEALING JIG
                         2   3   6   170, 171, 173, 177

     MUELLER,M.H.   SEE HEATON ET AL. (1970)

     MUELLER,J.M.   SEE HUFFMAN ET AL. (1973)

     MUKHIN,K.Y.   SEE BOIKO ET AL. (1973)

     MYASNIKOV,Y.G.   SEE BOIKO ET AL. (1972)

     MYERS,H.P.   SEE BUTTERS AND MYERS (1955)

339 NATTA,G. (1933)
     GAZZ. CHIM. ITAL. 63, 425-439
     STRUCTURE AND POLYMORPHISM OF HYDROHALIC ACIDS
                 323311   12, 133

340 NAUDON,A. AND JAULIN,M. (1968)
     REV. PHYS. APPL. FR. 3, 152-156
     DESCRIPTION OF A SMALL-ANGLE X-RAY SCATTERING
     LOW-TEMPERATURE APPARATUS
                 423487   101

341   NELMES,R.J. (1970)
      J. APPL. CRYST. 3, 422
      A METHOD OF ENCLOSING HIGHLY DELIQUESCENT OR SUBLIMING SPECIMENS
      FOR X-RAY SINGLE-CRYSTAL DATA COLLECTION
                              1   6   164, 176, 250

342   NEYNABER,R.H., BRAMMER,W.G., AND BEEMAN,W.W. (1959)
      J. APPL. PHYS. 30, 656-661
      MECHANISM OF THE SMALL-ANGLE X-RAY SCATTERING FROM
      COLD-WORKED METALS
                        423387

343   NIKOLAENKO,V.A. AND KARPUKHIN,V.I. (1967)
      INDUSTRIAL LAB. USSR 33, 135-136
      ATTACHMENT TO A DIFFRACTOMETER FOR OPERATION AT
      LIQUID-NITROGEN TEMPERATURE
                        321370   112

      NITTA,I.   SEE ODA ET AL. (1943)

344   NOLTE,M.J., SINGLETON,E., AND LAING,M. (1975)
      J. AM. CHEM. SOC. 97, 6396-6400
      REDETERMINATION OF THE STRUCTURE OF (IRO2(PH2PCH2CH2PPH2)2)(PF6)
                              7   20

345   NORDMAN,C.E. (1962)
      ACTA CRYST. 15, 18-23
      THE CRYSTAL STRUCTURE OF HYDRONIUM PERCHLORATE AT -80 D-C
                              4   182

346   NORDMAN,C.E. AND REIMANN,C. (1959)
      J. AM. CHEM. SOC. 81, 3538-3543
      MOLECULAR AND CRYSTAL STRUCTURES OF AMMONIA-TRIBORANE
                              4   182

347   NORTH,A.C.T., PHILLIPS,D.C., AND MATHEWS,F.S. (1968)
      ACTA CRYST. A24, 351-359
      A SEMI-EMPIRICAL METHOD OF ABSORPTION CORRECTIONS
                             18   220

      NOVICHKOV,V.P.   SEE MALYUSHITSKAYA ET AL. (1975)

348   OBSIEGER,R. (1963)
      ACTA PHYS. AUSTRIACA 16, 249-255
      THE EFFECT OF LOW TEMPERATURES ON THE INTENSITY OF
      X-RADIATION REFLECTED FROM CRYSTALS
                        123371   7   131

349   ODA,T., IIDA,T., AND NITTA,I. (1943)
      J. CHEM. SOC. JAPAN 64, 616-621
      CRYSTAL STRUCTURE OF TETRANITROMETHANE
                        382520   12, 158

350   ODOU,G. AND MORE,M. (1975)
      J. APPL. CRYST. 8, 684-686
      CONSTRUCTION OF A CRYOSTAT WITH ADJUSTABLE TEMPERATURE
      FOR SINGLE-CRYSTAL X-RAY DIFFRACTOMETRY
                        223171

351   IBID.
                        213171

352   OKAZAKI,A. AND KAWAMINAMI,M. (1973)
      JAPANESE J. APPL. PHYS. 12, 783-789
      ACCURATE MEASUREMENT OF LATTICE CONSTANTS
      IN A WIDE RANGE OF TEMPERATURES
                              7   201

OKAZAKI,A.   SEE SUGINO ET AL. (1973)

OKUDA,S.   SEE MAETA ET AL. (1975)

OLIVER,W.F.   SEE BURTON AND OLIVER (1936)

353 OLOVSSON,I.   (1960)
     ARK. F. KEMI 16, 437-458
     A LOW-TEMPERATURE STUDY OF SOME SIMPLE HYDROGEN-BONDED STRUCTURES
                    111306   2   3   4   16   64, 87, 177, 187, 199

     OLOVSSON,I.   SEE THOMAS ET AL. (1974)

354 OLOVSSON,I. AND TEMPLETON,D.H. (1959)
     ACTA CRYST. 12, 827-832
     THE CRYSTAL STRUCTURE OF AMMONIA MONOHYDRATE
                    212350   4   91, 177

     OMAROV,T.G.   SEE KOGAN AND OMAROV (1965)

     ONU,K.   SEE GOTO ET AL. (1969)

     ORCUTT,D.   SEE STAMMLER ET AL. (1963)

     OSKARSSON,A.   SEE DANIELSSON ET AL. (1976)

     OVCHARENKO,N.N.   SEE ARONOVA ET AL.(1959)

     OVCHINNIKOV,A.S. SEE BOIKO AND OVCHINNIKOV (1968)

355 OWEN,E.A. AND WILLIAMS,G.I. (1954)
     J. SCI. INSTRUM. 31, 49-54
     A LOW-TEMPERATURE X-RAY CAMERA
                    312313   14   64, 67, 154

356 OWSTON,P.G.   (1949)
     ACTA CRYST. 2, 222-228
     DIFFUSE SCATTERING OF X-RAYS BY ICE
                    185570   2   4   157, 162

     OWSTON,P.G.   SEE DAVIES ET AL.(1970)

     PACKARD,R.E.   SEE WILLIAMS AND PACKARD (1974)

     PARAK,F.   SEE THOMANEK ET AL. (1973)

357 PARKES,A.S. AND HUGHES,R.E. (1963A)
     ACTA CRYST 16, 734-736
     THE CRYSTAL STRUCTURE OF CYANOGEN
                    4   18   187

358 PARKES,A.S. AND HUGHES,R.E. (1963B)
     ACTA CRYST. 16, 1185-1187
     IRRADIATED VOLUME IN WEISSENBERG AND PRECESSION TECHNIQUES
                    18   219

     PARRY,D.   SEE COHEN ET AL. (1971)

     PARRY,G.S.   SEE CIMINO ET AL.(1959)

359 PAVLOVIC,A.S.   (1956)
     DISSERTATION ABSTRACTS 16, 1156 (PUBL. NO. 16725)
     DESIGN AND APPLICATION OF A WEISSENBERG X-RAY CAMERA FOR
     STUDIES NEAR LIQUID HELIUM TEMPERATURE
                    223150   116

360 PEARSON,W.B. (1954)
     CAN. J. PHYS. 32, 708-713
     THERMAL EXPANSION OF LITHIUM, 77 TO 300 O-K
              412311   99, 154

361 PEARSON,W.B. (1955)
     CAN. J. PHYS. 33, 473-475
     A METHOD OF EXAMINING STRUCTURAL CHANGES OF METALS ON
     DEFORMATION IN LIQUID HELIUM - EXAMINATION OF INDIUM
              373117   155

     PEARSON,W.B.   206

     PEARSON,W.B.   SEE BECK, ET AL. (1973)

     PEISER,H.S.   SEE BLACK ET AL. (1958)

     PEISER,H.S.   SEE BOLZ ET AL.(1962)

362 PEISL,H. AND WAIDELICH,W. (1959)
     Z. ANGEW. PHYS. 11, 474-477
     LOW-TEMPERATURE CAMERA FOR STRUCTURE INVESTIGATIONS WITH
     COUNTER-TUBE GONIOMETERS
              323371   132

     PEISL,H.   SEE HIMMLER ET AL. (1969)

363 PENFOLD,D.W. (1971)
     THESIS, IMPERIAL COLLEGE, LONDON
     SUPERCONDUCTIVITY AND STRUCTURE IN BETA-TUNGSTEN COMPOUNDS
              323171   115, 130, 135, 161

     PEPINSKY,R.   SEE FRAZER AND PEPINSKY (1950)

     PEPINSKY,R.   SEE KEELING ET AL. (1953)

     PERNOLET,R.   SEE GERARD AND PERNOLET (1973)

     PETER,M.   SEE JACCARD, ET AL. (1953)

364 PETERSON,O.G. AND SIMMONS,R.O. (1965)
     REV. SCI. INSTRUM. 36, 1316-1318
     RIGID-TAIL HELIUM CRYOSTAT FOR X-RAY DIFFRACTION
     STUDIES OF CRYSTALLIZED GASES
              323125   129, 136, 249

     PETROFF,J.F.   SEE MATHIOT AND PETROFF (1975)

365 PETSKO,G.A. (1975)
     J. MOL. BIOL. 96, 381-392
     PROTEIN CRYSTALLOGRAPHY AT SUB-ZERO TEMPERATURES,II.
              500000   1   20, 45, 84, 165, 172, 223

     PETSKO,G.A.   SEE MARSH AND PETSKO (1973)

366 PETZ,,J.I. (1963)
     NORELCO REPORTER 10, 131-132, 147
     A TEMPERATURE CONTROLLED SMALL ANGLE X-RAY SAMPLE HOLDER
              335572   145

     PHILLIPS, D.C.   SEE NORTH ET AL. (1968)

     PIERMARINI,G.J.   SEE SANTORO ET AL. (1968)

PINGS,C.J.   SEE HONEYWELL ET AL. (1964)

367 PINOT,M., SOUGI,M., AND TOURAND,G. (1965)
J. PHYS. (PARIS) 26, 529-530
X-RAY DIFFRACTION CAMERA WITH VERY LOW TEMPERATURES FOR POWDERS
        313110   82, 85

PLETCHER,J.   SEE LOW ET AL. (1966)

PODOLICH,V.B.   SEE BOIKO ET AL. (1973)

368 POHLAND,E. (1934)
Z. PHYSIK. CHEM. B26, 238-245
A HANDY X-RAY VACUUM CAMERA FOR ANY LOW TEMPERATURE
        323311
                                                                 133

369 POLLOCK,J.C. (1955)
ACTA CRYST. 8, 652-653
THE CALIBRATION OF MICROTHERMOSTATS FOR X-RAY MEASUREMENTS
ON CRYSTALS
                        14
                                                                 201

POOLE,D.   SEE AXON ET AL. (1953)

370 POST,B. (1964)
INTERNATIONAL UNION OF CRYSTALLOGRAPHY,
COMMISSION ON CRYSTALLOGRAPHIC APPARATUS,
BIBLIOGRAPHY 2 - LOW-TEMPERATURE X-RAY DIFFRACTION
                        20
                                                                 11

POST,B.   SEE STEINFINK ET AL. (1953)

POST,B.   SEE MIKSIC ET AL. (1959)

POST,B.   SEE RUDMAN AND POST (1968)

371 POST,B. AND FANKUCHEN,I. (1953)
ANAL. CHEM. 25, 736-737
LOW TEMPERATURE X-RAY CRYSTALLOGRAPHY
                        20

372 POST,B., SCHWARTZ,R.S., AND FANKUCHEN,I. (1951)
REV. SCI. INSTRUM. 22, 218-219
AN IMPROVED DEVICE FOR X-RAY DIFFRACTION STUDIES
AT LOW TEMPERATURES
        111300   2   3   4   5   6   44, 45, 46, 47, 49, 62, 67,
                                                      185, 187, 204, 209

373 POST,B., SCHWARTZ,R.S., AND FANKUCHEN,I. (1952)
ACTA CRYST. 5, 372-374
THE CRYSTAL STRUCTURE OF SULFUR DIOXIDE
                        3   4
                                                                 175

374 POTAPOV,L.P. (1962)
INSTRUM. EXPTL. TECH. (USSR) 5, 200-201
A CRYOSTAT FOR LOW-TEMPERATURE X-RAY INVESTIGATIONS
        323370
                                                                 132

POWELL,G.W.   SEE REEBER AND POWELL (1967)

POWELL,H.M.   SEE FORD AND POWELL (1954)

POWELL,R.L.   SEE SPARKS ET AL. (1972)

375   PRAKASH,A.   (1966)
      INDIAN J. PURE APPL. PHYS. 4, 362-363
      SIMPLE GAS-COOLING ATTACHMENT FOR A SINGLE CRYSTAL
      WEISSENBERG GONIOMETER
                    212350   68, 246

      PRANDL,W.   SEE IHRINGER ET AL. (1975)

      PREECE,C.M.   SEE COCKS ET AL. (1966)

      PREECE,C.M.   SEE KING AND PREECE (1967)

      PREECE,C.M.   SEE KING AND PREECE (1968)

376   PRESS,W. AND HÜLLER,A. (1973)
      ACTA CRYST. A29, 252-263
      ANALYSIS OF ORIENTATIONALLY DISORDERED STRUCTURES
                    19   221

      PROKHVATILOV,A.I.   SEE BOL'SHUTKIN ET AL. (1970)

      PROKHVATILOV,A.I.   SEE BOIKO ET AL. (1972)

      PROKHVATILOV,A.I.   SEE BOL'SHUTKIN ET AL. (1972)

377   PRZEDMOJSKI,J.   (1960)
      POSTEPY FIZYKI 2, 565-568
      X-RAY CAMERA FOR INVESTIGATIONS AT TEMPERATURES BETWEEN
      -180 AND +300 D-C
                    322312

378   PRZEDMOJSKI,J.   (1966)
      POMIARY,AUTOMAT., KONTR. 12, 436
      X-RAY APPARATUS FOR THE MEASUREMENT OF CRYSTALLINE
      PARAMETERS AND PHASE CHANGES AT HIGH AND LOW TEMPERATURES
      AND UNDER HIGH PRESSURES
                    312312   84

379   QUEISSER,H.J.   (1958)
      Z. PHYSIK 152, 495-506
      X-RAY INVESTIGATIONS ON BISMUTH LAYERS AT LOW TEMPERATURES
                    323110   130

      RABARDEL,L.   SEE ETOURNEAU ET AL. (1975)

380   RADOVICI,C. AND VLAHOVICI,N. (1971)
      LUCR. CONF. NAT. CHIM. ANAL. 3RD, 2, 109-115
      LOW-TEMPERATURE DEVICE FOR ANALYSIS BY X-RAY DIFFRACTION
                    322375

      RAMASESHAN,S.   SEE VISWAMITRA AND RAMASESHAN (1960)

      RAMASESHAN,S.   SEE VISWAMITRA AND RAMASESHAN (1963)

      RAMASESHAN,S.   SEE SINGH AND RAMASESHAN (1963)

      RAMASESHAN,S.   SEE SINGH AND RAMASESHAN (1964)

381   RAO,J.K.M. AND VISWAMITRA,M.A. (1972)
      ACTA CRYST. B28, 1484-1495
      CRYSTAL STRUCTURE OF GLYCINE SILVER(I)NITRATE
                    212356   216

382 RAY,SIDDHARTHA (1964)
     INDIAN J. PHYS. 38, 82-86
     A 19-CM DEBYE-SCHERRER CAMERA FOR WORKING
     BETWEEN 400 AND 106 D-K
                 323314   133

     RAYMOND,K.N.   SEE JURNAK AND RAYMOND (1974)

     RAYNE,J.A.   SEE FLINN ET AL. (1961)

383 REEBER,R.R. AND POWELL,G.W. (1967)
     J. APPL. PHYS. 38, 1531-1534
     THERMAL EXPANSION OF ZNS FROM 2 TO 317 D-K
                 322110   130

384 REED,T.B. AND LIPSCOMB,W.N. (1953)
     ACTA CRYST. 6, 45-48
     THE CRYSTAL AND MOLECULAR STRUCTURE OF
     1,2-DICHLOROETHANE AT -140 D-C
                 112300   2   3   4   37, 62, 170, 175, 246

     REED,T.B.   SEE ABRAHAMS, ET AL. (1950)

385 REEKIE,J. (1939)
     PROC. PHYS. SOC. (LONDON) 51, 683-688
     THE SENSITIVITY OF PHOTOGRAPHIC FILMS TO X-RADIATION
     AT VERY LOW TEMPERATURES
                             15   205

386 REES,B. AND COPPENS,P. (1973)
     ACTA CRYST. B29, 2516-2528
     ELECTRONIC STRUCTURE OF BENZENE CHROMIUM TRICARBONYL
     BY X-RAY AND NEUTRON DIFFRACTION AT 78 D-K
                 223371   16   206

     REES,B.   SEE COPPENS ET AL. (1974)

     REES,B.   SEE JUST ET AL. (1975)

387 REICHERT,J.F. (1972)
     REV. SCI. INSTRUM. 43, 1727-1728
     CEMENTING TEFLON AT LOW TEMPERATURES
                 21   249

388 REID,J.S. (1973)
     ACTA CRYST. A29, 248-251
     TEMPERATURE DEPENDENCE OF INTEGRATED THERMAL DIFFUSE SCATTERING
                 11   12   18, 198

     REIMANN,C.   SEE NORDMAN AND REIMANN (1959)

389 RENAUD,M. AND FOURME,R. (1966)
     BULL. SOC. FR. MINERAL. CRISTALLOGR. 89, 243-245
     TECHNIQUE FOR GROWING SINGLE CRYSTALS OF ORGANIC LIQUIDS
     ON A GONIOMETER HEAD
                             4   189, 191

390 RENAUD,M. AND FOURME,R. (1967)
     ACTA CRYST. 22, 695-698
     APPARATUS AND TECHNIQUES FOR ROUTINE STUDY OF CRYSTALLINE ORGANIC
     MATERIALS AT LOW TEMPERATURES
                 211300   2   14   35, 36, 50, 72, 91, 199, 204

RENAUD,M.   SEE ANDRE ET AL. (1971)

RENAUD,M.   SEE ANDRE ET AL. (1972)

RENAUD,M.   SEE KAHN ET AL. (1973)

391   RENTON,J.J. AND BAUN,W.L.   (1963)
      WRIGHT-PATTERSON AIR FORCE BASE, REPORT NO. ASD-TDR-63-469
      SOME SPECIALIZED ATTACHMENTS FOR THE SIEMENS X-RAY DIFFRACTOMETER
                  123371   112, 126, 127, 132, 139, 223

RENTON,J.J.   SEE BAUN AND RENTON (1963)

RENTON,J.J.   SEE BAUN AND RENTON (1964)

REUTH,E.C.,VAN   SEE BRADY AND REUTH (1966)

392   RHODES,R.G.   (1951)
      ACTA CRYST.   4, 105-110
      BARIUM TITANATE TWINNING AT LOW TEMPERATURES
                  112342   44

RICHARDS,J.P.G.   SEE WOODLEY ET AL. (1971)

RICHARDS,J.P.G.   SEE HINE ET AL. (1975)

RICHARDS,S.M.   SEE CARPENTER AND RICHARDS (1962)

RICHARDSON,H.W.   SEE JAUNCEY AND RICHARDSON (1934)

RILEY,D.P.   SEE FIGGINS ET AL. (1956)

RINFRET,A.P.   SEE DOWELL AND RINFRET (1960)

393   RINNE,F.   (1914)
      CENTR. MIN. GEOL.   1914, 705-718
      CHANGES IN CRYSTAL ANGLE OF RELATED SUBSTANCES
      WITH CHANGE OF TEMPERATURE   3

394   RINNE,F.   (1917)
      BER. VERH. GES. WISS. MATH-PHYS.   69, 57-62
      THE CRYSTALLINE SYSTEM AND THE AXIAL RATIO OF ICE
                  282520   3, 142, 158

395   ROBERGE,R.   (1975)
      J. LESS-COMMON METALS   40, 161-164
      LATTICE PARAMETER OF NIOBIUM BETWEEN 4.2 AND 300K
                  323170   136

ROBERTSON,B.E.   SEE GUTTORMSON AND ROBERTSON (1973)

396   ROBERTSON, J.H.   (1960)
      J. SCI. INSTRUM.   37, 41-45
      X-RAY DIFFRACTION BY SINGLE CRYSTALS AT LOW TEMPERATURES -
      A CRYOSTAT FOR USE WITH LIQUID HYDROGEN
                  211152   31, 32, 33, 36, 80, 81

397   ROBERTSON,J.H.   (1965)
      ACTA CRYST.   18, 410-417
      AMMONIUM OXALATE MONOHYDRATE - STRUCTURE REFINEMENT AT 30 D-K
                  212252   14   80, 201

ROBERTSON,J.H.   SEE CRUICKSHANK ET AL. (1956)

ROBINSON,W.   SEE FRENZ ET AL. (1969)

398   ROESSLER,B. AND BOLLING,G.F. (1964)
REV. SCI. INSTRUM. 35, 230-231
SIMPLE SPECIMEN HOLDER FOR LOW-TEMPERATURE
X-RAY DIFFRACTION STUDIES
361371   132, 155, 156, 157

399   ROSE-INNES,A.C. (1973)
LOW TEMPERATURE LABORATORY TECHNIQUES
UNIVERSITY OF LONDON PRESS, LONDON
21   233, 234

400   ROSS,F.K. AND WILLIAMS,J.M. (1974)
ACA MEETING ABSTR. SER. 2, 2, 246
LOW TEMPERATURE SINGLE-CRYSTAL NEUTRON DIFFRACTION USING
THE CT-38 CRYOREFRIGERATOR
14   17   122, 208

ROSS,F.K.   SEE COPPENS ET AL. (1974)

ROSS,F.K.   SEE WANG ET AL. (1976)

ROSSMANN,M.G.   SEE HAAS AND ROSSMANN (1970)

401   RUDMAN,R. (1966A)
THESIS - POLYTECHNIC INSTITUTE OF BROOKLYN
LOW TEMPERATURE PHASE TRANSITIONS IN MOLECULAR CRYSTALS
111306   2   4   5   62, 67, 88, 101, 191

402   RUDMAN,R. (1966B)
NOKELCO REPORTER 13, 61-62
LOW-TEMPERATURE X-RAY DIFFRACTION POWDER TECHNIQUES
311310   5   62, 67, 101

403   RUDMAN,R. (1967)
J. CHEM. EDUC. 44, 331-334
LABORATORY EXPERIMENTS IN LOW-TEMPERATURE X-RAY DIFFRACTION
311310   5   48, 62, 101

404   RUDMAN,R. (1968)
J. APPL. CRYST. 1, 126-127
SINGLE-CRYSTAL ALIGNMENT WITH A POLAROID ADAPTER FOR
THE WEISSENBERG GONIOMETER
16   90, 92, 210, 212, 213

405   RUDMAN,R. (1970A)
J. CRYST. GROWTH 6, 163-166
PREPARATION OF SINGLE CRYSTALS OF THE METHYLCHLOROMETHANE COMPOUNDS
FOR X-RAY DIFFRACTION ANALYSIS
2   4   195

406   RUDMAN,R. (1970B)
MOL. CRYST. AND LIQ. CRYST. 6, 427-429
POLYMORPHISM OF THE CRYSTALLINE
METHYLCHLOROMETHANE COMPOUNDS, II.
2   179, 191

407   RUDMAN,R. (1972)
J. APPL. CRYST. 5, 143
MECHANICALLY REFRIGERATED GAS STREAMS FOR STUDIES TO -70 0-C
111300   64, 65

RUDMAN,R.   SEE LIPPMAN AND RUDMAN (1976)

RUDMAN,R.   SEE SILVER AND RUDMAN (1971)

RUDMAN,R.   SEE SILVER AND RUDMAN (1972)

408   RUDMAN,R. AND GÖDEL,J. (1969)
      J. APPL. CRYST. 2, 109-112
      AN AUTOMATIC LOW-TEMPERATURE APPARATUS FOR
      SINGLE-CRYSTAL DIFFRACTOMETRY
                111300
                                                   36, 37, 38, 54, 55, 56, 77, 78

409   RUDMAN,R. AND POST,B. (1968)
      MOL. CRYST. 5, 95-110
      POLYMORPHISM OF THE CRYSTALLINE METHYLCHLOROMETHANE COMPOUNDS
                        2   4
                                                                       191

410   RUHEMANN,B. (1935)
      PHYSIK Z. SOWJETUNION 7, 572-582
      NEW CAMERA FOR LOW-TEMPERATURE X-RAY DIFFRACTION WORK
                323110

      RUHEMANN,B.   SEE RUHEMANN AND RUHEMANN (1937)

411   RUHEMANN,B. AND SIMON,F. (1931)
      Z. PHYSIK. CHEM. B15, 389-413
      THE CRYSTAL STRUCTURE OF KR, XE, HI, AND HBR
                323210   3
                                                              133, 135, 175

412   RUHEMANN,M. (1932)
      Z. PHYSIK 76, 368-385
      X-RAY INVESTIGATION OF SOLID NITROGEN AND HYDROGEN
                323110   3
                                                                   130, 175

413   RUHEMANN,M. AND RUHEMANN,B. (1937)
      LOW TEMPERATURE PHYSICS
      CAMBRIDGE UNIVERSITY PRESS, PAGES 103-113
                        20
                                                                       11

414   RÜHL,W. (1954)
      Z. PHYSIK 138, 121-135
      X-RAY INVESTIGATION OF CONDENSED THIN FILMS
      AT LOW TEMPERATURES
                323111
                                                                   130

      RUSAKOVA-LUKOVSKAYA,N.Y.   SEE BAGARYATSKII ET AL. (1951)

      SABINE,T.M.   SEE COPPENS ET AL. (1967)

      SAINI,H.   SEE WEIGLE AND SAINI (1936)

415   SAKURAI,T. AND SUZUKI,T. (1959)
      NORELCO REPORTER 6, 122,128
      A LOW TEMPERATURE ATTACHMENT FOR THE NORELCO X-RAY DIFFRACTOMETER
                312372
                                                                   101

      SAKURAI,T.   SEE ITO AND SAKURAI (1973)

416   SALJE,E. AND VISWANATHAN,K. (1975)
      ACTA CRYST. A31, 356-359
      PHYSICAL PROPERTIES AND PHASE TRANSITIONS IN WO3
                323330

417   SAMPSON,C.F.   (1970)
      U.K.AT. ENERGY RES. GROUP, REP. 1970,AERE-R 6532
      LOW-TEMPERATURE CRYOSTAT FOR SINGLE-CRYSTAL AND
      POWDER X-RAY DIFFRACTION
                       323171   *112, 113, 129*

418   SANDLER,N.I. AND AKHMECHET,M.N.   (1958)
      INDUSTRIAL LAB. USSR 24, 731
      A LOW-TEMPERATURE BACK-REFLECTION CAMERA WITH INDEPENDANT STANDARD
                       365323   *157*

      SANGER,P.L.   SEE LEVY ET AL. (1975)

419   SANTORO,A., WEIR,C.E., BLOCK,S. AND PIERMARINI,G.J.   (1968)
      J. APPL. CRYST. 1, 101-107
      ABSORPTION CORRECTIONS IN COMPLEX CASES
                       18   *218*

420   SANTOS,J.A. AND WEST,J.   (1933)
      J. SCI. INSTRUM. 10, 219-221
      A METHOD OF TAKING X-RAY PHOTOGRAPHS OF CRYSTALLINE POWDERS AT THE
      TEMPERATURE OF LIQUID AIR
                       362311   *152*

      SASS,R.L.   SEE SCHEUERMAN AND SASS (1962)

421   SAYETAT,F. (1975A)
      ACTA CRYST. A31, S239
      PRESENT STATUS AND FUTURE TRENDS OF X-RAY POWDER DIFFRACTOMETRY
      AT LOW AND VERY LOW TEMPERATURES
                       20

422   SAYETAT,F. (1975B)
      J. APPL. PHYS. 46, 3619-3625
      X-RAY POWDER DIFFRACTION AT LOW TEMPERATURE APPLIED TO
      THE DETERMINATION OF MAGNETOELASTIC PROPERTIES
      IN TERBIUM IRON GARNET
                       323171   *130*

      SCHAR,R.   SEE GRANICHER ET AL. (1959)

423   SCHEUERMAN,R.F. AND SASS,R.L.   (1962)
      ACTA CRYST. 15, 1244-1247
      THE CRYSTAL STRUCTURE OF VALERIC ACID
                       4   *182*

      SCHIRBER,J.E.   SEE ATOJI ET AL. (1959)

      SCHIRBER,J.E.   SEE MOROSIN AND SCHIRBER (1974)

      SCHMIDT,G.M.J.   SEE HIRSHFELD AND SCHMIDT (1956)

      SCHMIDT,W.   SEE BRUEHL AND SCHMIDT (1971)

      SCHROEDER,L.W.   SEE FRENZ ET AL. (1969)

      SCHROEDER,L.W.   SEE COYLE AND SCHROEDER (1971)

424   SCHUCH,A.F. (1958)
      PROC. INTL. CONF. ON LOW-TEMPERATURE PHYSICS AND CHEMISTRY,LT5
      ED., J.R. DILLINGER, UNIV. OF WISCONSIN PRESS, PPS. 79-81
      THE STRUCTURE OF SOLID HELIUM
                       323115   *130*

SCHUCH,A.F.   SEE MILLS AND SCHUCH (1974)

425 SCHUCH,A.F. AND MILLS,R.L. (1962)
PHYS. REV. LETTERS 8, 469-470
STRUCTURE OF THE GAMMA FORM OF SOLID HELIUM-4
              223125   12, 130

SCHUMAKER,N.E.   SEE BONILLA ET AL. (1970)

SCHWAGER,P.   SEE THOMANEK ET AL. (1973)

SCHWARTZ,R.D.   SEE ASHBY AND SCHWARTZ (1974)

SCHWARTZ,R.S.   SEE POST ET AL. (1951)

SCHWARTZ,R.S.   SEE POST ET AL. (1952)

426 SCOTT,R.B. (1959)
CRYOGENIC ENGINEERING
D.VAN NOSTRAND, N.Y.
              21   229, 233

427 SEARS,D.R. AND KLUG,H.P. (1962)
J. CHEM. PHYS. 37, 3002-3006
DENSITY AND EXPANSIVITY OF SOLID XENON
              323171   5   129, 178

428 SEARS,W.M. (1974)
THESIS, MCMASTER UNIVERSITY
(ALSO PUBLISHED IN J.CHEM. PHYS. 62, 2736-2739 (1975))
SOLID SIH4 - STRUCTURE AND ORIENTATIONAL ORDER
              323272   129

SEGERMAN,E.   SEE MIKSIC ET AL. (1959)

429 SEGMÜLLER,A., MELCHER,R.L., AND KINDER,H. (1974)
SOLID STATE COMM. 15, 101-104
X-RAY DIFFRACTION MEASUREMENT OF THE JAHN-TELLER
DISTORTION IN TMVO4
              364170   154, 157, 200

SELLA,C.   SEE GERVAIS ET AL. (1966)

SELTE,K.   SEE HOPE ET AL. (1973)

SEN,S.N.   SEE KRISHNA MURTI AND SEN (1956)

SENEGAS,J.   SEE TRUT ET AL. (1973)

SEPP,A.   SEE HIMMLER ET AL. (1969)

SHABELNIKOV,L.G. SEE AKNAZAROV ET AL. (1974)

430 SHAH,J.S. AND STRAUMANIS,M.E. (1971)
J. APPL. PHYS. 42, 3288-3289
THERMAL EXPANSION OF TUNGSTEN AT LOW TEMPERATURES
              7   14   201

431 SHAH,J.S. AND STRAUMANIS,M.E. (1972)
SOLID STATE COMM. 10, 159-162
THERMAL EXPANSION BEHAVIOR OF SILICON AT LOW TEMPERATURES
              7   14   201

SHAH,J.S.   SEE JOHNSON ET AL. (1974)

432 SHALLCROSS,F.V. AND CARPENTER,G.B. (1958)
ACTA CRYST. 11, 490-496
THE CRYSTAL STRUCTURE OF CYANOACETYLENE
                    2   4
                                                            170

SHARMA,B.D.   SEE GRUBER ET AL. (1969)

433 SHARON,B. (1975)
PERSONAL COMMUNICATION
                223360
                                                95, 126, 127, 223

SHCHERBAKOV,G.N.   SEE GRUSHKO ET AL. (1970)

SHEKHTMAN,V.S.   SEE AKNAZAROV ET AL. (1974)

SHEVCHENKO,N.F.   SEE GRUSHKO ET AL. (1970)

434 SHEVELEV,A.K. AND BALAKINA,L.M. (1960)
SOVIET PHYSICS - CRYSTALLOGRAPHY 4, 225-226
AN X-RAY DIFFRACTION CAMERA FOR USE AT LOW AND HIGH TEMPERATURES
                323314
                                                            133

435 SHIMURA,Y. (1960)
ACTA CRYST. 13, 986
NEW CONTINUOUS RECORDING HIGH AND LOW TEMPERATURE
X-RAY DIFFRACTOMETERS
                323371
                                                            132

SHMYTKO,I.M.   SEE BOIKO AND SHMYTKO (1974)

436 SIEMONS,W.J. AND TEMPLETON,D.H. (1954)
ACTA CRYST. 7, 194-198
THE CRYSTAL STRUCTURE OF AMMONIUM OXIDE
                115356   3   4
                                                        175, 177

437 SILVER,L. AND RUDMAN,R. (1971)
REV. SCI. INSTRUM. 42, 671-673
A STABLE LOW-TEMPERATURE GAS STREAM SYSTEM WITH VARIABLE
TEMPERATURE CONTROL
                111300
                                            36, 54, 73, 74, 75, 76, 77

438 SILVER,L. AND RUDMAN, R. (1972)
J. CHEM. PHYS. 57, 210-216
POLYMORPHISM OF THE CRYSTALLINE METHYLCHLOROMETHANE COMPOUNDS,IV.
                    7
                                                            12, 22

439 SIMMONS,R.O. (1976)
PERSONAL COMMUNICATION
                243175
                                                    4, 107, 143

SIMMONS,R.O.   SEE PETERSON AND SIMMONS (1965)

SIMMONS,R.O.   SEE LOSEE AND SIMMONS (1968)

SIMMONS,R.O.   SEE BALZER AND SIMMONS (1974)

440 SIMON,A. (1971)
J. APPL. CRYST. 4, 138-145
DESCRIPTION OF A NEW X-RAY CAMERA
                311334
                                                            25, 101

SIMON,A.   SEE DEISEROTH ET AL. (1975)

441 SIMON,F. AND SIMSON,C. (1924A)
     Z. PHYSIK 21, 168-177
     THE CRYSTAL STRUCTURE OF HYDROGEN CHLORIDE
          323310   3   4, 115, 133, 135, 175

442 SIMON,F. AND SIMSON,C. (1924B)
     Z. PHYSIK 25, 160-164
     THE CRYSTAL STRUCTURE OF ARGON
                         7   115, 133, 135, 175

443 SIMON,F. AND VOHSEN,E. (1928)
     Z. PHYSIK. CHEM. 133, 165-187
     CRYSTAL STRUCTURE INVESTIGATION OF THE ALKALI METALS
          323313   134

     SIMON,F.   SEE RUHEMANN AND SIMON (1931)

     SIMONETTA,M.   SEE FILIPPINI ET AL. (1974)

     SIMSON,C.   SEE SIMON AND SIMSON (1924A)

     SIMSON,C.   SEE SIMON AND SIMSON (1924B)

444 SINGH,A.K. AND RAMASESHAN,S. (1963)
     CRYSTALLOGRAPHY AND CRYSTAL PERFECTION
     ED., G.N. RAMACHANDRAN
     ACADEMIC PRESS, NY PAGES 309-315
          211350   90

445 SINGH,A.K. AND RAMASESHAN,S. (1964)
     PROC. INDIAN ACAD. SCI. 60A, 20-24
     A LOW-TEMPERATURE ATTACHMENT FOR THE WEISSENBERG GONIOMETER
          212350   4   86, 188

     SINGLETON,E.   SEE NOLTE ET AL. (1975)

     SINGMAN,L.   SEE LOW ET AL. (1966)

     SINGMAN,L.   SEE CUCKA ET AL. (1970)

446 SKABA,V. AND KRIVY,I. (1970)
     USTAV. JAD. VYZK. NO.2549-A, 23 PPS.
     LOW-TEMPERATURE ATTACHMENT FOR AN X-RAY DIFFRACTOMETER
          323371   6   177

447 SMIRNOV,Y.N. AND FINKEL,V.A. (1965)
     SOVIET PHYSICS JETP 20, 315-317
     THE CRYSTAL STRUCTURE OF CHROMIUM AT 113-373 K
          323370

     SMITH,B.L.   SEE HONEYWELL ET AL. (1964)

448 SMITH,D.K. (1961)
     NORELCO REPORTER 8, 11-12
     A SIMPLE LOW TEMPERATURE SPECIMEN HOLDER
     FOR THE NORELCO DIFFRACTOMETER
          322372   131, 139

449 SMITH,D.K. AND LEIDER,H.R. (1968)
     J. APPL. CRYST. 1, 246-249
     LOW-TEMPERATURE THERMAL EXPANSION OF (LI)H, (MG)O, AND (CA)O
          223171   16   129, 206, 249

SMITH,H.   SEE LONSDALE AND SMITH (1941)

450 SMITH,H.M. AND HEADY,H.H. (1955)
ANAL. CHEM 27, 883-888
IDENTIFICATION OF FROZEN LIQUID SAMPLES WITH THE
X-RAY DIFFRACTOMETER
          322372   7   13   *131*

451 SMITH,H.W. (1966)
THESIS, HARVARD UNIVERSITY
CRYSTALLOGRAPHIC STUDIES AT LOW TEMPERATURES
          123140   3   4   18   19   20   *108, 123, 174, 175, 188,*
                                                      206, 207, 209, 219

452 SMITH,H.W. AND LIPSCOMB,W.N. (1965)
J. CHEM. PHYS. 43, 1060-1064
SINGLE-CRYSTAL X-RAY DIFFRACTION STUDY OF BETA-DIBORANE
          123170   *12, 108, 112, 122*

SMITH,L.B.   SEE HORNE ET AL. (1959)

SOANE,C.M.   SEE CHEESMAN AND SOANE (1957)

453 SOLENTE,P. (1965)
CRYOGENICS 5, 112
AN APPARATUS FOR X-RAY STUDIES AT TEMPERATURES IN THE
LIQUID HELIUM RANGE
          312100   *82*

SOLOVEICHIK,M.B.   SEE BOIKO ET AL. (1972)

SOUGI,M.   SEE PINOT ET AL. (1965)

454 SPARKS,L.L., POWELL,R.L., AND HALL,W.J. (1972)
REFERENCE TABLES FOR LOW-TEMPERATURE THERMOCOUPLES
NBS MONOGRAPH 124
                    21   *229*

455 SPENCER,J.B. AND LUNDGREN,J. (1973)
ACTA CRYST. B29, 1923-1928
HYDROGEN BOND STUDIES. LXXIII.
                    7   *14, 15*

SPRITZER,C.   SEE GERVAIS ET AL. (1966)

456 STAMMLER,M., ORCUTT,D., AND COLODNY,P.G. (1963)
ADV. X-RAY ANALYSIS 6, 202-209
LOW-TEMPERATURE TRANSITIONS OF SOME AMMONIUM SALTS
          322371   *132, 142*

457 STEINFINK,H., LADELL,J., POST,B., AND FANKUCHEN,I. (1953)
REV. SCI. INSTRUM. 24, 882-883
A LOW-TEMPERATURE WEISSENBERG X-RAY CAMERA
          211356   *88*

458 STEWARD,E.G. (1960)
X-RAY DIFFRACTION BY POLYCRYSTALLINE MATERIALS,
REINHOLD PUBLISHING CO., N.Y.
EDS., H.S. PEISER, H.P. ROOKSBY AND A.J.C. WILSON
PPS. 265-277 - LOW-TEMPERATURE METHODS
                    20   *153*

STEWARD,E.G.   SEE KELLETT AND STEWARD (1962)

459   ST. JOHN,A. (1918)
      PROC. NAT. ACAD. SCI. (U.S.) 4, 193-197
      THE CRYSTAL STRUCTURE OF ICE
                  285500   3, 6, 156, 157

460   STOCHL,C.A. AND ULLMAN,S.G. (1963)
      REV. SCI. INSTRUM. 34, 1134-1138
      LIQUID HELIUM DEWAR ATTACHMENT FOR AN X-RAY DIFFRACTOMETER
                  323172   129, 137

      STOKHUYZEN,R.   SEE BECK, ET AL. (1973)

461   STRAUMANIS,M.E. AND WOODARD,C.L. (1971)
      ACTA CRYST. A27, 549-551
      LATTICE PARAMETERS AND THERMAL EXPANSION COEFFICIENTS OF
      AL, AG, AND MO AT LOW TEMPERATURES
                         7   14   24, 201, 202

      STRAUMANIS,M.E.   SEE SHAH AND STRAUMANIS (1971)

      STRAUMANIS,M.E.   SEE WOODARD AND STRAUMANIS (1971)

      STRAUMANIS,M.E.   SEE SHAH AND STRAUMANIS (1972)

      STRAWBRIDGE,D.J. SEE HUME-ROTHERY AND STRAWBRIDGE (1947)

462   STREIB,W.E. AND LIPSCOMB,W.N. (1962)
      PROC. NAT. ACAD. SCI. (U.S.) 48, 911-913
      GROWTH, ORIENTATION, AND X-RAY DIFFRACTION OF SINGLE CRYSTALS NEAR
      LIQUID HELIUM TEMPERATURES
                  223140   4   16   108, 123

      STREIB,W.E.   SEE HUFFMAN ET AL. (1973)

463   STRONG,S.L. (1957)
      U.S. DEPT. COMM., OFFICE TECH. SERV., PB REPT. 133,875   23 PPS.
      STRUCTURE PARAMETER CHANGES IN (MN)O2 BELOW 85 D-K
                  323112

      STROUD,R.M.   SEE KRIEGER ET AL. (1974)

464   STROUSE,C.E. (1975)
      PERSONAL COMMUNICATION (PUBLISHED: REV. SCI. INSTRUM. 47, 871-876, 1976)
      A VARIABLE TEMPERATURE ACCESSORY FOR USE WITH A
      FOUR-CIRCLE X-RAY DIFFRACTOMETER
                  211371   69, 96, 208

      STRUNZ,H.   SEE GUNTHER ET AL. (1939)

      STRYLAND,J.C.   SEE BRONSVELD ET AL.(1973)

465   STURTEVANT,J.M. (1971)
      TECHNIQUES OF CHEMISTRY, VOL 1, PART V
      EDS., A. WEISSBERGER AND B.W. ROSSITER
      WILEY-INTERSCIENCE, N.Y.
      PPS. 1-22 TEMPERATURE MEASUREMENT
                         21   229

      SUFFRITTI,G.B.   SEE FILIPPINI ET AL. (1974)

466   SUGINO,K., HIDAKA,M., AND OKAZAKI,A. (1973)
      JAPANESE J. APPL. PHYS. 12, 1124-1129
      A LOW-TEMPERATURE ATTACHMENT FOR WEISSENBERG GONIOMETERS
                  212154   45, 46, 55, 80

SUVOROV,E.V.   SEE BOIKO ET AL. (1973)

SUZUKI,T.   SEE SAKURAI AND SUZUKI (1959)

SWENSON,C.A.   SEE ATOJI ET AL. (1959)

467 SYNTEX ANALYTICAL INSTRUMENTS (1974)
MANUAL FOR LOW TEMPERATURE ATTACHMENT LT-1
          111300   50, 51, 68, 209

468 TACHEZ,M. AND THÉOBALD,F. (1975)
J. APPL. CRYST. 8, 496
A VERY INEXPENSIVE DEVICE FOR COLLECTING X-RAY DATA
BELOW THE AMBIENT TEMPERATURE
          182500   158

TACONIS,K.W.   SEE KEESOM AND TACONIS (1935)

TACONIS,K.W.   SEE KEESOM AND TACONIS (1938)

469 TALLMAN,R.L., WAMPLER,D.L., AND MARGRAVES,J.L. (1961)
J. INORG. NUCL. CHEM. 21, 38-39
THE X-RAY POWDER DIFFRACTION PATTERN AND DENSITY OF
SOLID PERCHLORYL FLUORIDE
          364310

TAKAGI,S.   SEE LENHERT AND TAKAGI (1974)

470 TANAKA,J. AND AMMA,E.L. (1964)
REV. SCI. INSTRUM. 35, 634
THIN-WALLED PYREX CAPILLARIES FOR SINGLE-CRYSTAL X-RAY DIFFRACTION
          2   3   6   170, 172

TATKIN,L.Z.   SEE BOIKO ET AL. (1972)

471 TAYLOR,A. (1960)
APPL. SPECT. 14, 116-118
(ALSO PUBLISHED IN ADV. X-RAY ANALYSIS 3, 41-47)
A VERSATILE 19-CM DIAMETER LOW-TEMPERATURE DEBYE-SCHERRER CAMERA
          311310   99

TAYLOR,J.C.   SEE LEVY ET AL. (1975)

TAYLOR,L.S.   SEE CIOFFI AND TAYLOR (1922)

472 TAYLOR,N.W. (1931)
REV. SCI. INSTRUM. 2, 751-755
AN X-RAY CAMERA FOR POWDER DIAGRAMS AT ANY TEMPERATURE
          362314   3   153, 175

TELLGREN,R.   SEE THOMAS ET AL. (1974)

TEMPLETON,D.H.   SEE SIEMONS AND TEMPLETON (1954)

TEMPLETON,D.H.   SEE OLOVSSON AND TEMPLETON (1959)

TERMINASOV,Y.S.   SEE MINDUKSHEEV AND TERMINASOV (1958)

473 TERREY,H. AND WRIGHT,C.M. (1928)
PHIL. MAG. 6, 1055-1069
STRUCTURE OF MERCURY, COPPER, AND COPPER AMALGAM
          323310   133

TESK,J.A.   SEE GRUBER ET AL. (1969)

TESTARD,O.   SEE BONFIGLIOLI AND TESTARD (1964)

474   THAXTON,C.B. AND JACOBSON,R.A. (1970)
      J. PHYS. E., SCI. INSTRUM. 3, 245-246
      A TECHNIQUE FOR COOLING SINGLE CRYSTALS BELOW 90 K
      FOR X-RAY DIFFRACTION
                    114300
                                                                    36, 50, 72, 73, 223

THÉOBALD,F.   SEE TACHEZ AND THEOBALD (1975)

475   THEWLIS,J. AND DAVEY,A.R. (1955)
      J. SCI. INSTRUM. 32, 79
      ADAPTATION OF A STANDARD X-RAY POWDER CAMERA FOR WORK
      AT LOW TEMPERATURES
                    312310
                                                                        39, 46, 98

476   THOMANEK,U.F., PARAK,F., MÖSSBAUER,R.L., FORMANEK,H.,
      SCHWAGER,P. AND HOPPE,W. (1973)
      ACTA CRYST. A29, 263-265
      FREEZING OF MYOGLOBIN CRYSTALS AT HIGH PRESSURE
                    565365   1
                                                                          165

477   THOMAS,J.O. (1972)
      J. APPL. CRYST. 5, 102-106
      A CRYOGENIC ATTACHMENT FOR SINGLE CRYSTAL X-RAY STUDIES
                    222152
                                                                   86, 123, 125, 249

478   THOMAS,J.O., TELLGREN,R. AND OLOVSSON,I. (1974)
      ACTA CRYST. B30, 1155-1166
      HYDROGEN BOND STUDIES. LXXXIV. AN X-RAY DIFFRACTION STUDY OF THE
      STRUCTURES OF KHCO3 AND KDCO3 AT 298, 219, AND 95 D-K
                         7
                                                                         198, 207

THOMAS,M. (1975)   PERSONAL COMMUNICATION   57

479   TICHY,K. (1962)
      CESK. CAS. FYS. A12, 240-243
      X-RAY DIFFRACTION MEASUREMENTS ON SAMPLES COOLED TO
      TEMPERATURES CLOSE TO 100 D-K
                    112371

TICHY,K.   SEE HINE ET AL. (1975)

480   TIEN,C.L. AND CUNNINGTON,G.R. (1972)
      CRYOGENICS 12, 419-421
      RECENT ADVANCES IN HIGH-PERFORMANCE CRYOGENIC THERMAL INSULATION
                         21   245

TIPPE,A.   SEE HOHNE ET AL. (1968)

TIPPE,A.   SEE CLAUDET ET AL. (1976)

TIUSANEN,K.   SEE HOVI ET AL. (1966)

481   TOMBS,N.C. (1952)
      J. SCI. INSTRUM. 29, 364
      A LOW-TEMPERATURE X-RAY POWDER CAMERA
                    362311
                                                                         152

TOMIZUKA,C.T.   SEE EUBIG AND TOMIZUKA (1972)

482 TOMPA,A.S. (1968)
     APPL. SPECT. <u>22</u>, 491-493
     LOW-TEMPERATURE GRINDING TECHNIQUE FOR INFRA-RED ANALYSIS
                              21   *178*

     TOURAND,G.   SEE PINOT ET AL. (1965)

     TRAUTZ,O.R.   SEE BARRETT AND TRAUTZ (1948)

483 TROITSKAYA,Z., VERESHCHAGIN,L., AND KABALKINA,S. (1969)
     APP. METODY RENTGENOVSK. ANAL. <u>4</u>, 26-29
     (ALSO PUBLISHED IN INSTRUM EXPTL. TECH (USSR)
              <u>12</u>, 1017-1019 (1969))
     HIGH-PRESSURE X-RAY CAMERA FOR STUDIES AT TEMPERATURES TO -150 D-C
              322315

484 TROPP,J. (1975)
     PRIVATE COMMUNICATION
              115500   *66*

485 TROTTER,J. (1959)
     ACTA CRYST. <u>12</u>, 884-888
     THE CRYSTAL STRUCTURE OF NITROBENZENE AT -30 D-C
              211356   *87*

     TRUS,B.L.   SEE KRIEGER ET AL. (1974)

486 TRUT,,L., SENEGAS,J., AND GALY,J. (1973)
     REV. PHYS. APPL. <u>8</u>, 99-100
     A SIMPLE LOW TEMPERATURE DIFFRACTOMETER SAMPLE HOLDER
              323271   *132, 139*

     TUCK,D.G.   SEE EINSTEIN ET AL.(1972)

487 TWEET,A.G. (1954)
     PHYS. REV. <u>93</u>, 15-20
     SMALL ANGLE X-RAY SCATTERING FROM LIQUID HELIUM I AND
     LIQUID HELIUM II
              323110

488 UBBELOHDE,A.R. AND WOODWARD,I. (1946)
     PROC. ROY. SOC. (LONDON) A<u>185</u>, 448-465
     STRUCTURE AND THERMAL PROPERTIES OF CRYSTALS
              111300   *43, 44, 45, 55*

489 UBBELOHDE,A.R. AND WOODWARD,I. (1947)
     PROC. ROY. SOC. (LONDON) A<u>188</u>, 358-371
     BEHAVIOUR OF KH2PO4 AND KH2(AS)O4 ON COOLING
              121343   *74, 126*

     UBBELOHDE,A.R.   SEE CIMINO ET AL.(1959)

     UCHIDA,T.   SEE GOTO ET AL. (1969)

490 ULICKY,L. (1965)
     CHEM. ZVESTI <u>19</u>, 655-659
     TEMPERATURE-CONTROLLED VACUUM CHAMBER FOR
     X-RAY ANALYSIS OF POLYMERS
              323410

     ULLMAN,S.G.   SEE STOCHL AND ULLMAN (1963)

     UMANSKII,M.M.   SEE ZUBENKO AND UMANSKII (1957)

UMANSKII,M.M.   SEE BOIKO AND UMANSKII (1968)

UMANSKII,M.M.   SEE BOIKO ET AL. (1972)

UMANSKII,Y.S.   SEE KAGAN AND UMANSKII (1960)

491 UNDERWOOD,F.A. AND CHAPMAN,B.F. (1976)
J. APPL. CRYST. 9, 258
USE OF HIGH AND LOW TEMPERATURE ATTACHMENT
FOR WEISSENBERG CAMERA
          211350
                                                                         61

VALLE,G.   SEE DEL PRA ET AL. (1967)

492 VANCE,R.W. AND DUKE,W.M. (1962)
APPLIED CRYOGENIC ENGINEERING
J. WILEY AND SONS, N.Y.
          21
                                                                         233

VAHTEVA,M.   SEE HOVI ET AL. (1964)

493 VAVRA,F. (1956)
CESK. CAS. FYS. 6, 548-549
A DEVICE FOR STRUCTURE INVESTIGATIONS AT LOW TEMPERATURES
          312312

494 VEGARD,L. (1931)
Z. PHYSIK 68, 184-203
STRUCTURE OF SOLID N2O4 AT LIQUID AIR TEMPERATURE
          322311   3
                                                                         133, 175

495 VEITH,M. (1975)
ACTA CRYST. A31, S191
TRIMETHYL-SILYL-, GERMYL-, AND STANNYL- HYDRAZINES   -
A PLASTIC CRYSTAL CLASS
          324230   2   6
                                                                         103, 177

496 VENKATESWARAN,C.S. (1941)
PROC. INDIAN ACAD. SCI. 14A, 387-394
LOW-TEMPERATURE STUDIES OF THE RAMAN X-RAY REFLECTIONS IN CRYSTALS
          223320

VENTURINO,A.S.   SEE KRAMER ET AL. (1963)

VERESHCHAGIN,L.   SEE TRIOTSKAYA ET AL. (1969)

497 VERSCHOOR,G.C. AND KEULEN,E. (1971)
ACTA CRYST. B27, 134-145
ELECTRON DENSITY DISTRIBUTION IN CYANURIC ACID, I.
          211370   1   14
                                              15, 46, 54, 163, 208, 218

498 VISWAMITRA,M.A. (1962)
J. SCI. INSTRUM. 39, 381-383
A LOW-TEMPERATURE WEISSENBERG CAMERA
          212356   1
                                                     154, 162, 246

VISWAMITRA,M.A.   SEE RAO AND VISWAMITRA (1972)

499 VISWAMITRA,M.A. AND KANNAN,K.K. (1962)
J. SCI. INSTRUM. 39, 318-319
SIMPLE GAS-COOLING DEVICE FOR LOW-TEMPERATURE INVESTIGATIONS
WITH WEISSENBERG CAMERAS
          212350
                                                                         89

500   VISWAMITRA,M.A. AND RAMASESHAN,S. (1960)
      REV. SCI. INSTRUM. 31, 456-457
      SIMPLE DEVICE FOR GROWING CRYSTALS AT LOW TEMPERATURES
      IN X-RAY CAMERAS
                                  4   188, 191

501   VISWAMITRA,M.A. AND RAMASESHAN,S. (1963)
      Z.F. KRISTALLOGR. 119, 79-89
      A BACK-REFLECTION X-RAY CAMERA FOR THERMAL EXPANSION STUDIES
      AND THE THERMAL EXPANSION OF NACL FROM -180 D-C TO +200 D-C
                    222323   134

      VISWANATHAN,K.   SEE SALJE AND VISWANATHAN (1975)

      VLAHOVICI,N.   SEE RADOVICI AND VLAHOVICI (1971)

      VOHSEN,E.   SEE SIMON AND VOHSEN (1928)

502   VONNEGUT,B. AND WARREN,B.E. (1936)
      J. AM. CHEM. SOC. 58, 2459-2461
      THE STRUCTURE OF CRYSTALLINE BROMINE
                    115340   2   4   162, 167, 182

      VORA,P.M.   SEE JOHNSON ET AL. (1974)

      VOS,A.   SEE COPPENS AND VOS (1971)

      VOUSDEN,P.   SEE KAY AND VOUSDEN (1949)

      WAGENFELD,H.   SEE LING AND WAGENFELD (1965)

503   WAGNER,C.N.J. (1960)
      Z. METALLK. 51, 259-264
      STACKING FAULTS IN GOLD AFTER A COLD-WORK DISTORTION
      AT LOW TEMPERATURE
                    312372

504   WAHL,W. (1913)
      Z. PHYSIK. CHEM. 84, 101-111
      OPTICAL STUDY OF CRYSTALLINE NITROGEN, ARGON, METHANE AND
      ASSORTED ORGANIC MATERIALS

      WAIDELICH,W.   SEE PEISL AND WAIDELICH (1959)

      WAIDELICH,W.   SEE HIMMLER ET AL. (1969)   3

505   WALLWORK, S.C. AND HARDING,T.T. (1954)
      J. SCI. INSTRUM. 31, 163-164
      A SIMPLE METHOD OF OBTAINING LOW-TEMPERATURE X-RAY
      DIFFRACTION PHOTOGRAPHS
                    112342   62

      WALLWORK,S.C.   SEE BROWN AND WALLWORK (1962)

      WALTER,H.   SEE HOHNE ET AL. (1968)

      WAMPLER,D.L.   SEE TALLMAN ET AL. (1961)

506   WANG,Y., BLESSING,R.H., ROSS,F.K., AND COPPENS,P. (1976)
      ACTA CRYST. B32, 572-578
      CHARGE DENSITY STUDIES BELOW LN2 TEMPERATURE-
      X-RAY ANALYSIS OF P-NITROPYRIDINE-N-OXIDE AT 30K
                          7   14, 16

507   WANG,Y. AND COPPENS,P. (1976)
      INORG. CHEM. 15, 1122-1127
      ELECTRON DISTRIBUTION IN MU-ACETYLENE
      BIS(CYCLOPENTADIENYLNICKEL) BY LOW-TEMPERATURE
      X-RAY DIFFRACTION
                                          7
                                                                          19

      WARREN,B.E.   SEE VONNEGUT AND WARREN (1936)

      WASSERMAN,J.   SEE BARRETT ET AL. (1967)

508   WATKIN,D.J. (1975)
      J. APPL. CRYST. 8, 491-492
      ABSORPTION CORRECTION FOR CRYSTAL IN CAPILLARY TUBES
                                         18
                                                                          220

509   WEIGLE,J. AND SAINI,M. (1936)
      HELV. PHYS. ACTA 9, 515-519
      THE STRUCTURE OF AMMONIUM BROMIDE AT LOW TEMPERATURES
                              323313
                                                                     116, 134

      WEINER,K.L.   SEE IHRINGER ET AL. (1975)

      WEIR,C.E.   SEE SANTORO ET AL. (1968)

510   WELLS,M. (1960)
      ACTA CRYST. 13, 722-726
      COMPUTATION OF ABSORPTION CORRECTIONS
                                 18
                                                                          218

511   WELTMAN,H.J. (1962)
      ADV. X-RAY ANALYSIS 5, 48-56
      LOW-TEMPERATURE X-RAY DIFFRACTION OF FROZEN ELECTROLYTES
                        322372   2   5   16   132, 139, 142, 168, 207

      WEST,J.   SEE SANTOS AND WEST (1933)

      WESTERBECK,E.   SEE DEISEROTH ET AL. (1975)

      WHALLEY,E.   SEE BERTIE ET AL. (1963)

512   WHEATLEY,P.J. (1960)
      ACTA CRYST. 13, 80-85
      THE CRYSTAL AND MOLECULAR STRUCTURE OF PYRIMIDINE
                          185500
                                                                          157

513   WHEELER,D. (1968)
      THESIS, UNIVERSITY OF NEW MEXICO
      AN X-RAY DIFFRACTION STUDY OF THE PLASTIC CRYSTAL PHASE
                          312312   5
                                                                        63, 178

      WHEELER,D.A.   SEE EBDON AND WHEELER (1971)

514   WHEELER,G.L. AND COLSON,S.D. (1975)
      ACTA CRYST. B31, 911-913
      GAMMA-PHASE P-DICHLOROBENZENE AT 100 D-K
                          253370   1
                                                                     145, 148, 168

515   WHITE,G.K. (1968)
      EXPERIMENTAL TECHNIQUES IN LOW-TEMPERATURE PHYSICS
      SECOND EDITION
      OXFORD,U.K.
                                 21
                                                                          248

WIEBENGA,E.H.   SEE DRENTH AND WIEBENGA (1955)

WILHELM,J.O.   SEE MCLENNAN AND WILHELM (1925)

516   WILLIAMS,G.A. AND PACKARD,R.E.   (1974)
REV. SCI. INSTRUM. 45, 1029-1030
EXTRACTION OF IMAGES FROM A VERY LOW TEMPERATURE CRYOSTAT
USING FIBER OPTICS
                              21   111, 224

WILLIAMS,G.I.   SEE OWEN AND WILLIAMS (1954)

WILLIAMS,J.M.   SEE ROSS AND WILLIAMS (1974)

517   WILLIAMS,P.S.   (1933)
REV. SCI. INSTRUM. 4, 334-336
THE COOLING OF CRYSTALS FOR X-RAY SCATTERING MEASUREMENTS
                  222371   112, 125

WILLIAMS,R.O.   SEE JETTER ET AL. (1957)

WILSON,P.W.   SEE LEVY ET AL. (1975)

WILSON,S.A.   SEE BACON,ET AL. (1964)

518   WITTSTADT,W. (1940)
Z. ELEKTROCHEM. 46, 521-527
X-RAY PHOTOGRAPHY UNDER EXTREME CONDITIONS
                  20

519   WOOD, E.A. (1953)
REV. SCI. INSTRUM. 24, 325-326
SIMPLE ATTACHMENT FOR LOW-TEMPERATURE USE OF AN X-RAY
DIFFRACTION CAMERA
                  312310   100

520   WOODARD,C.L. AND STRAUMANIS,M.E. (1971)
J. APPL. CRYST. 4, 201-204
PRECISION DETERMINATION OF LATTICE PARAMETERS AT LOW TEMPERATURES
WITHOUT THE USE OF LIQUID GASES
              343213   14   143, 201

WOODARD,C.L.   SEE STRAUMANIS AND WOODARD (1971)

521   WOODLEY,M.J.A., HINE,R., AND RICHARDS,J.P.G. (1971)
J. APPL. CRYST. 4, 9-12
A HELIUM-COOLED, THREE CIRCLE X-RAY DIFFRACTOMETER
                  223171   125, 206, 209, 217

WOODWARD,I.   SEE UBBELOHDE AND WOODWARD (1946)

WOODWARD,I.   SEE UBBELOHDE AND WOODWARD (1947)

WRIGHT,C.M.   SEE TERREY AND WRIGHT (1928)

YAKEL,H.L.,JR.   SEE JETTER ET AL. (1957)

YAKIMENKO,L.F.   SEE KOGAN ET AL. (1964)

YELON,W.B.   SEE JUST ET AL. (1975)

YELON,W.B.   SEE CLAUDET ET AL. (1976)

522   YOUNG,R.A. (1966)
      J. SCI. INSTRUM. 43, 449-453
      X-RAY SPECIMEN TEMPERATURE CONTROL WITH GAS STREAMS
                  111300   20   32, 39, 49, 54, 232

523   ZABETAKIS,M.G. (1967)
      SAFETY WITH CRYOGENIC FLUIDS
      PLENUM PRESS,N.Y.
                                  21   233, 234

524   ZAKHAROV,A.I. (1961)
      INSTRUM. EXPTL. TECH. (USSR) 4, 731-734
      A LOW-TEMPERATURE X-RAY DIFFRACTOMETER
                  323374

525   ZAKHAROV,A.I. (1969)
      INSTRUM. EXPTL. TECH. (USSR) 12, 498-499
      CRYOGENICS 10, 65-67 (1970)
      HELIUM CRYOSTAT FOR X-RAY STUDIES
                  423170   129

      ZECHMEISTER,K.   SEE ANDRE ET AL. (1972)

526   ZEVIN,L.S. AND KHEIKER,D.M. (1958)
      INDUSTRIAL LAB. USSR 24, 716-718
      HIGH- AND LOW-TEMPERATURE ATTACHMENTS FOR THE URS-501
      X-RAY DIFFRACTOMETER
                  365374   154

527   ZHURAVLEV,N.N. AND KATSENEL'SON,A.A. (1959)
      SOVIET PHYSICS - CRYSTALLOGRAPHY 3, 639-641
      AN X-RAY CAMERA FOR THE STUDY OF CRYSTALS OVER THE RANGE
      OF -175 TO +300 D-C
                  312310

528   ZICKERT,K. AND GUTTZEIT,R. (1966)
      ACTA CRYST. 21, A226
      A LOW TEMPERATURE THERMOELECTRIC SAMPLE HOLDER
                  133600   145

      ZINOVIEV,M.V.   SEE BOL'SHUTKIN ET AL. (1969)

529   ZUBENKO,V.V. AND UMANSKII,M.M. (1957)
      SOVIET PHYSICS - CRYSTALLOGRAPHY 2, 505-509
      THE X-RAY DIFFRACTION DETERMINATION OF THE THERMAL
      EXPANSION OF SINGLE CRYSTALS
                  222442   115, 127

      ZUBENKO,V.V.   SEE BOIKO ET AL. (1972)

B.3. Apparatus Code-Number Listing

EACH TYPE OF APPARATUS IS CLASSIFIED BY A CODE NUMBER CONSISTING
OF SIX(6) DIGITS THE MEANING OF EACH OF THESE DIGITS IS AS FOLLOWS •

FIRST DIGIT • TYPE OF SAMPLE THAT CAN BE STUDIED
•••
    1   ANY TYPE
    2   SINGLE CRYSTAL
    3   POWDER
    4   METAL
    5   PROTEINS
    6   NEUTRON DIFFRACTION

SECOND DIGIT • TYPE OF COOLING USED
•••••••••••••••••••••••••••••••••••
    1   COLD-GAS STREAM
    2   CONDUCTION (CRYOGENIC FLUID AS COOLANT)
    3   CONDUCTION (THERMOELECTRIC COOLING)
    4   CONDUCTION (MECHANICAL REFRIGERATION)
    5   JOULE-THOMSON EXPANSION
    6   IMMERSION OF SAMPLE
    7   IMMERSION OF CAMERA
    8   USE OF COLD ROOM

THIRD DIGIT • METHOD OF FROST PREVENTION
••
    1   DRY GAS STREAM
    2   DRY CHAMBER
    3   EVACUATED CHAMBER
    4   NOT GIVEN
    5   NONE

FOURTH DIGIT • MINIMUM TEMPERATURE ATTAINABLE (DEGREES KELVIN)
•••
    1   LESS THAN 20
    2   20-78
    3   78-200
    4   200-260
    5   GREATER THAN 260
    6   NOT AVAILABLE

FIFTH DIGIT • TYPE OF X-RAY INSTRUMENT MENTIONED
••
    0   ANY TYPE
    1   DEBYE-SCHERRER CAMERA (INCLUDES BACK-REFLECTION)
    2   FLAT-CASSETTE AND LAUE CAMERAS
    3   GUINIER CAMERA
    4   OSCILLATION-ROTATION CAMERA
    5   WEISSENBERG GONIOMETER
    6   PRECESSION CAMERA
    7   DIFFRACTOMETER
    8   SMALL-ANGLE
    9   TOPOGRAPHIC STUDIES

SIXTH DIGIT - SPECIAL CHARACTERISTICS
----
    0   NONE
    1   HORIZONTAL
    2   VERTICAL
    3   BACK-REFLECTION
    4   HIGH-TEMPERATURE ALSO
    5   HIGH-PRESSURE ALSO
    6   WEISSENBERG GONIOMETER ACCESSORIES
    7   COLD-WORKING AT LOW TEMPERATURES

   FOR EXAMPLE, CODE NUMBER 111310 REFERS TO A DEVICE
THAT CAN BE USED WITH ANY TYPE OF SAMPLE, COOLS WITH A
COLD-GAS STREAM, PREVENTS ICING WITH A CONCENTRIC
DRY GAS STREAM,CAN REACH A MINIMUM TEMPERATURE BETWEEN
78 AND 200 D-K AND WAS DESIGNED FOR USE WITH A DEBYE-SCHERRER
CAMERA. IT HAS NO SPECIAL CHARACTERISTICS.

111300	-	522	112342	-	505	123371	-	391
		488			392			348
		467						48
		437	112356	-	6			47
		408						
		407	112371	-	479	133600	-	528
		372						
		307	112390	-	65	153100	-	76
		305			9			
		297				153200	-	8
		243	113600	-	12			
		240				162350	-	225
		231	114300	-	474			
		221			296	165300	-	309
		212			213			302
		180						192
		177	115300	-	237			108
		162			104			
		145				165320	-	303
		133						
		114	115340	-	502	182400	-	188
		86	115356	-	436			
		69				182500	-	468
111306	-	401	115500	-	484			
						185450	-	53
		353	121343	-	489			
		144				185500	-	512
111343	-	265	121370	-	251			
						165570	-	356
		103	123140	-	451			
111400	-	197				211152	-	395
			123170	-	452			
112300	-	384				211300	-	390
		327	123171	-	101			241
		158						234
		4	123190	-	323			175
		2				211350	-	491
			123320	-	236			444
112306	-	117						143
			123345	-	80			
112340	-	264						

211356	▪	485	215571	▪	132	253371	▪	161
		457						153
			222152	▪	477			
211360	▪	89				262320	▪	311
		88	222175	▪	31			
						263320	▪	203
211364	▪	191	222323	▪	501			
						272150	▪	49
211370	▪	497	222340	▪	32			
						282520	▪	394
211371	▪	464	222370	▪	107			29
		232						
		62	222371	▪	517	285500	▪	459
211374	▪	15	222440	▪	124	300070	▪	183
212100	▪	226	222442	▪	529	300310	▪	319
212140	▪	174	223125	▪	425	311310	▪	471
								403
212154	▪	466	223140	▪	462			402
								291
212171	▪	140	223145	▪	275			235
								127
212252	▪	397	223150	▪	359	311334	▪	440
212340	▪	166	223170	▪	202	311372	▪	332
212342	▪	64	223171	▪	521	312100	▪	453
					449			
212350	▪	499			350	312110	▪	227
		445			336			
		375			299	312111	▪	300
		354			121			
		245				312130	▪	228
		182	223190	▪	44			
		152				312310	▪	527
		134	223300	▪	179			519
		99						475
		83	223320	▪	496			329
		82						242
		59	223340	▪	219			178
		14						165
		13	223350	▪	274			
212354	▪	16	223352	▪	266	312311	▪	250
212356	▪	498	223360	▪	433	312312	▪	513
		381						493
		292	223370	▪	215			378
		138						
			223371	▪	386	312313	▪	355
212360	▪	66			268			
						312315	▪	96
212365	▪	337	223390	▪	17			
						312370	▪	290
212370	▪	27	224390	▪	123			
						312371	▪	109
212371	▪	135	243175	▪	439			68
213141	▪	50	253370	▪	514	312372	▪	503
					249			415
213171	▪	351			248			43
					206			
214356	▪	105			181			

332470	=	255	364310	=	469	423487	=	340
		214	365311	=	21	463107	=	150
335572	=	366	365323	=	418	465323	=	195
		233	365374	=	526	500000	=	365
343111	=	256	372310	=	217			293
343213	=	520	373117	=	361			283
361371	=	398	373120	=	325			208
362310	=	335	382520	=	349			207
		34	385510	=	81	515400	=	320
362311	=	481	400000	=	169	531560	=	113
		420	412311	=	360	565365	=	476
		285	422371	=	164	613171	=	106
362312	=	294	423170	=	525	623170	=	24
362314	=	472	423177	=	36	623171	=	218
362320	=	220	423217	=	278			3
362372	=	42	423320	=	30	633470	=	7
363311	=	55	423387	=	342	633471	=	277
363312	=	271				653371	=	159
364170	=	429						120

REFERENCES ARE SORTED ON THE BASIS OF
DIGIT NUMBER   1 OF THE APPARATUS-TYPE CODE

FIRST DIGIT - TYPE OF SAMPLE THAT CAN BE STUDIED
----
            1   ANY TYPE
            2   SINGLE CRYSTAL
            3   POWDER
            4   METAL
            5   PROTEINS
            6   NEUTRON DIFFRACTION

1 - 528	522	512	505	502	489	488	484
	479	474	468	467	452	451	437
	436	408	407	401	392	391	384
	372	356	353	348	327	323	309
	307	305	303	302	297	296	265
	264	251	243	240	237	236	231
	225	221	213	212	197	192	188
	180	177	162	158	145	144	133
	117	114	108	104	103	101	86
	80	76	69	65	53	48	47
	12	9	8	6	4	2	
2 - 529	521	517	514	501	499	498	497
	496	491	485	477	466	464	462
	459	457	449	445	444	439	433
	425	397	396	394	390	386	381
	375	359	354	351	350	337	336
	311	299	292	275	274	268	266
	249	248	245	241	234	232	226
	219	215	206	203	202	191	182
	181	179	175	174	166	161	153
	152	143	140	138	135	134	132
	124	123	121	107	105	99	89
	88	83	82	66	64	62	59
	50	49	44	32	31	29	27
	17	16	15	14	13		

3 - 527	526	524	520	519	513	511	509
	503	495	494	493	490	487	486
	483	481	475	473	472	471	469
	463	460	456	453	450	448	447
	446	443	441	440	435	434	429
	428	427	424	422	420	418	417
	416	415	414	412	411	410	403
	402	398	395	383	382	380	379
	378	377	374	368	367	366	364
	363	362	361	355	349	343	339
	335	334	332	331	329	328	326
	325	324	322	321	319	318	316
	310	304	300	295	294	291	290
	289	288	287	286	285	282	281
	280	279	276	273	271	267	263
	262	261	256	255	253	250	247
	246	244	242	235	233	229	228
	227	223	220	217	214	211	205
	204	200	199	198	196	193	190
	189	186	185	184	183	178	171
	167	165	163	160	157	156	151
	146	142	141	139	136	127	115
	110	109	102	100	98	96	95
	94	92	91	81	79	78	75
	74	71	70	68	63	61	60
	58	57	56	55	46	45	43
	42	41	40	35	34	33	26
	21	11	5	1			
4 - 525	360	342	340	278	195	169	164
	150	36	30				
5 - 476	365	320	293	283	208	207	113
6 - 277	218	159	120	106	24	7	3

REFERENCES ARE SORTED ON THE BASIS OF
DIGIT NUMBER   2 OF THE APPARATUS-TYPE CODE

SECOND DIGIT - TYPE OF COOLING USED
----
            1   COLD-GAS STREAM
            2   CONDUCTION (CRYOGENIC FLUID AS COOLANT)
            3   CONDUCTION (THERMOELECTRIC COOLING)
            4   CONDUCTION (MECHANICAL REFRIGERATION)
            5   JOULE-THOMSON EXPANSION
            6   IMMERSION OF SAMPLE
            7   IMMERSION OF CAMERA
            8   USE OF COLD ROOM

1 - 527	522	519	513	505	503	502	499
	498	497	493	491	488	485	484
	479	475	474	471	467	466	464
	457	453	445	444	440	437	436
	415	408	407	403	402	401	397
	396	392	390	384	381	378	375
	372	367	360	355	354	353	351
	337	332	329	327	320	307	305
	300	297	296	292	291	290	276
	265	264	250	245	243	242	241
	240	237	235	234	232	231	228
	227	226	221	213	212	200	197
	191	182	180	178	177	175	174
	166	165	162	158	157	152	145
	144	143	140	138	135	134	133
	132	127	117	115	114	109	106
	105	104	103	99	96	89	88
	86	83	82	78	75	69	68
	66	65	64	63	62	61	59
	50	43	27	16	15	14	13
	12	9	6	4	2		
2 - 529	525	524	521	517	511	509	501
	496	495	494	490	489	487	486
	483	477	473	463	462	460	456
	452	451	450	449	448	447	446

```
            443     441     435     434     433     428     427
            425     424     422     417     416     414     412
            411     410     395     391     386     383     382
            380     379     377     374     368     364     363
            362     359     350     348     343     342     340
            339     336     334     331     328     326     324
            323     322     321     318     316     310     304
            299     295     289     288     287     286     282
            281     280     279     278     275     274     273
            268     267     266     263     262     261     253
            251     247     246     244     236     229     223
            219     218     215     211     205     204     202
            199     198     196     193     190     189     186
            185     184     179     171     167     164     163
            160     156     151     146     142     141     139
            136     124     123     121     110     107     102
            101     100      98      95      94      92      91
             80      79      74      71      70      60      58
             57      56      48      47      46      45      44
             41      40      36      35      33      32      31
             30      26      24      17      11       5       3
              1

3 -  528    366     277     255     233     214     113       7
4 -  520    439     256
5 -  514    249     248     206     181     161     159     153
            120      76       8
6 -  526    481     476     472     469     429     420     418
            398     335     311     309     303     302     294
            285     271     225     220     203     195     192
            150     108      55      42      34      21
7 -  361    325     217      49
8 -  512    468     459     394     356     349     188      81
             53      29
```

REFERENCES ARE SORTED ON THE BASIS OF
DIGIT NUMBER   3 OF THE APPARATUS-TYPE CODE

THIRD DIGIT - METHOD OF FROST PREVENTION
----
                1   DRY GAS STREAM
                2   DRY CHAMBER
                3   EVACUATED CHAMBER
                4   NOT GIVEN
                5   NONE

1 - 522	497	491	489	488	485	471	467
	464	457	444	440	437	408	407
	403	402	401	398	396	390	372
	353	343	332	307	305	297	291
	265	251	243	241	240	235	234
	232	231	221	212	197	191	186
	185	180	177	175	162	156	145
	144	143	133	127	114	113	103
	89	88	86	69	62	15	
2 - 529	527	519	517	513	511	505	503
	501	499	498	494	493	483	481
	479	477	475	472	468	466	456
	453	450	448	445	420	415	397
	394	392	384	383	381	380	378
	377	375	360	355	354	349	337
	335	329	328	327	326	322	318
	311	300	294	292	290	287	285
	273	264	255	250	245	242	228
	227	226	225	220	217	214	211
	205	204	199	188	184	182	178
	174	166	165	164	158	152	142
	141	140	138	135	134	124	117
	115	110	109	107	99	96	94
	83	82	68	66	65	64	59

	49	43	42	35	34	33	32
	31	29	27	16	14	13	9
	6	5	4	2			
3 • 528	525	524	521	520	514	509	496
	490	487	486	473	463	462	460
	452	451	449	447	446	443	441
	439	435	434	433	428	427	425
	424	422	417	416	414	412	411
	410	395	391	386	382	379	374
	368	367	364	363	362	361	359
	351	350	348	342	340	339	336
	334	331	325	324	323	321	316
	310	304	299	295	289	288	286
	282	281	280	279	278	277	275
	274	271	268	267	266	263	261
	256	253	249	248	247	246	244
	236	229	223	219	218	215	206
	203	202	200	196	193	190	189
	181	179	171	167	163	161	160
	159	157	153	151	150	146	136
	121	120	106	101	100	98	95
	92	91	80	79	78	76	75
	74	71	70	63	61	58	57
	56	55	50	48	47	46	44
	41	40	36	30	26	24	17
	12	11	8	7	3	1	
4 • 495	474	469	429	296	262	213	139
	123	105	102	60			
5 • 526	512	502	484	476	459	436	418
	366	356	320	309	303	302	276
	237	233	198	195	192	132	108
	104	81	53	45	21		

REFERENCES ARE SORTED ON THE BASIS OF
DIGIT NUMBER   4 OF THE APPARATUS-TYPE CODE

FOURTH DIGIT - MINIMUM TEMPERATURE ATTAINABLE (DEGREES KELVIN)
•••
                 1   LESS THAN 20
                 2   20-78
                 3   78-200
                 4   200-260
                 5   GREATER THAN 260
                 6   NOT AVAILABLE

1 - 525	521	487	477	466	463	462	460
	453	452	451	449	439	429	427
	425	424	422	417	414	412	410
	396	395	383	379	367	364	363
	361	359	351	350	336	334	331
	325	324	323	316	310	304	300
	299	288	287	286	282	281	280
	279	275	273	256	246	228	227
	226	223	218	202	200	190	189
	174	171	167	163	160	150	142
	140	121	106	101	95	76	74
	71	70	63	61	58	56	50
	49	44	41	40	36	31	26
	24	3	1				
2 - 520	495	486	428	411	397	278	98
	8						
3 - 527	526	524	522	519	517	514	513
	511	509	505	503	502	501	499
	498	497	496	494	493	491	489
	488	485	483	481	479	476	475
	474	473	472	471	469	467	464
	457	456	450	448	447	446	445
	444	443	441	440	437	436	435
	434	433	420	418	416	415	408
	407	403	402	401	398	392	391

			390	386	384	382	381	380	378
			377	375	374	372	368	362	360
			355	354	353	348	343	342	339
			337	335	332	329	328	327	326
			322	321	319	318	311	309	307
			305	303	302	297	296	295	294
			292	291	290	289	285	276	274
			271	268	267	266	265	264	263
			261	253	251	250	249	248	247
			245	244	243	242	241	240	237
			236	235	234	232	231	229	225
			221	220	219	217	215	213	212
			206	205	203	199	196	195	193
			192	191	186	185	182	181	180
			179	178	177	175	166	165	164
			162	161	159	158	157	156	153
			152	151	146	145	144	143	141
			139	138	136	135	134	133	127
			123	120	117	115	114	110	109
			108	107	105	104	103	100	99
			96	94	92	91	89	88	86
			83	82	80	79	78	75	69
			68	66	65	64	62	60	59
			57	55	48	47	46	45	43
			42	34	33	32	30	27	21
			17	16	15	14	13	11	9
			6	5	4	2			
4	•	529	490	340	320	277	262	255	214
			211	204	198	197	188	184	124
			53	35	7				
5	•	512	484	468	459	394	366	356	349
			233	132	113	81	29		
6	•	528	102	12					

REFERENCES ARE SORTED ON THE BASIS OF
DIGIT NUMBER   5 OF THE APPARATUS-TYPE CODE

FIFTH DIGIT - TYPE OF X-RAY INSTRUMENT MENTIONED
••
               0   ANY TYPE
               1   DEBYE-SCHERRER CAMERA (INCLUDES BACK-REFLECTION)
               2   FLAT-CASSETTE AND LAUE CAMERAS
               3   GUINIER CAMERA
               4   OSCILLATION-ROTATION CAMERA
               5   WEISSENBERG GONIOMETER
               6   PRECESSION CAMERA
               7   DIFFRACTOMETER
               8   SMALL-ANGLE
               9   TOPOGRAPHIC STUDIES

0 - 528   522   512   488   484   474   468   467
          459   453   437   408   407   401   390
          384   372   365   353   327   320   309
          307   305   302   297   296   293   283
          243   241   240   237   234   231   226
          221   213   212   208   207   197   192
          188   185   180   179   177   175   169
          162   158   150   145   144   133   117
          114   108   104   86   76   69   12
            0   4   2

1 - 527   520   519   513   509   494   493   490
          487   483   481   475   473   472   471
          469   463   443   441   434   424   420
          414   412   411   410   403   402   383
          382   379   378   377   368   367   361
          360   355   339   335   329   328   326
          321   319   318   300   295   294   291
          289   288   287   286   285   278   273
          271   267   263   256   250   244   242
          235   227   217   205   204   200   196
          190   189   178   167   165   160   157
          146   142   141   139   127   100   96
           91   81   78   55   35   34   33
           26   21   11

2 - 501   496   425   418   394   364   349   334

		325	311	310	303	236	220	211
		203	195	156	102	58	30	29
3	•	495	440	416	246	228		
4	•	529	505	502	489	462	451	392	275
		265	264	219	174	166	124	103
		80	64	50	32			
5	•	499	498	491	485	477	466	457	445
		444	436	397	396	381	375	359
		354	292	274	266	245	225	182
		152	143	138	134	105	99	83
		82	59	53	49	16	14	13
		6						
6	•	476	433	337	191	113	89	88	66
7	•	526	525	524	521	517	514	511	503
		497	486	479	464	460	456	452
		450	449	448	447	446	439	435
		429	428	427	422	417	415	398
		395	391	386	380	374	366	363
		362	356	351	350	348	343	336
		332	331	324	322	316	304	299
		290	282	281	280	279	277	276
		268	262	261	255	253	251	249
		248	247	233	232	229	223	218
		215	214	206	202	199	198	193
		186	184	183	181	171	164	163
		161	159	153	151	140	136	135
		132	121	120	115	110	109	107
		106	101	98	95	94	92	79
		75	74	71	70	68	63	62
		61	60	57	56	48	47	46
		45	43	42	41	40	36	31
		27	24	15	7	5	3	1
8	•	342	340					
9	•	323	123	65	44	17	9	

REFERENCES ARE SORTED ON THE BASIS OF
DIGIT NUMBER   6 OF THE APPARATUS-TYPE CODE

SIXTH DIGIT - SPECIAL CHARACTERISTICS
----
        0   NONE
        1   HORIZONTAL
        2   VERTICAL
        3   BACK-REFLECTION
        4   HIGH-TEMPERATURE ALSO
        5   HIGH-PRESSURE ALSO
        6   WEISSENBERG GONIOMETER ACCESSORIES
        7   COLD-WORKING AT LOW TEMPERATURES

        CATEGORY -0- NOT LISTED

1 - 521	517	494	486	481	479	464	456
	449	446	435	427	422	420	417
	414	398	391	386	368	363	362
	360	351	350	348	339	336	324
	322	316	304	300	299	285	282
	281	279	277	268	267	256	250
	232	218	200	164	161	159	153
	140	135	132	121	120	110	109
	106	101	98	95	79	75	74
	71	70	68	63	62	61	56
	55	50	48	47	21	3	
2 - 529	513	511	505	503	493	477	463
	460	450	448	428	415	397	396
	392	378	377	366	332	328	294
	276	271	266	233	229	198	186
	163	160	157	151	146	141	136
	94	92	78	64	45	43	42
	5						
3 - 520	509	501	489	443	418	355	321

			265	195	167	103			
4	=	526	524	472	466	440	434	382	223
			199	193	191	171	115	57	46
			16	15	11				
5	=	483	476	439	425	424	380	364	337
			334	318	310	275	262	204	184
			102	96	80	58	31		
6	=	498	485	457	436	401	381	353	292
			144	138	117	105	6		
7	=	361	342	340	331	278	253	150	41
			40	36	1				

B.4. Techniques and Applications Code-Number Listing

SAMPLE PREPARATION

```
        1   SAMPLE PREPARATION IF SOLID AT ROOM TEMPERATURE
        2   SAMPLE PREPARATION IF LIQUID AT ROOM TEMPERATURE
        3   SAMPLE PREPARATION IF GAS AT ROOM TEMPERATURE
        4   CRYSTAL GROWTH IF LIQUID OR GAS AT ROOM TEMPERATURE
        5   TECHNIQUES FOR RANDOMLY ALIGNING POWDER SAMPLE
        6   REACTIVE OR RADIOACTIVE SAMPLE
```

APPLICATIONS

```
        7   APPLICATIONS
```

DATA COLLECTION AND REDUCTION

```
       11   PURPOSE OF LOW-TEMPERATURE METHODS
       12   CHOICE OF TEMPERATURE
       13   CHOICE OF COOLING METHOD
       14   TEMPERATURE CALIBRATION
       15   CHARACTERISTICS OF X-RAY FILM AT LOW TEMPERATURES
       16   ALIGNMENT OF CRYSTAL AND/OR APPARATUS
       17   SPECIAL PROGRAMMING OF AUTOMATIC DIFFRACTOMETER
       18   CORRECTION FOR ABSORPTION DUE TO SAMPLE HOLDER
       19   REFINEMENT OF DISORDERED MODELS
       20   REVIEW ARTICLE
```

OTHER

```
       21   GENERAL LOW-TEMPERATURE TECHNIQUES
```

1	-	22	36	54	111	112	143	164
			207	208	283	298	326	330
			333	341	365	476	497	498
			514					
2	-	4	52	53	86	117	143	175
			240	283	314	330	332	338
			353	356	372	384	390	401
			405	406	409	432	470	495
			502	511				

3	-	4	25	40	74	100	116	142
			205	240	274	287	288	310
			324	338	353	372	373	384
			411	412	436	441	451	470
			472	494				
4	-	4	20	53	73	77	80	86
			87	90	97	117	137	149
			154	155	175	176	203	221
			231	240	257	260	264	292
			302	310	312	313	314	345
			346	353	354	356	357	372
			373	384	389	401	405	409
			423	432	436	445	451	462
			500	502				
5	-	52	72	100	111	112	139	217
			328	372	401	402	403	427
			511	513				
6	-	25	41	53	86	143	173	214
			254	288	289	298	324	330
			338	341	372	440	470	495
7	-	10	37	38	119	168	189	211
			258	269	272	344	348	352
			430	431	438	442	450	455
			461	478	506	507		
11	-	85	118	122	130	152	222	225
			239	309	388			
12	-	118	129	209	210	239	388	
13	-	86	118	450				
14	-	2	28	118	164	187	248	281
			282	304	355	369	390	397
			400	430	431	461	497	520

Table B–1. Code Numbers Used in Classifying Low-Temperature X-Ray Diffraction Apparatus and Techniques

a. Type of Apparatus

First digit: Type of sample that can be Studied
1. Any type
2. Single crystal
3. Powder
4. Metal
5. Protein
6. Neutron diffraction

Second digit: Type of cooling used
1. Cold gas stream
2. Conduction (cryogenic fluid as coolant)
3. Conduction (thermoelectric cooling)
4. Conduction (mechanical refrigeration)
5. Joule–Thomson expansion
6. Immersion of sample
7. Immersion of camera
8. Use of cold room

Third digit: Method of frost prevention
1. Dry gas stream
2. Dry chamber
3. Evacuated chamber
4. Not given
5. None

Fourth digit: Minimum temperature attainable (Kelvin)
1. Less than 20
2. 20–78
3. 78–200
4. 200–260
5. Greater than 260
6. Not available

Fifth digit: Type of X-ray instrument mentioned
0. Any type
1. Debye–Scherrer camera (includes back-reflection)
2. Flat-cassette and Laue cameras
3. Guinier camera
4. Oscillation–rotation camera
5. Weissenberg goniometer
6. Precession camera
7. Diffractometer
8. Small-angle instrument
9. Topographic studies

Sixth digit: Special characteristics
0. None
1. Horizontal
2. Vertical
3. Back-reflection
4. High-temperature capability also
5. High-pressure capability also
6. Weissenberg goniometer accessories
7. Cold-working at low temperatures

b. Techniques and Applications

1. Sample preparation if solid at room temperature
2. Sample preparation if liquid at room temperature
3. Sample preparation if gas at room temperature
4. Crystal growth if liquid or gas at room temperature
5. Techniques for randomly aligning powder sample
6. Reactive or radioactive sample
7. Applications
11. Purpose of low-temperature method
12. Choice of temperature
13. Choice of cooling method
14. Temperature calibration
15. Characteristics of X-ray film at low temperatures
16. Alignment of crystal
17. Special programming of automatic diffractometer
18. Correction for absorption due to sample holder
19. Refinement of disordered models
20. Review article
21. General low-temperature techniques

Index